Insect Metamorphosis

From Natural History to Regulation of Development and Evolution

Insect Metamorphosis

From Natural History to Regulation of Development and Evolution

Xavier Belles
Institute of Evolutionary Biology (CSIC-Pompeu Fabra University),
Barcelona, Spain

Academic Press is an imprint of Elsevier
125 London Wall, London EC2Y 5AS, United Kingdom
525 B Street, Suite 1650, San Diego, CA 92101, United States
50 Hampshire Street, 5th Floor, Cambridge, MA 02139, United States
The Boulevard, Langford Lane, Kidlington, Oxford OX5 1GB, United Kingdom

British Library Cataloguing-in-Publication Data
A catalogue record for this book is available from the British Library

Library of Congress Cataloging-in-Publication Data
A catalog record for this book is available from the Library of Congress

ISBN: 978-0-12-813020-9

For Information on all Academic Press publications
visit our website at https://www.elsevier.com/books-and-journals

Publisher: Charlotte Cockle
Acquisitions Editor: Anna Valutkevich
Editorial Project Manager: Pat Gonzalez
Production Project Manager: Punithavathy Govindaradjane
Cover Designer: Christian Bilbow

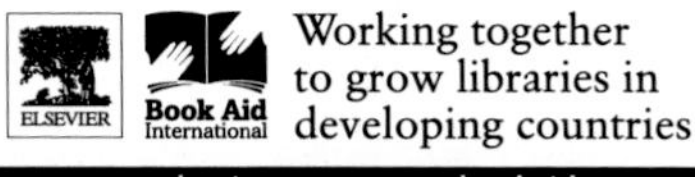

Typeset by MPS Limited, Chennai, India

Contents

Preface

How does a caterpillar transform into a butterfly? That question, which sums up the wonder and mystery of insect metamorphosis, has built a timeless enigma, which has fascinated humans since the earliest times. Thanks to the systematic observations initiated in the Renaissance and followed during the Enlightenment, today we know the details of metamorphosis in numerous species of insects of the most diverse groups. Scientific research began in the 19th century and in the years since has taught us what are the main factors that regulate it, which turned out to be mainly hormonal. The most notable progress was made in the field of experimental physiology, in which the most important hormones, juvenile hormone and molting hormone, were discovered and their essential functions were revealed, and in the field of chemistry, with the structural elucidation of these hormones. The most important developments of the last quarter of the 20th century came from molecular scale studies, which unveiled the essential mechanisms underlying the molting hormone action.

Only those mechanisms underlying the action of the juvenile hormone, arguably the most important hormone in the regulation of insect metamorphosis, were still pending. These mechanisms, at least the most essential ones, have been reported in the 21st century. Their importance goes beyond their strictly biochemical and molecular interest, since they allow envisaging the regulatory aspects from an integral point of view, leading to the reconnaissance of general rules and to the analysis of the evolution of insect metamorphosis in a more robust way. These recent achievements suggest that it may be a good time to prepare a book like this. Certainly the issues that clearly need an update are those related to molecular regulation and the evolutionary aspects. However, this book aims to explain insect metamorphosis in a comprehensive way, from natural history to regulatory mechanisms and evolution. And it goes without saying that in all fields there have been important advances in recent years that this book also tries to cover.

As for acknowledgments, I have kept my research on insect metamorphosis active for more than 30 years thanks to the uninterrupted financial support of the Spanish Ministry of Education and Science (now Ministry of Economy and Competitiveness), the Catalan Government, and the European Fund for Economic and Regional Development.

The significant diversity of topics discussed has led me to ask for help from colleagues who critically read different parts of the manuscript of which they were experts. The colleagues involved are Javier Alba-Tercedor (Chapter 4), Jordi Bernués and Montserrat Corominas (Chapter 8 focusing on DNA methylation and histone modifications), Malcolm Davies (classical Greece, in Chapter 1), Ryo Futahashi (Odonata, in Chapter 4), Miquel Gaju-Ricart (Chapter 3), Arturo Goldarazena (Thysanoptera, in Chapter 4), Klaus Hartfelder (Chapter 12), Marek Jindra (Chapter 7 focusing on the transduction of the juvenile hormone signal, Chapter 10, and Chapter 12), Jeyaraney Kathirithamby (Chapter 1 and Strepsiptera, in Chapter 5), Pierre Léopold (regulation of right allometric growth, in Chapter 9), Jesus Lozano (Chapter 2), Jose-Luis Maestro (Chapter 6 and Chapter 7 focusing on peptide hormones), Chieka Minakuchi (Chapter 10), Gerald B. Moritz (Thysanoptera, in Chapter 4), Michael O'Connor (the metamorphic molt, in Chapter 9), Genta Okude (Odonata, in Chapter 4), Subba Reddy Palli (Chapter 8 focusing on DNA methylation and histone modifications), John D. Pinto (hypermetamorphosis, in Chapter 5), Lynn M. Riddiford (Chapter 12), Carl Thummel (transduction of the 20-hydroxyecdysone signal, in Chapter 7), Jose Manuel Tierno de Figueroa (Plecoptera, in Chapter 4), Yoshinori Tomoyasu (the wings, a crucial innovation, in Chapter 2), James W. Truman (Chapter 12), Jozef Vanden Broeck (Chapter 6 and Chapter 7 focusing on peptide hormones), and Isabelle Vea (Coccomorpha, in Chapter 4).

Moreover, I have discussed the topics of the book with many people, especially with Maria-Dolors Piulachs and Jose-Luis Maestro, my closest colleagues in the Institute of Evolutionary Biology (CSIC-Pompeu Fabra University, Barcelona), and with Takaaki Daimon, from Kyoto University, who critically read the entire manuscript. I would also highlight the discussions about paleontology, a subject on which I have relied more heavily on expert opinion. Thus I thank Jarmila Kukalová-Peck for her hospitality and advice during the week I spent in Ottawa in 2017, especially examining the juvenile stages of fossil mayflies, and André Nel for the rich discussions on the paleontology of insect metamorphosis at the National Museum of Natural History in Paris.

Needless to say, the fact that all these colleagues helped me to improve the contents of the book does not mean that they necessarily share the ideas presented in it. In this context, the author is aware that he alone remains responsible the opinions expressed in the book and for any imperfections that may have remained. Finally, the author would like to acknowledge the assistance of the Elsevier's team, mainly Pat Gonzalez, Anna Valutkevich, Narmatha Mohan and Punitha Govindaradjane, who has efficiently solved the logistics and technical issues involved in the preparation of the book.

Xavier Belles

Institute of Evolutionary Biology (CSIC-Pompeu Fabra University), Barcelona, Spain

Chapter 1

The evolution of ideas on insect metamorphosis

In an unusual moment of modesty, Isaac Newton coined a celebrated phrase: "If I can see further than anyone else, it is only because I am standing on the shoulders of giants." By this, he meant with beautiful words that present knowledge is the product of a long chain of partial progress achieved by successive thinkers. It is never superfluous to know how the ideas that led to the concepts that we handle today have evolved over time. Especially because knowing how concepts on a subject have been modified by new knowledge significantly helps us to fully understand the subject itself.

The present chapter briefly describes how insect metamorphosis has been understood from the historic period to modern times. We know that in the Egypt of the Pharaohs, insect metamorphosis aroused admiration and awe by its parallelism with the resurrection. Very precise data on the metamorphosis of the sacred scarab beetle came to be known, as drawings in papyri, bas reliefs, paintings, and mummified beetles shown us today. However, it seems reasonable to begin the scientific history in classical Greece, with Aristotle, who was the first to go from mystical conceptions to directly interrogate nature and provided the first formal descriptions of insect transformations.

CLASSICAL GREECE

In classical Greece, the word *psychê* had two disparate meanings: soul and butterfly. Not in vain, the Greek word for the chrysalis, *nekydallos*, also means "little corpse." Similarly, the Latin word *anim(ul)a* is also used to denote the soul and butterflies. The symbolism is obvious, as from the inanimate chrysalis arises the vivacious butterfly as if it is resurrected from death. The famous sarcophagus of Prometheus, preserved in the Capitoline Museum of Rome, shows the goddess Athena holding a butterfly-shaped soul. Christian writers, such as St. Basil and other Church Fathers, abundantly used this ambiguity as an allegory of the resurrection, a tradition that was preserved even to the moralizing medieval bestiaries (Davies and Kathirithamby, 1986).

In spite of the deep roots of the ancient popular culture, which sees animals more as symbols than as living organisms, in the classical times of the

Insect Metamorphosis. DOI: https://doi.org/10.1016/B978-0-12-813020-9.00001-6

Western civilization, there were observers of the real nature and among them stands the gigantic figure of Aristotle. He lived immersed in the culture of the duality of *psychê*: soul—butterfly, but far from conforming to the metaphor, he tirelessly studied the book of nature. In his *Historia Animalium*, Aristotle remarked about the metamorphosis of butterflies and other insects in a very precise way. The writings do not make it clear whether he distinguished the continuity between egg and larva since he considered the pupa as the initial egg from which the "perfect" insect emerges (Reynolds, 2019). Nevertheless, Aristotle perfectly described the molts, the detachment of the exuvia, that is, the metamorphosis in all its more significant details.

The correspondence of the larva, pupa, and butterfly stages, however, was clear even among laymen, at least for some species, as shown by various graphical documents of the time, such as a gem engraved in the Hellenic region in the first century (Fig. 1.1). The Hellenic jewel probably represents a species of silk-producing lepidopteran that would have been well known in areas of Greece where silk was exploited and manufactured, like in Kos island, for example. In his *Historia Animalium*, Aristotle described a large larva with prominent horns that, in 6 months, transform into a pupa inside a cocoon. He also reported that Greek women untangled the cocoon and made fabrics with the thread, using the procedure invented by Pamphila, a woman from Kos island, the daughter of Plateus (Aristotle, 1991). Of course, Aristotle is not referring to the famous silkworm, *Bombyx mori*, since the introduction of this species into Europe from the East occurred shortly before the reign of Justinian, about 550 AD. Aristotle probably referred to the species *Pachypasa otus*, a large moth, whose larva presents a kind of horns and produces a cocoon with silk of mediocre quality, which is presently distributed across Southeast Europe, including Greece.

Aristotle not only described only the metamorphosis of moths and butterflies but also dealt with mosquitoes, "which come from small worms that live in the bottom of wells and ponds and, after a few days, rise to the surface

FIGURE 1.1 Gem engraved in the Hellenic region in the first century showing the larva, pupa, and adult of a lepidopteran species. From Davies and Kathirithamby (1986), with permission.

of the water, become immobile, harden to form a carcass, of which emerges the mosquito that still stays immobile at the beginning, until the sun and the wind make him move." Similarly, Aristotle described the metamorphosis of flies, cicadas, wasps, and mayflies and always carried out in precise terms, often specifying the time span of each stage of development and interspersing details that cannot be but the result of direct observation. The report on the metamorphosis of the cicada, for example, suggests that the information is firsthand. Aristotle said that cicadas live in places where trees do not give much shade (e.g., in olive fields), which lay eggs on fallow land, and that juveniles live underground but appear in large numbers after the first rains. He added that the larva is smooth until the skin is broken due to the transformation, and that by the time of the summer solstice, adult cicadas emerge at night, darken their color, harden, and (the males) begin to sing (Aristotle, 1991).

Aristotle observed the metamorphosis of mayflies in the Hypanis River, in the Cimmerian Bosporus region. He reported that in the summer solstice, there appears a kind of rigid object from which emerged winged and four-legged (sic) creatures that fly until twilight, when they die (Aristotle, 1991). If we ignore the misconception of the four legs, Aristotle was certainly referring to the species *Palingenia longicauda*, which is still very common in southern Russia, where the river Hypanis (now Kuban River) is located. Today, the description of the behavior of mayflies, like so many other naturalistic descriptions of Aristotle, seems superficial and impregnated with certain ingenuity. But we must not forget that we are talking about observations made in 350 BCE. The giant step taken by Aristotle toward the direct observation of nature will forever be an unavoidable reference.

FROM CLASSICAL GREECE UNTIL THE RENAISSANCE

The detailed observations and high level of knowledge achieved by Aristotle succumbed to oblivion in the following centuries. The writings of the authors of classical Rome who dealt with natural history, such as Pliny the Elder and Elian, were often secondhand. What most resembles the description of insect metamorphosis is Pliny the Elder's dissertation on the life cycle of bees. The author of *Naturalis Historia* writes that "the king (sic) of bees mates with common bees, and the resulting eggs are incubated by the bees in the same way as the hens do, and a small worm emerges from the egg, although the king is born directly with wings." About the mealybugs that were used in Hispania to obtain dyes, Pliny explained us an imaginary metamorphosis in reverse, that is, the adult animal is transformed into a small worm, which ends up being an egg. The nonsense is anthological, but the situation will not become much better in medieval times.

In the West, the medieval period did not contribute anything relevant to the knowledge of insect metamorphosis. The Bestiary, the famous medieval book that explains the qualities of various animals, inferring moral lessons from them in the context of the Christian imaginary, contains much more fantasy than reality. When reading the Bestiary, one immediately has the impression that the description of animal qualities has been written to suit the moral lesson that comes afterward. Whether what is said on the animal qualities is real or not is irrelevant. Insects appear little and are almost exclusively represented by social insects, bees, and ants, with which the Bestiary praises how perfect their organization is and establishes parallelisms with the ordered life that must pursue a good Christian (Belles, 2004).

As in different subjects of science and culture, the knowledge of insect metamorphosis recovers with the Renaissance. The *Historia Animalium* of Aristotle is rediscovered and first printed in Venice in 1495, and the new encyclopedias of animals that begin to appear are based more on Aristotle than on the medieval Bestiary. Those of Edward Wotton, Conrad Gesner, and Ulises Aldrovandi are the most famous, the latter with a whole volume dedicated specifically to insects entitled *De animalibus insectis*, although Aldrovandi does not deal only with insects but includes practically any type of invertebrate known at the time, from other arthropods (scorpions, spiders, centipedes) to worms, slugs, tapeworms, and even seahorses (Aldrovandi, 1602). The book still retains the extensive compilation style that was so popular in the Renaissance. For example, for each type of insect, Aldrovandi collected all the available information, not only biological but also related to history, symbols, numismatics, proverbs, mysticism, medicinal uses, etc. However, especially when dealing with lepidopterans, he described the different stages of the life cycle of several species and separately illustrated the adult, the caterpillar, and the pupa using woodcuts. These woodcuts (108 adults, 43 caterpillars, and 6 pupae are represented), although rather crude, show the morphological diversity in each stage. In that of the pupae, for example, the author separated the two types, which are presently known as adecticous exarate and obtect (see Chapter 5: The holometabolan development) (Fig. 1.2).

It is also worth mentioning the book *Insectorum Theatrum*, published in London in 1634 but written successively by Conrad Gesner, Thomas Penny, and Thomas Moufet (the three died in 1565, 1588, and 1607, respectively, without being able to finish the manuscript) (Moufet, 1634). The organization and style of the book denote that it was written by different authors, which gives the whole a rather heterogeneous character, from the erudite passages of Gesner to the most vivid descriptions of Moufet, through the lyrical pages of Penny. But the basic ordering of concepts is as correct as a work of those characteristics can be when written in the middle of the 16th century. Faithful to Pliny the Elder, some nonsense is still mentioned, such as the assertion that bees are ruled by a king, but it is not uncommon to recognize the identity of the species, especially when dealing with moths and butterflies, which are depicted, like in the book of Aldrovandi, with hundreds

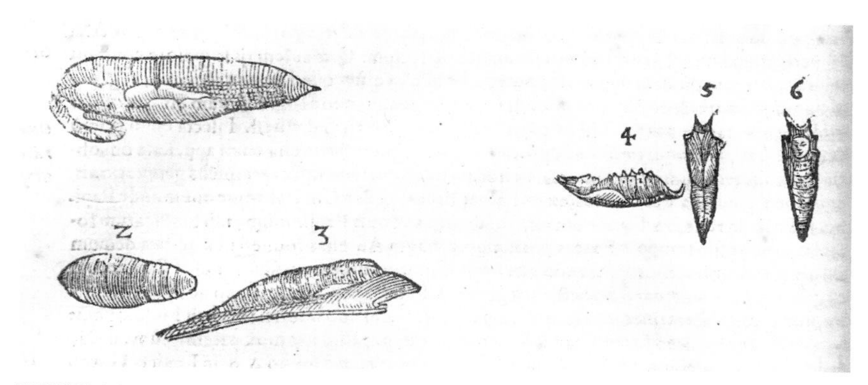

FIGURE 1.2 Types of lepidopteran pupae drawn by Ulisses Aldrovandi at the end of the 16th century. Aldrovandi recognized different pupae presently known as adecticous obtect. See, however, the naive representation of the human-faced head of the pupae on the right. From Aldrovandi (1602), reproduced from the author's copy of the book.

of woodcuts. In a significant number of cases, the adult, the caterpillar, and the pupa of the same species are represented side by side (Fig. 1.3). However, larvae and pupae orphans of the corresponding adult also appear quite often and are placed in the most peregrine locations or in an appendix at the end of the book, where, like in the book of Aldrovandi, a gallery of noninsect invertebrates is also included.

A notable person in the 16th and 17th centuries is William Harvey. His most famous work is *Exercitationes anatomicae, motu cordis et sanguinis circulatione*, published in 1628, which describes the circulation of blood driven by the heart. However, Harvey (1651) published another important book, *Exercitationes de generatione animalium*, in which he studied the embryonic development of some 50 species of animals, including a number of insects. This book is an interesting reference in the history of the theories about the origin of insect metamorphosis. On the one hand, Harvey denies Aristotle by stating that vermiform animals do not arise spontaneously from putrefaction and maintains that worms have an oviparous origin. Moreover, Harvey contemplates the worms as imperfect eggs, crawling eggs that are still developing until arriving at the stage of pupa, which he considers a sort of definitive egg, from which the adult animal arises. This theory of the imperfect egg, although much modified according to the progress of the knowledge, has survived until the present time (Erezyilmaz, 2006), as we will see later.

THE 17TH CENTURY

The first work specifically dedicated to insect metamorphosis was written by a painter who was at the same time a great observer. Jan Goedart was born

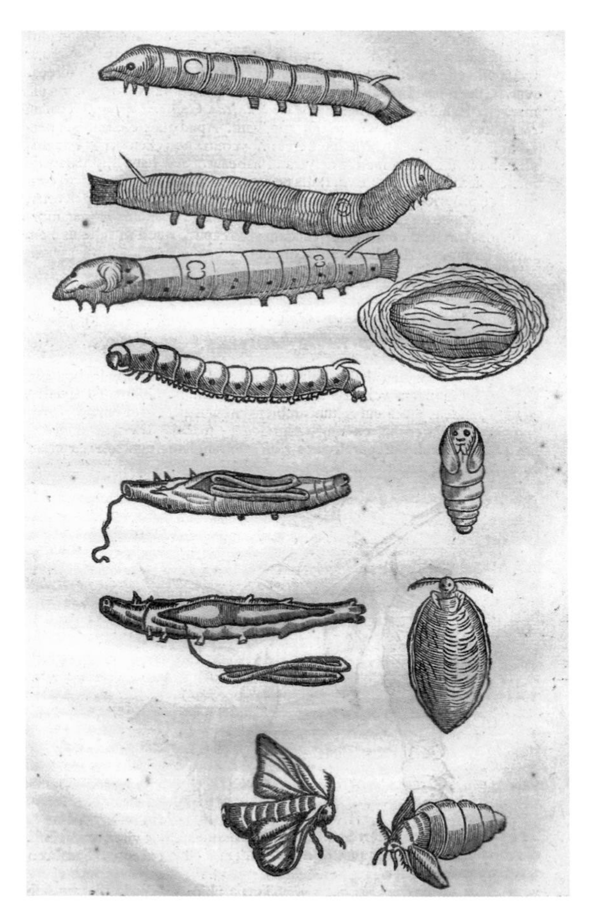

FIGURE 1.3 Life cycle of the silkworm, *Bombyx mori*, as appear in the book *Insectorum Theatrum*, published in 1634. The drawings were borrowed from Aldrovandi (1602). Note the naive human-faced head of the pupae and the emerging adult. From Moufet (1634), reproduced from the author's copy of the book.

and died in Middleburg and dedicated his life to painting insects, especially butterflies. He did so in a way so faithful that the species that served as models can be identified today without any difficulty. We should indeed refer to lepidopterans and not to butterflies, since Goedart was not content with just painting the adult, but represented all stages of the life cycle. If he could not observe the entire life cycle in nature, he bred the species in captivity and painted the successive stages as they were produced. All these observations, accompanied by Goedart's beautiful watercolor illustrations, were published in three volumes under the title *Metamorphosis Naturalis*, which started to appear in Middleburg in 1662 (Goedart, 1662–1667). In these volumes, and under the epigraphs entitled "Experiments," "History," or "Transformation," Goedart reported their observations, with the original dates of each one and with a simple language that breathes descriptive freshness. He was the first to use the word "metamorphosis" with the entomological meaning it has today, most probably inspired by the reading of Ovid. A phenomenon that caused Goedart a great surprise and confusion was to observe small wasps emerging from the body of larvae or pupae of various lepidopteran species (Fig. 1.4), concluding that it seemed unnatural that more than one species could develop from the same animal. He did not understand that this was a case of parasitism, which today we would consider trivial. Understanding this phenomenon would wait until the extremely careful dissections and observations that Jan Swammerdam performed about 10 years later.

From Goedart, the path widens considerably as it enters the 17th century, in which the most outstanding person is the Dutchman Jan Swammerdam. In 1669, he published the memoir *Historia Insectorum Generalis*, with numerous observations on the cycle and the transformations of several insects and in 1675, he published a monograph on mayflies. Swammerdam's life, however, was unfortunate as he died at the age of 43 practically in misery, leaving most of his work unpublished. He had studied the life and external morphology and internal anatomy of bees, mosquitoes, mayflies, ants, dragonflies, butterflies, moths, beetles, gall wasps, hermit crabs, water fleas, terrestrial and aquatic mollusks, and amphibians. In all his studies, the precision of the observations and the accuracy of his drawings are extraordinary. At his death, the abundant materials that Swammerdam left

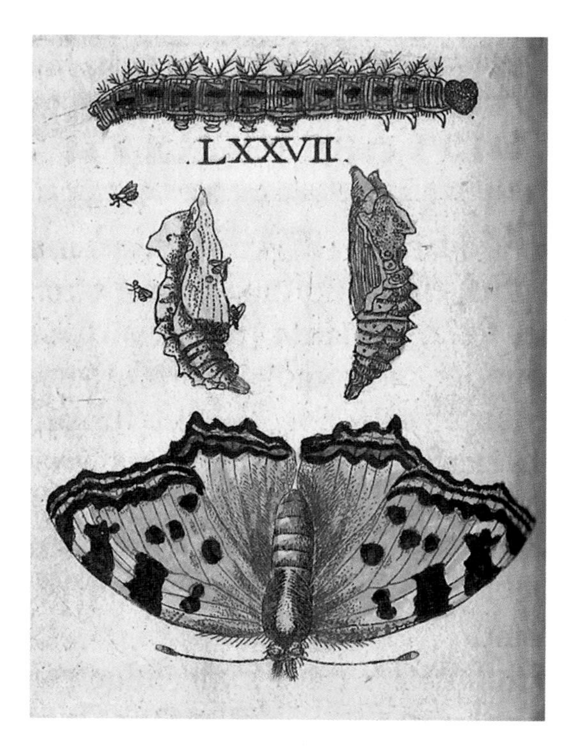

FIGURE 1.4 Caterpillar, pupae and adult of the large tortoiseshell (*Nymphalis polychloros*) as appear in the first volume of *Metamorphosis naturalis*, published by Goedart in 1662. Note that the pupa on the left has been parasitized by a wasp. From Goedart (1662), reproduced from the author's copy of the book.

unpublished passed from hand to hand until reaching the Dutch physician and anatomist Hermann Boerhaave, who published them in a work with the suggestive title *Bybel der natuure*, in Leiden in 1737 (Swammerdam, 1737). Among Swammerdam's prodigious observations stand out the details about the life of bees and ants and the studies on the process of metamorphosis, in which he compared insects with amphibians. In insects, he recognized four fundamental types of transformations. The first is represented by species that grow without changes (using the lice as an example); a second includes the species that gradually develop the wings and pass to adult directly, without any intermediate quiescent stage (such as cockroaches and crickets); a third type accommodates species whose wings develop under the larval cuticle and which pass through a quiescent pupal stage before transforming into an adult (such as butterflies, beetles, or ants) (Fig. 1.5); and finally, a fourth type includes those that undertake the pupal stage under the skin of the last larval instar (represented by flies). Swammerdam relentlessly fought Harvey's theory of the imperfect egg and demonstrated with careful dissections the continuity of the insect's life cycle, especially for those that include a pupal stage. The fact that the types of transformations recognized by Swammerdam have remained virtually unchanged to the present day demonstrates the quality of their studies.

THE ENLIGHTENMENT

In the transition between the 17th and 18th centuries, Maria Sibylla Merian, the daughter of the famous engraver and publisher Matthaeus Merian, shines with her own light. Passionate about the study of insects, this tenacious lady put her life to the service of that passion. Maria Sibylla was born in Frankfurt am Main in 1647. Although her father died when she was only 3 years old, the family environment soon led her to draw, paint, and engrave plants and animals. Her first collection of drawings of insect caterpillars originally painted on parchment was printed in Nuremberg in two parts, one in 1669 and the other in 1683, under the title *Der Raupen wunderbare Verwandlung*. In 1684, she moved to the Netherlands, and in Amsterdam, she visited the cabinet of Nicolaas Witsen, who besides being the burgomaster of the city was the president of the Company of the West Indies. Maria Sibylla was impressed by the size and beauty of the insects from Suriname that Witsen had in his collection. The impression must have been very strong, as she subsequently made the decision to go to Suriname to draw the insects in their natural environment. Thus in June 1699 and accompanied by her eldest daughter, Johanna Helena, Maria Sibylla Merian embarked for Suriname, where she would remain for 2 years, until June 1701. From her stay, she brought back a collection of splendid drawings of insects painted on parchment, 60 of which were published in the famous work *Metamorphosis Insectorum Surinamensium*, published in Amsterdam

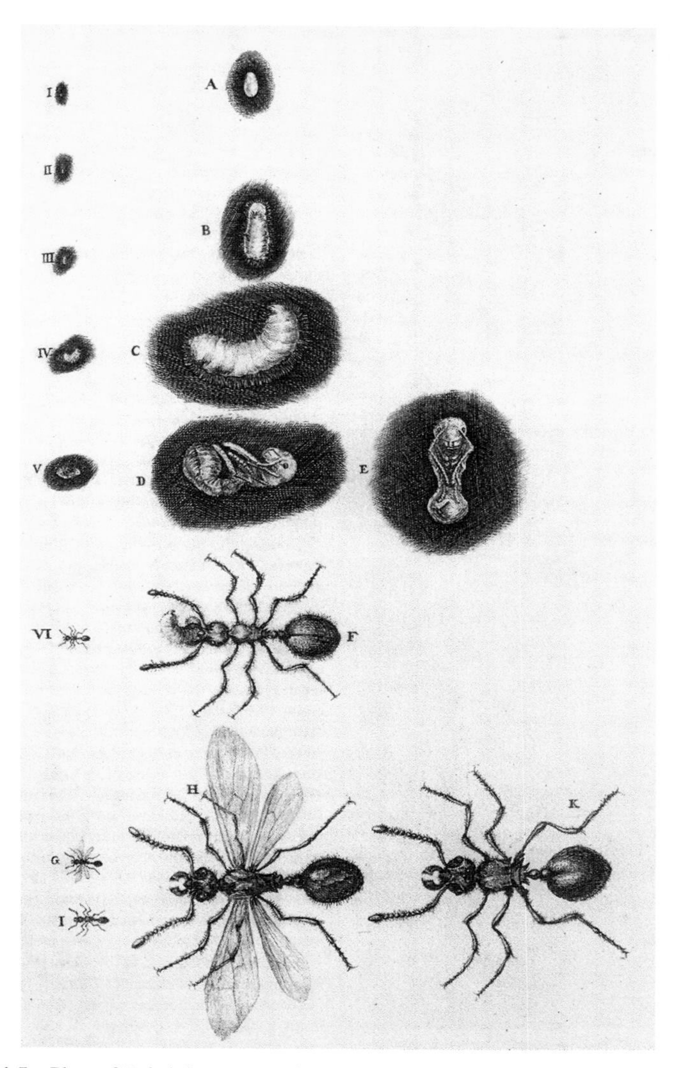

FIGURE 1.5 Plate of *Bybel der natuure* showing the ant's life cycle and metamorphosis. From Swammerdam (1737), reproduced from the author's copy of the book.

(Merian, 1705). Most of the drawings represent lepidopteran species, including all stages of the life cycle and the feeding plant. Each drawing, with its corresponding explanation, is an elegant lesson of entomology and ecology based on beautifully realistic images, which contrast with the static "portraits" of insects drawn by the entomologists of the time. For example, in her drawing of the lepidopteran *Arsenura armida* perched on a branch of the "palisade tree" (*Erythrina fusca*) (Fig. 1.6), Maria Sibylla Merian explained

FIGURE 1.6 Plate of *Metamorphosis Insectorum Surinamensium* showing the different life cycle stages of the Giant Silk Moth (*Arsenura armida*) perched on a branch of the "palisade tree" (*Erythrina fusca*). From Merian (1705), reproduced from a copy of the book belonging to the Smithsonian library (http://www.biodiversitylibrary.org/page/41398732#page/42/mode/1up).

that "Each year this kind of caterpillar comes three times to this tree; it is yellow with black stripes and decorated with six black spines. When they have reached a third of their final size, they shed their skin and changes the color to orange-yellow with black spots on their limbs... Several days later they shed the skin once more; on April 1700 they transformed into chrysalides; on 12 June moths like those showed in the drawing emerged. The one at the bottom, smaller, is the male, the larger at the top is the female." With this succinct text and the magnificent drawings, it is not necessary to add anything else.

The 18th century witnessed an unparalleled advance in the observation, inventory, and ordering of natural objects. Natural history cabinets proliferated at the hands of outstanding naturalists, such as Buffon or Linnaeus. Knowledge was systematized, and information was ordered. Linnaeus' effort to classify plants and animals in a rational way is a clear example of this trend. The success of some of the proposed solutions, like the *Systema Naturae* of Linnaeus, was so great that they are still in use today. To Linnaeus, moreover, we owe the introduction of the terms larva, pupa, and imago, widely used afterward to describe the life cycle and metamorphosis of insects.

It is also worth mentioning the achievements of René-Antoine Ferchault de Réaumur, partially compiled in his *Mémoires pour servir à l'histoire des insectes*, published in six large volumes (Réaumur, 1734–1742), in which he described the life cycle, metamorphosis, and behavior of numerous insect species. Bees are treated in great detail, as observations were based on glass hives made by Réumur, which allowed him to observe what was happening in the interior. Practically half of the second volume of his *Mémoires* deals with the transformation of caterpillars into pupae and then into adults, in several species of moths and butterflies. His experiments on the influence of temperature on the speed of changes are remarkable, as well as the descriptions of these changes. In the case of the metamorphosis of flies, among many other observations, Réaumur describes the ptilinum, an inward fold of the cuticle in the frontal part of the head that is able to be projected outward or retracted. The ptilinum enables the fly to emerge from the puparium and escape the substrate, often hard, where it is buried.

The school founded by Réaumur was prodigal in entomologists who would become famous. One of the most outstanding was Pierre Lyonet, a contemporary of Réaumur, a man of law who was seduced by the world of insects, to which he dedicated a large part of his life. In his splendid monograph on the goat moth, *Cossus cossus* (Lyonet, 1762), the internal anatomy of the larva is represented with drawings and engravings made by Lyonet himself, with an extraordinary degree of detail. To Lyonet we owe, for example, the discovery of the imaginal discs, which are crucial structures in holometabolan metamorphosis (see Chapter 5: The holometabolan development), and the corpora allata, tiny glands that produce the juvenile hormone, which we will deal with in Chapter 6 (Hormones involved in the regulation of metamorphosis, as a crucial metamorphosis regulator).

DARWIN'S TIMES

Swammerdam's conclusions on the different modes of postembryonic development are consolidated into a classification of three main types, which are now called ametabolan, hemimetabolan, and holometabolan. Although using different names for the last two (metamorphosis incomplete or gradual for the hemimetabolan mode or complete or perfect for the holometabolan), the first manuals of modern entomology popularized the classification of metamorphosis into these three types. William Kirby and William Spence's *An Introduction to Entomology*, published in 1815, was tremendously popular worldwide. Among Anglo-Saxon readers, the *Manual of Entomology*, published by Hermann Burmeister in 1836, or those by John O. Westwood—*The Entomologist's Text Book*, of 1838, and *An Introduction to the Modern Classification of Insects*, of 1839—were also well known. For French readers, the most popular was *Les métamorphoses des insectes*, published by Maurice Girard in 1866.

From the robust descriptive bases established by Swammerdam and Réaumur, the processes underlying the metamorphic changes began to be studied in some detail in the 19th century. In 1813, Cesare Majoli artificially triggered the formation of intermediates between larva, pupa, and imago in the silkworm, *B. mori*, by submitting the insects to high temperatures, although he was unable to explain the underlying mechanisms. In 1823, Auguste Odier treated the elytra of the common cockchafer, *Melolontha melolontha*, with a solution of potassium hydroxide and isolated an insoluble residue. He established its protein nature and gave it the name of chitin, from the Greek *khitōn*, meaning tunic or covering. Odier identified chitin from the demineralized shell of crabs and suggested that it could be the base material of the exoskeleton of all arthropods. In 1857, Ernst Haeckel was one of the first to recognize that the cuticle of arthropods is a product generated by the epidermal cell layer and defined them as chitinogenic cells. Réaumur had already observed the role of the ptilinum in the emergence of flies from the puparium, but in 1864 August Weismann showed that the ptilinum was not projected outward by the injection of air, but by the pressure of the hemolymph due to the action of the thoracic and abdominal muscles. This was also demonstrated by Philippe Alexandre Jules Kunkel d'Herculais in 1890, using the locust *Locusta migratoria* as a model insect. In the 19th century, all entomologists recognized that for an animal with a rigid exoskeleton of chitin, the molts were necessary to grow, but in 1898 Joseph Pantel pointed out that molts were also an essential process for postembryonic development to produce new morphologies and new cuticular structures. As stated above, in the 18th century, Pierre Lyonet had discovered the imaginal discs in the moth *C. cossus*, but it was Weismann who unveiled their function and gave them that name in 1864. The number and location of imaginal discs led Weismann to suspect that they could be a sort of primordial wing, and he observed that during metamorphosis of flesh flies they grew rapidly, while most of the tissues disintegrated.

At the end of the 19th century, John Lubbock, a scholar in many fields of knowledge, became interested in entomology, especially in social insects such as ants and bees, and was deeply involved in the study of metamorphosis. A close friend of Charles Darwin and a convinced follower of his theories of evolution by natural selection, Lubbock opposed the generalized anti-Darwinist thoughts of the time and attempted to find a scientific explanation for the metamorphosis of insects. This is explicitly stated in his book *On the Origin and Metamorphoses of Insects* (Lubbock, 1873), when he wonders: "Why do insects pass through metamorphoses? Messrs Kirby and Spence tell us they can only answer that such is the will of the Creator; this, however, is a general confession of faith, not an explanation of metamorphoses." Lubbock described the life cycle of different species with gradual metamorphosis (hemimetabolan) and with complete metamorphosis (holometabolan), considering that the distinct larval types are associated with different types of life styles, which would imply that natural selection operates in juvenile stages. Regarding the evolution of metamorphosis, although without mentioning Harvey, Lubbock adhered to his imperfect egg theory, assuming that "insects leave the egg at different stages of embryonal development." It seems that the theory of "de-embryonization" was popular in the 19th century, as shown by the various quotations of authors of the time who adhered to it. This is the case, for example, for Armand de Quatrefages who in his book *Métamorphoses de l'homme et des animaux*, published in Paris in 1862, resolutely stated that "the larva is only an embryo with independent life." In his monograph, Lubbock continued his arguments comparing the embryonic development of various species and finished with the conclusion that "the metamorphoses of insects depend then primarily on the fact that the young quit the egg at a more or less early stage of development" (Lubbock, 1873).

THE 20TH CENTURY

At the beginning of the 20th century, the French entomologist Charles Pérez considered Lubbock's hypothesis to be an extravagance. On the basis of the morphological and histological changes that occur between the last nymphal instar and the adult in hemimetabolan species and between the last larval instar, the pupa, and the adult in holometabolans, Pérez argued that the pupa would be equivalent to the last nymphal instar (Pérez, 1902, 1910). Nevertheless, taking into account the information on insect embryogenesis of holometabolan and hemimetabolan species, the Italian Antonio Berlese reworked the de-embryonization theory, proposing that the larva of holometabolans hatched in an earlier embryogenic stage to that of the hemimetabolans (Fig. 1.7) (Berlese, 1913). Berlese's conceptualizations received the support of Augustus Daniel Imms, who in his influential book *Recent advances in entomology* (Imms, 1931) diffused these proposals widely. According to Berlese and Imms, the embryonic development takes place through three successive stages: protopod, polypod, and oligopod. The hemimetabolan embryo would pass through the three stages, thus hatching in a postoligopod stage, that is, the nymph. In contrast, the holometabolan embryo would hatch in the protopod or oligopod stage and would resume the development in the pupal stage. Coleopteran and neuropteran larvae would

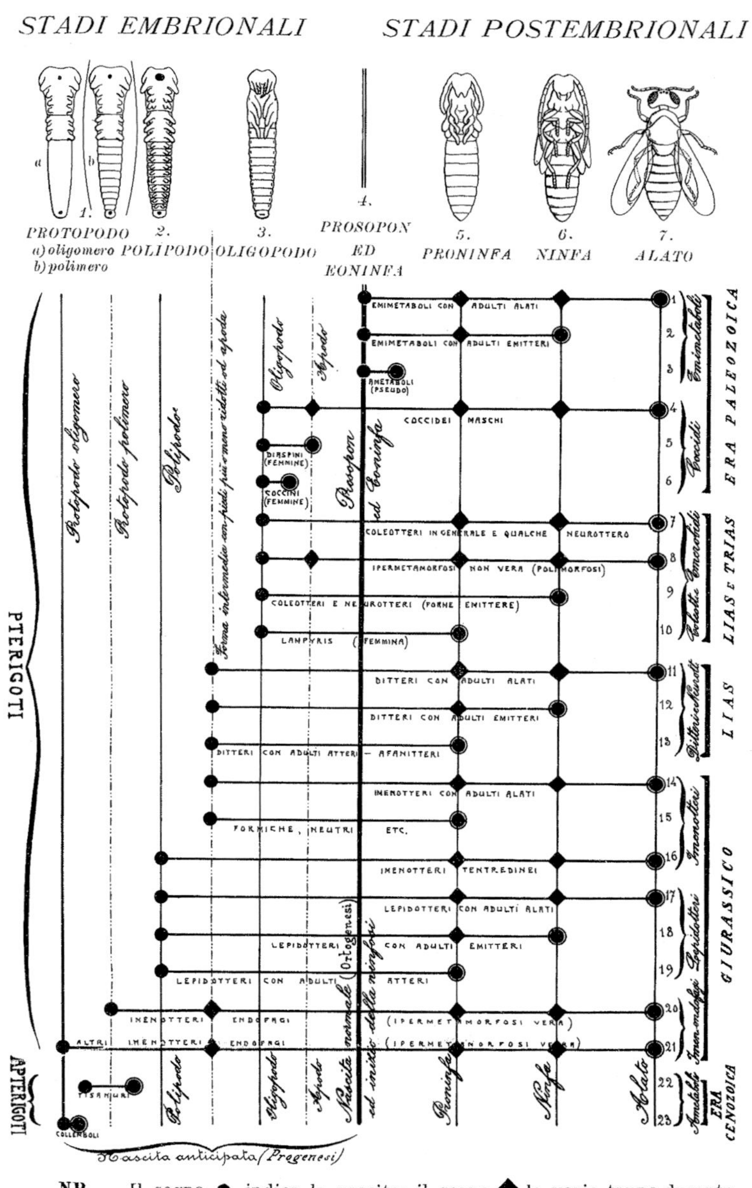

NB. — Il segno ● indica la nascita; il segno ◆ le varie tappe durante lo sviluppo postembrionale; il segno ◉ indica la forma matura sessualmente.

FIGURE 1.7 Plate of the article *Intorno alle metamorfosi degli insetti* that summarizes the Berlese theory about the evolution of metamorphosis by premature hatching of the embryo. From Berlese (1913).

be examples of insects that hatch in the oligopod stage. Dipteran larvae were considered to be oligopods that had secondarily lost their appendages. As an example of a protopod larva, Imms suggested the endoparasitic larvae of certain hymenoptera, which have a very simple morphology, with a differentiated head and thorax but with a rudimentary abdomen. Today we know that the simplicity of these larvae is due to an adaptive process of secondary simplification, associated with the parasitic way of life.

The theory of Berlese–Imms was followed by other ideas that agreed or disagreed with it. For example, Poyarkoff opposed this theory on the basis of his studies on the musculature changes during the transition to pupa and adult. He proposed that the pupal stage arises from the unfolding of the adult stage in two phases (Poyarkoff, 1914). At the same time, Ježikov (1929) followed Berlese, contending that the larva is a free-living continuation of the embryo, and that the nymphal stages of its ancestors condensed into the pupal stage. In the 1940s, Henson conceived the curious theory of considering that the pupal stage is a true repetition of embryonic development rather than its continuation, which reminds one of Harvey's ideas of the 17th century. Henson proposed his theory after studying the process of formation of the digestive tract during metamorphosis, which presents some aspects reminiscent of the same process in embryogenesis (Henson, 1946). About 10 years later, Heslop-Harrison presented a view similar to that of Berlese, but considering that the ancestral insect that gave rise to holometabolan and hemimetabolan groups hatched as a polypodan larva, which, after a number of molts in that stage, transformed into an oligopod larva. From that ancestor, the hemimetabolan insects would originate by the transposition of the larval instars to the process of embryogenesis, whereas the holometabolans would arise by the compression of all the nymphal instars into a single pupal stage (Heslop-Harrison, 1958). In this period, the British entomologist Howard E. Hinton, who initially favored Poyarkoff's theory (Hinton, 1948), studied in depth the changes of the musculature in the pupa and in the adult and came to the conclusion that the juvenile stages of the holometabolan species were equivalent to those of the hemimetabolans, including the pupal stage, which would be homologous to the last nymphal instar (Hinton, 1963). In the 1990s, Berlese's de-embryonization theory was reformulated in a modern endocrine context (Truman and Riddiford, 1999) into the so-called pronymph theory, championed by James Truman and Lynn Riddiford. Nowadays, the pronymph theory competes with the theory of the homologous juvenile stages, formalized by Hinton (1948), a controversy that will be discussed in Chapter 12 (The evolution of metamorphosis).

Concerning the regulatory aspects of insect metamorphosis, numerous experiments of castration and gonadal transplantation performed in the 19th and early 20th centuries always yielded negative results, which seemed to discard any hypothesis of hormonal regulation, at least regarding the sex organs. Moreover, the administration of vertebrate hormones to insects did not produce any positive results either. All this evidence led to the idea that insects did not have hormones, which discouraged any research of the possible endocrine regulation of metamorphosis. The paradigm of the absence of hormones in insects was denied by the Polish entomologist Stephan Kopeć, who, at the beginning of the 20th century, demonstrated the existence of brain hormones in the gypsy moth, *Lymantria dispar* (Kopeć, 1922). With ligature and transplantation experiments, Kopeć showed that insects do have hormones, thus founding the field of endocrinology of metamorphosis, a field that was brilliantly cultivated shortly after by the British researcher Vincent B. Wigglesworth, as we shall see in Chapter 6 (Hormones involved in the regulation of metamorphosis).

REFERENCES

Aldrovandi U., 1602. De animalibus insectis libri septem, cum singulorum iconibus ad viuum expressis. Ioan. Bapt. Bellagambam, Bononiae.

Aristotle, 1991. History of Animals. Harvard University Press, Cambridge, Three volumes.

Belles, X., 2004. Els bestiaris medievals: llibres d'animals i de símbols, Dalmau Editors, Barcelona. Berlese, A., 1913. Intorno alle metamorfosi degli insetti. Redia 9, 121–136.

Davies, M., Kathirithamby, J., 1986. Greek Insects. Oxford University Press.

Erezyilmaz, D.F., 2006. Imperfect eggs and oviform nymphs: a history of ideas about the origins of insect metamorphosis. *Integr. Comp. Biol.* ***46***. pp. 795–807. Available from: https://doi.org/10.1093/icb/icl033.

Goedart, J., 1662. Metamorphosis Naturalis. Jaques Fierens, Boeck-verkooper, inde Globe, Middelburg, Three volumes.

Harvey, W., 1651. Exercitationes de Generatione Animalium, Typis Du-Gardianis, Impensis Octav. Pulleyn, London.

Henson, H., 1946. The theoretical aspect of insect metamorphosis. Biol. Rev. 21, 1–14. Available from: https://doi.org/10.1111/j.1469-185X.1946.tb00449.x.

Heslop-Harrison, G., 1958. On the origin and function of the pupal stadia in Holometabolous Insecta. Proc. Univ. Durham Phil. Soc. 13, 57–79.

Hinton, H.E., 1948. On the origin and function of the pupal stage. Trans. R. Entomol. Soc. Lond. 99, 395–409.

Hinton, H.E., 1963. The origin and function of the pupal stage. Proc. R. Entomol. Soc. Lond. 38, 77–85.

Imms, A.D., 1931. Recent Advances in Entomology. J. & A. Churchill, London.

Ježikov, J., 1929. Zur Frage über die Entstehung der vollkommenen Verwandlung. Zool. Jahrb. Anat. 50, 601–650.

Kopeć, S., 1922. Studies on the necessity of the brain for the inception of insect metamorphosis. Biol. Bull. 42, 223–242.

Lubbock, J., 1873. On the Origin and Metamorphoses of Insects. Macmillan, London.

Lyonet, P., 1762. Traité anatomique de la chenille, qui ronge le bois de saule, Pierre Gosse Jr., Daniel Pinet, Marc Michel Rey, The Hague and Amsterdam.

Merian, M.S., 1705. Metamorphosis of de Verandering der Surinaamsche insecten, Koninklijke Bibliotheek. Voor den Auteur, Amsterdam.

Moufet, T., 1634. Insectorum sive minimorum animalium theatrum, Ex Officina Thom. Cotes, Londini.

Pérez, C., 1902. Contribution à l'étude des métamorphoses. Imprimerie L. Danel, Lille.
Pérez, C., 1910. Signification phylétique de la nymphe chez les insectes métaboles. Bull. Sci. Fr. Belg. 44, 221–233.
Poyarkoff, E., 1914. Essai d'une théorie de la nymphe des insectes holométaboles. Arch. Zool. Exp. Gén. 54, 221–265.
Réaumur, R.-A.F.D., 1734. Memoires pour servir a l'histoire des Insectes. Imprimerie Royal, Paris, Six volumes.
Reynolds, S., 2019. Cooking up the perfect insect: Aristotle's transformational idea about the complete metamorphosis of insects. Philos. Trans. R. Soc. B: Biol. Sci 374, 20190074. Available from: https://doi.org/10.1098/rstb.2019.0074.
Swammerdam, J., 1737. Bybel der natuure door Jan Swammerdam, Amsteldammer, of Historie der insecten, Isaak Severinus, Boudewyn Van der Aa, Pieter Van der Aa, Leyden.
Truman, J.W., Riddiford, L.M., 1999. The origins of insect metamorphosis. Nature 401, 447–452. Available from: https://doi.org/10.1038/46737.

Chapter 2

A spectacular diversity of forms and developmental modes

Insects, with about 1 million species formally described, are the most diverse animal group on earth, collectively comprising over half of all extant species. Insects inhabit virtually all land places, and they tend to dominate the small fauna in the ecosystems that insects occupy. Insects are surprisingly well represented in arid deserts and can be found in areas relatively close to the two poles: in the Antarctica, a few species of flies, springtails, and parasitic lice have been recorded, and in the Arctic region, more than 300 species of insects, especially flies, are known from the Canadian islands north of the 75th parallel. Insects have been found in hot springs, salt lakes, deep caves, and even pools of petroleum. A modest percentage (c.3%) are adapted to live in freshwater, and there are five species of water striders belonging to the genus *Halobates* that permanently live in the open ocean. In other words, insects occupy all major terrestrial and freshwater habitats, and only special insects have colonized the marine milieu. Moreover, insects can exploit almost every organic resource, from dead plants and animal matters, to all parts of green plants. They also can feed on other kinds of animals as predators or parasitoids. These ecological specializations involve the corresponding adaptations, which led to a formidable diversity in terms of morphology, physiology, and life cycles. Besides all, the extraordinary success of insects is due in great part to their long evolutionary history, as they emerged early in the history of life. This gave them enough time to evolve the above adaptations, as well as a series of key innovations that acted as drivers of expansion and diversification (Grimaldi and Engel, 2005).

A LONG EVOLUTIONARY HISTORY

Despite some challenging opinions (Haug and Haug, 2017), the earliest fossil remains that can be attributed to an insect are a pair of mandibles preserved in Lower Devonian materials from Scotland. The species was described under the name *Rhyniognatha hirsti* (Engel and Grimaldi, 2004) and corresponds to an insect with double-jointed mandibles (dicondylic), a feature that emerged some 400 million years ago (Mya) (Fig. 2.1). However, the first

Insect Metamorphosis. DOI: https://doi.org/10.1016/B978-0-12-813020-9.00002-8

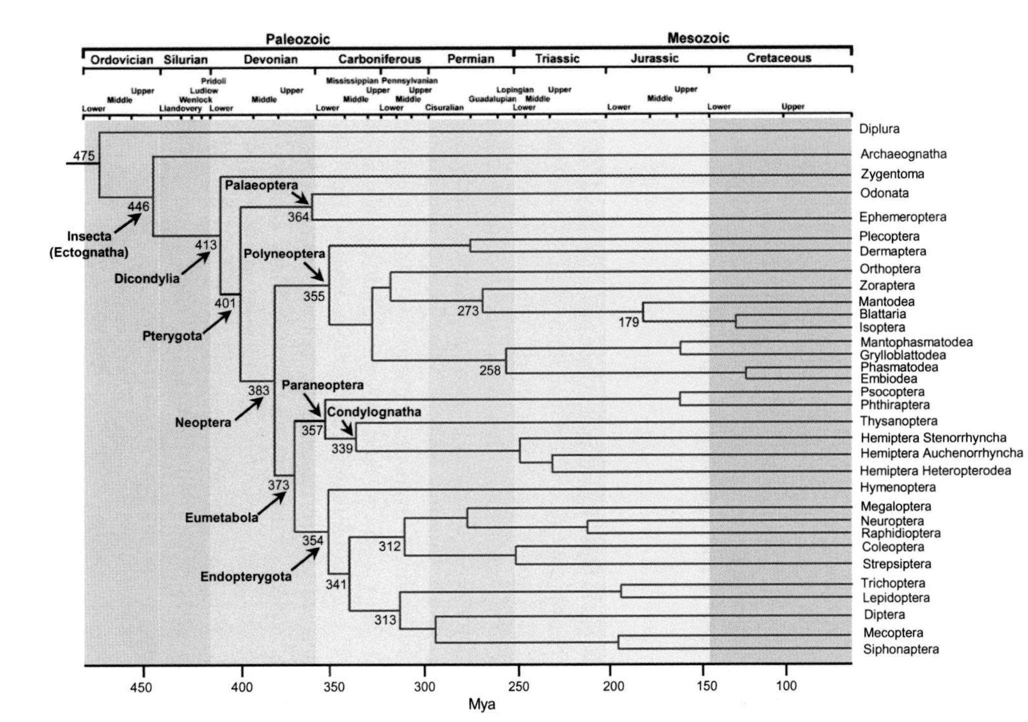

FIGURE 2.1 Cladogenesis of the main insect groups in a chronological context. The phylogenetical reconstruction is based on Misof et al. (2014) and Wang et al. (2016). The main discrepancy between these two proposals is the situation of Paraneoptera, monophyletic for Wang et al. and polyphyletic for Misof et al. Divergence times are generally similar in both proposals. Those indicated here are based on the average values reported by Wang et al. (2016).

insects appeared earlier, probably in the Ordovician, when a stock of entognathous (mouthparts enclosed in a cavity formed by the labium fused with pleural folds of the cephalic capsule) hexapods had already diversified into distinct groups, including Protura, Collembola, and Diplura. Divergence times inferred from phylogenetic studies (Wang et al., 2016) suggest that in the Lower Ordovician, approximately 475 Mya, the first ectognathous (mouthparts not enclosed in a cavity) and wingless true insects, today represented by the Archaeognatha and Zygentoma, diverged from entognathous hexapodan ancestors (Fig. 2.1).

Archeognatha and Zygentoma still share a number of characteristics with entognathous hexapods, such as the three thoracic segments well differentiated from each other, the presence of vestigial appendages at least on some distal abdominal segments, and spiracles without muscular closing devices. The amniotic cavity can be open, incompletely closed, or closed, depending on the species. Characteristically, molting continues throughout life, also after attaining reproductive capabilities, and fertilization is achieved by indirect transfer of the sperm.

Pterygota, which originated in the Lower Devonian, some 410 Mya (Misof et al., 2014; Wang et al., 2016) (Fig. 2.1), show mesothoracic and metathoracic wings as the most obvious feature. However, pterygotes are distinguished by other characteristics such as having the mesothorax and metathorax structurally and functionally coupled (forming what is called the pterothorax), the abdominal segments devoid of appendages, and the spiracles with muscular closing devices (except in the Ephemeroptera and Diptera, in the latter case due to secondary loss). The amniotic cavity of Pterygota is always completely closed, molting ceases after attaining the reproductively competent adult stage, and fertilization is achieved by direct transfer of sperm through copulation. The adult stage is reached through postembryonic development and metamorphosis.

Shortly after their emergence, Pterygota insects began to radiate. Palaeoptera and Neoptera diverged in the Lower Devonian, some 400 Mya, whereas Polyneoptera and Eumetabola diverged from neopteran ancestors around the Middle and Upper Devonian boundary, some 380 Mya. Finally, Paraneoptera and Endopterygota (=Holometabola) diverged from eumetabolan ancestors around the Upper Devonian, some 370 Mya. It is very likely that early radiations in each of the three superorders Polyneoptera, Paraneoptera, and Endopterygota occurred during the Lower Carboniferous, between 360 and 350 Mya (Misof et al., 2014; Wang et al., 2016) (Fig. 2.1).

NEW RESOURCES TO BE EXPLOITED, NEW ECOLOGICAL OPPORTUNITIES

Phyogenetic reconstructions suggest that hexapods first diversified around the Lower Ordovician, which is long before the first traces of terrestrial animal life are found in the fossil record. Thus, the early hexapodan lineages, basal to Entognatha + Insecta, were probably marine (Wang et al., 2016). Phylogenetic studies place land colonization by hexapods in the Ordovician (Lozano-Fernandez et al., 2016), and after terrestrialization, true insects diverged from basal entognathous hexapods and developed flight. This did not take long, given that the origin of the winged insects (Pterygota) is dated at about 413 Mya (Fig. 2.1), only slightly prior to the age of *R. hirsti*. This points to a terrestrial origin for Pterygota, with subsequent independent invasions of freshwater systems only by two palaeopteran orders (Ephemeroptera and Odonata) and by a few groups of Neoptera, particularly because freshwater was not abundant in the Lower Devonian (Wang et al., 2016).

Subsequently, waves of diversification of winged insects coincide with important emergences and changes in terrestrial floras during the Devonian, which can be divided into three main stages. First, the short, riparian rhyniopsid-dominated habitats in the Lower Devonian, then the period of arborescent plants evolution until the Middle Devonian, and the apparition of medium-sized to giant tree fern forests in the Upper Devonian. Finally, around the time of the Devonian and Carboniferous boundary, the first seed plants emerged and radiated (Field et al., 2012; Morris et al., 2018; Puttick et al., 2018) (Fig. 2.2).

Most probably, insects were largely benefited from the nutritious resources and the ecologically diverse structure afforded by these new plant lineages. Significantly, an important portion of Paraneoptera and Endopterygota diversification is explained in part by mouthpart specializations that follow the dramatic increase of diverse vegetal resources available. It is clear that the groundplans for many insect lineages, including early saprophagous, mycophagous, predacious, and omnivorous groups, reflect an increased diversity of diets. In the case of Paraneoptera, the diversification timescale is consistent with the origin and initial diversification of seed plants, which probably promoted the evolution of the mouthpart adaptations. A number of evolutionary changes can be distinguished in the mouthparts of Paraneoptera species, from the chewing structures of Psocoptera to the probing and puncturing devices of Condylognatha, then to the piercing–sucking rostrum, associated to a suppression of the mandibular and maxillary palps, in the Hemiptera (Farrell, 1998; Grimaldi and Engel, 2005; Wang et al., 2016).

In addition to new resources and ecological niches provided by plant evolution, spatial heterogeneity became significantly modified toward the latter half of the Devonian. With the increasing height and diversity of sciophilous plants, concealed spaces would have become more varied and numerous, which might have facilitated appropriate shelters for Polyneoptera and Paraneoptera species, whose preference for cryptic habitats is typical of these groups (Wang et al., 2016).

THE WINGS, A CRUCIAL INNOVATION

Winged insects emerged about 410 Mya (Fig. 2.1). Significantly, the innovation of powered flight coincides with a period approximately 408 Mya

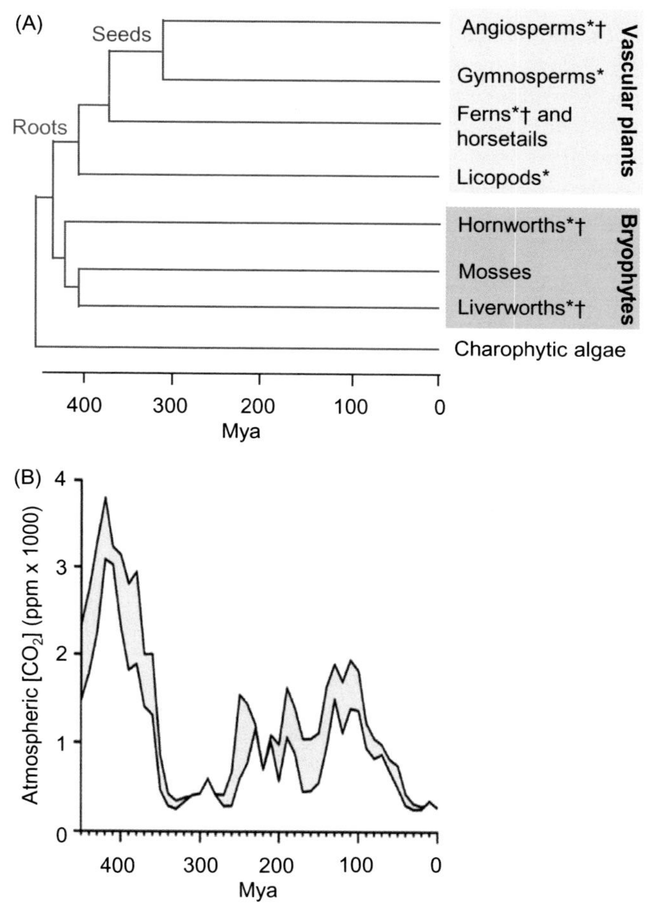

FIGURE 2.2 Coevolution of land plant life cycles and atmospheric CO_2 concentrations. (A) Land plant phylogeny. (B) Modeled global [CO_2] history for the past 500 million years. Mapped onto the phylogeny is the occurrence of arbuscular mycorrhizal (AM) fungal associations across plant taxa (*), and fossil evidence of AM fungi (†). Modified from Field et al. (2012), especially the topology of the node Mosses-Liverworths-Hornworths, which reflects the results of Morris et al. (2018) and Puttick et al. (2018).

characterized by a hyperoxic atmosphere (Fig. 2.2). Given that flight muscles demand oxygen at a fast rate, an oxygen concentration as high as approximately 24%–25% had to facilitate acquisition of flight (Ward, 2006). Although insect flight is one of the key innovations in the history of insects, which has been the object of continuous interest, the origin of insect wings is still controversial. Throughout the history of the ideas, two basic theories have been proposed. One theory postulates that wings originated from an extension of the thoracic tergum, first forming paranotal lobes and then fully articulated wings. The tergal origin theory was initially suggested by F. Müller in 1875, formalized by G. Crampton in 1916 and supported by R.E.S. Snodgrass in 1935 and by K.G.A. Hamilton in 1971. The theory of tergal origin has also been invoked in modern developmental studies based on the molecular approaches (see Clark-Hachtel and Tomoyasu, 2016).

The second theory considers that the wing derives from pleural structures and also has very ancient origins, dating back to the works of L. Oken of 1809–1811 and those of C.W. Woodworth of 1906. In modern times, it has been championed by Jarmila Kukalová-Peck, who proposed that the wing derives from articulated exites located on the proximal leg segments of ancestral insects, which migrated dorsally and finally formed the articulated and flying wing. An influential paper of M. Averof and S.M. Cohen, published in 1997, reported that insect wing-related genes are expressed in dorsal gills of crustacean branched limbs, providing support for the pleural origin theory. Toward 2015, morphological and molecular investigations of J.F. Coulcher and coworkers showed that the embryonic subcoxa contribute to form pleural sclerites of the adult and provided significant arguments in favor of the pleural origin theory (see Clark-Hachtel and Tomoyasu, 2016).

Until recently, the theory of tergal origin has proved the most popular, possibly because it appears coherent with the flat form and position of modern wings; however, it does not account for the complex musculature and articulations needed for flight functions. Conversely, the theory of pleural origin explains the origin of muscles and articulations as they are already contained in the leg branch limb. However, this hypothesis is counterintuitive when trying to imagine how a leg branch can transform into a flattened and flexible structure like the wing of a flying insect. A balanced alternative that came to solve this disparity has been to merge both theories, thus explaining the origin of wings according to a dual contribution of tergal and pleural structures (Fig. 2.3). Indeed this eclectic approach has a historical background as it was first suggested by G. Crampton in 1916, although he clearly favored the tergal origin theory. However, convincing data pointing to this dual origin have been produced more recently following molecular and functional genomics approaches (see Clark-Hachtel et al., 2013; Clark-Hachtel and Tomoyasu, 2016). Recently, a fossil of a Carboniferous species of the extinct order Palaeodictyoptera has provided direct evidence that supports the dual model for insect wing origins. It shows the occurrence of three pairs of nymphal wing pads that were medially articulated to the thorax by the sclerites and also markedly fused anteriorly and posteriorly to the notum (Prokop et al., 2017).

A subsequent innovation in winged insects was wing flexion. There are two main situations that are characteristic of two major pterygote lineages. One is represented by Palaeoptera, in which the simple articulation is not modified and thus the wings can just flap up and down. The other situation is represented by the Neoptera, which evolved rotation of wings around the tergum sclerite where the wing articulates, thus permitting the wing to flex

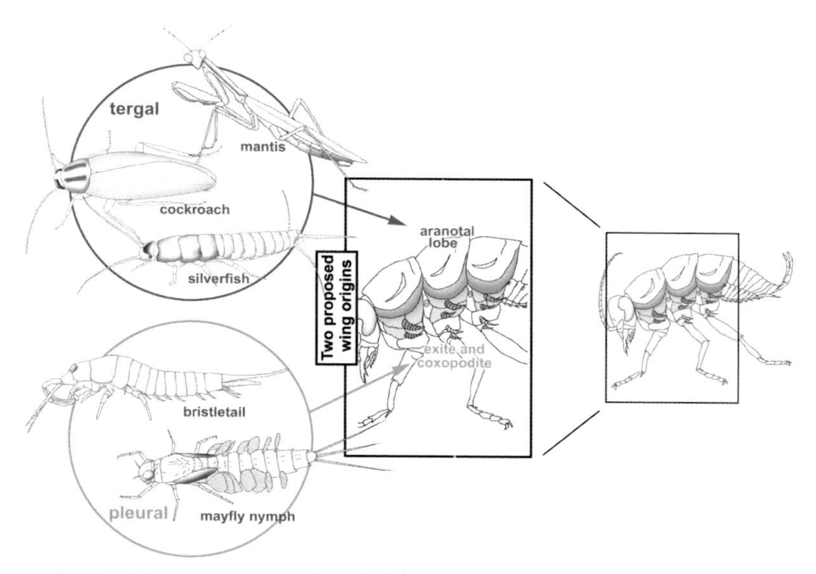

FIGURE 2.3 Schematic representation of the dual theory explaining the innovation of insect wings, from pronotal (tergal theory) and pleural (pleural theory) origin. Images courtesy of Yoshinori Tomoyasu, slightly modified from Clark-Hachtel and Tomoyasu (2016), with permission.

over the dorsal body side. A similar solution for wing flexion was independently acquired by the Diaphanopterodea, an extinct order (Kukalová-Peck, 1978). Arguably, wing flexion can be considered a key innovation, as it allows winged insects to radiate into concealed and architecturally complex microhabitats (Mayhew, 2002).

THE INNOVATION OF METAMORPHOSIS AND THEIR DIFFERENT TYPES

In wingless groups, that is, Archaeognatha and Zygentoma, the insect attains the reproductive capabilities, in terms of body size and genital structures, through gradual growth with minor changes in form. This practically ametamorphic postembryonic development is commonly known as ametabolan. Characteristically, the insect continues molting after attaining adulthood. In contrast, pterygote insects undertake metamorphosis, that is, the insect changes the morphology during postembryonic development, until reaching the adult, reproductively competent stage. The adult is a final stage, as there are no further molts. Metamorphosis appeared with the emergence of the pterygote insects, some 410 Mya.

In the Pterygota, postembryonic morphological changes can be more or less dramatic depending on the group. In Palaeoptera, Polyneoptera, and

Paraneoptera (i.e., exopterygotes), the developmental cycle comprises three characteristic stages, the embryo, the juvenile instars, conveniently called nymphs, and the adult. The nymphs are morphologically similar to the adult and develop gradually until the adult stage; the main differences of the adult are the fully developed wings and genital structures. This gradual metamorphosis of the exopterygotes has been collectively categorized as hemimetaboly, a well-established term in the modern entomological literature. However, the postembryonic development of exopterygotes is far from uniform, as a quite dramatic diversity of forms and lifestyles has been described for different insect groups (see Chapter 4: The hemimetabolan development).

In the Endopterygota, which emerged some 370 Mya, the developmental cycle comprises four characteristic stages, the embryo, the juvenile instars, conveniently called larvae, the pupa, and the adult. The larvae are morphologically more or less divergent with respect to the adult, and the pupal stage, normally a nonfeeding, immobile stage, bridges the morphological gap between the larvae and the adult. This postembryonic development has been generally categorized as holometaboly, which is conserved in all endopterygotes. However, there is a considerable morphological diversity in the larvae of different groups, from body structures relatively similar to that of the adult, like those exhibited by the larvae of Neuropterida (Neuroptera, Megaloptera, and Raphidioptera), to the vermiform shapes, like those of Diptera Brachycera. Moreover, in a few endopterygote groups, notably the Strepsiptera and some endoparasitoid species of different orders, postembryonic development passes through different types of larvae that are morphologically distinct, a phenomenon called hypermetamorphosis. We will further explore more details about larval types and hypermetamorphosis in Chapter 5 (The holometabolan development).

This simple classification of insect postembryonic development into ametabolan, hemimetabolan, and holometabolan is the most popular, and the one that will be used in this book, although other more detailed classifications have been proposed. In general, these classifications place the emphasis on naming particular types of metamorphosis but without a clear phylogenetic contextualization. One of the most popular was that of Berlese (1913), who classified the types of metamorphosis into three major categories: hemimetabola, neometabola, and holometabola. In turn, the hemimetabola were divided into pseudoametabola, paurometabola, and heterometabola, according to whether the adult is more or less similar to the nymph. Neometaboly was considered by Berlese as an intermediate type between hemimetaboly and holometaboly, which would be characterized by having completely immobile nymphal stages. The examples are provided by species of whiteflies, scale insects, and mealybugs. Finally, holometaboly is further divided into current holometaboly and hypermetamorphosis, exemplified by some endoparasitic hymenopterans.

Some 35 years later, Weber (1949) proposed more complicated categorizations. In his classification, the ametabola are called epimetabola, and the

Ephemeroptera are called prometabola, as they show a gradual development during the nymphal period and have two postmetamorphic stages: the subimago and the imago. Groups showing a gradual development of adult features in nymphs that are aquatic (like Odonata and Plecoptera) are categorized as archimetabola, whereas completely terrestrial groups with gradual development, such as most exopterygotes, are classified as paurometabola. Four further categories are proposed as a function of the more or less precocious externalization of the wing primordia. The metamorphosis observed in winged females of Hemiptera Phylloxeroidea, where the wing primordia appear in the last nymphal instar, is classified as homometabola. That of males of Hemiptera Coccomorpha, where two or three feeding wingless stages are followed by one or two nonfeeding stages with external wing primordia, is termed parametabola. The Hemiptera Aleyrodoidea, where the transformation to winged adults occurs within the last wingless and immobile nymphal instar (only the first of four nymphal instars is mobile), is called allometabola. Finally, the Thysanoptera, where two wingless nymphal instars are followed by one mobile and then by one, two, or three immobile instars showing wing primordia, are named remetabola. Within the holometabola the most common and well-known type is named euholometabola, whereas the Megaloptera, with their larvae aquatic but similar to the adult, and the fully motile pupae, are distinguished with the name eoholometabola. Then, different types of hypermetamorphosis are categorized as polymetabola (parasitic Coleoptera and Strepsiptera) and hypermetabola (Coleoptera Meloidae), while viviparous Phoridae (Diptera), like *Termitoxenia* spp., where a third larval instar emerges from the egg and it pupates within a few minutes, are categorized as cryptometabola. While the classification of Weber has the value of recognizing the great diversity of metamorphosis variants, in both hemimetabolan and holometabolan species, it is formed by unequal groupings with very limited phylogenetic sense and is excessively detailed and complicated. Nowadays practically nobody uses it and its interest has become only historical.

A more recent classification of insect metamorphosis is that of Nüesch (1987) who, based on the degree of morphological change during the postembryonic development, proposed the following four categories of insects: ametamorphs, paurometamorphs, hemimetamorphs, and holometamorphs. The ametamorphs show direct development, and thus they do not possess juvenile structures that degenerate or disappear in the adult molt. According to Nüesch, this would be the commonest category, represented by the Archaeognatha and Zygentoma, practically all Polyneoptera, the Psocoptera and Phthiraptera, and most Hemiptera. Roughly, it would correspond to the traditional ametabola (Archaeognatha and Zygentoma) plus the hemimetabola. The paurometamorphs are defined by showing a few specific juvenile structures, like particular muscles or sensilla, which degenerate or disappear in the adult molt. Paurometamorphs would have as representatives the

Orthoptera, Dermaptera, several groups of Hemiptera (Psylloidea, Cicadomorpha, Fulgoromorpha, Coleorrhyncha, and Heteroptera), and Psocoptera. Nüesch recognizes the hemimetamorphs because they show extensive replacement of juvenile by adult structures, especially in the last molt, although they do not possess a proper pupal stage. The following groups are considered hemimetamorphs: Ephemeroptera, Odonata, Plecoptera, and Hemiptera Aleyrodoidea and Heteroptera. Finally, the holometamorphs also show extensive replacement of juvenile by adult structures, which takes place within one, two, or three quiescent pupal or pupal-like instars. The endopterygote orders, the Thysanoptera, and the females of Hemiptera Coccomorpha, would belong to this category (Nüesch, 1987). Nüesch's classification, based on exclusively morphological and very general criteria, is disconnected from any phylogenetic context. Thus it is not very useful from an evolutionary point of view, which is the perspective followed in this book.

In summary, we will follow the most general classification in three types of postembryonic development: ametabolan, hemimetabolan, and holometabolan. In a phylogenetic context, holometaboly would be a synapomorphy that defines one clade, the Endopterygota. The rest of the insect clades that possess wings follow the hemimetabolan type of metamorphosis, whereas those that do not have wings (apterygota) are all ametabolan.

UNITY AND DIVERSITY OF EMBRYONIC DEVELOPMENT

We have been dealing with postembryonic development, but the ontogeny of insects begins with the embryogenesis, by which a nymph (in ametabolan and hemimetabolan species) or a larva (in holometabolans) is formed from an egg. Through an early cleavage, the unique cell of the fertilized egg produces hundreds and thousands of cells that later develop and differentiate into different cell types that form the tissues and organs. The embryogenesis itself can be very different in different insect groups, even between closely related species. What follows is thus a description of general traits. Detailed descriptions of the embryo development of ametabolan, hemimetabolan, and holometabolan species can be found in the comprehensive reviews of Johannsen and Butt (1941), Jura (1972), and Anderson (1972a,b).

Generally, the early cleavages involve nuclear subdivisions that are not accompanied by cleavage of the cytoplasm. The process is known as syncytial cleavage and the result is the formation of a general compartment or syncytium formed by up to 6000 nuclei. Syncytial nuclei are surrounded by associated portions of cytoplasm that separate one nucleus from one another. The nuclei and associated cytoplasm portions are known as energids, which migrate to the periphery of the egg, where they can continue dividing and subsequently acquire a cell membrane. The layer of cells ending at the periphery of the egg forms the blastoderm, which will coalesce forming the germ anlage that later develops into the germ band. The cells that do not contribute to the germ anlage form an extraembryonic membrane called serosa. Moreover, a second membrane, called amnion, forms later from the cells adjacent to the germ anlage. Ultimately, the amnion cells meet in the middle of the embryo and form a single cell layer that, lying between the embryo and the serosa, covers the embryo leaving a space known as the amniotic cavity (Fig. 2.4).

In most holometabolan species, the germ anlage forms from almost the entire blastoderm, whereas in the great majority of hemimetabolan species, after the formation of the syncytial blastoderm, nuclei migrate to the posterior pole region and form there the germ anlage from a relatively small proportion of the blastoderm. In the former case, known as long germ band type, the complete body (head, gnathal, thoracic, and abdominal segments) is configured at the blastoderm stage, and all segments are formed at once. In the second case, called short germ band type, the head lobes, the most anterior trunk segments, and the posterior terminus are configured first. Then new segments are added progressively through proliferative growth (Fig. 2.5A). Between the two extreme short and long germ band types, a number of species start embryo development with intermediate germ band types (Davis and Patel, 2002). Less modified groups, mainly among the

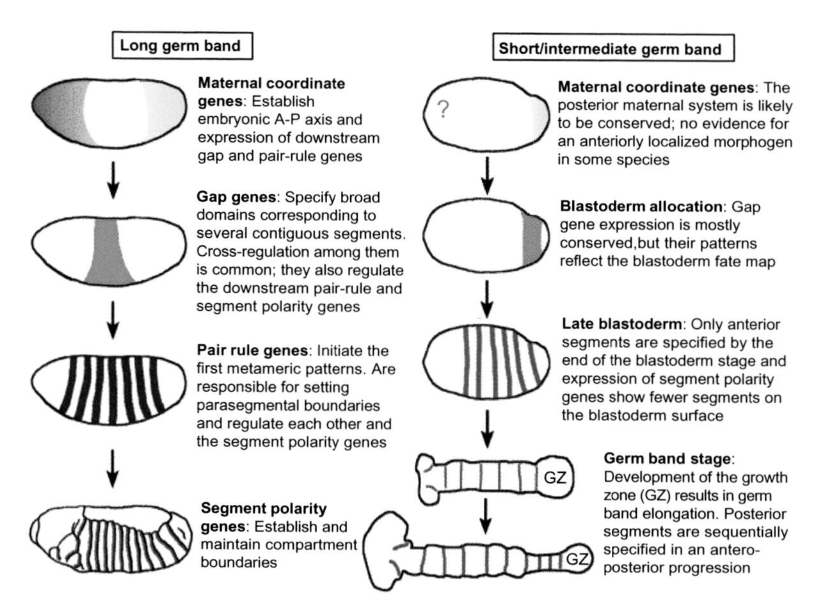

FIGURE 2.4 Comparison of long and short/intermediate germ band segmentation, exemplified by the fly *Drosophila melanogaster* (left) and the bug *Oncopeltus fasciatus* (right). The genes involved and the main processes are indicated in each germ band type. From Liu and Kaufman (2005), with permission.

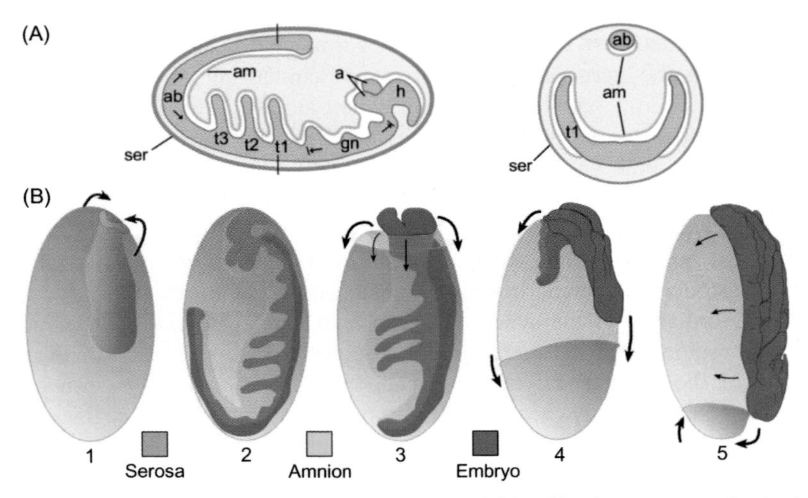

FIGURE 2.5 The amnion and serosa membranes and blastokinesis movements in the bug *Oncopeltus fasciatus*. (A) Germ band stage embryo in midsagittal section view (left) and transverse section (right), showing the serosa (blue) and amnion (orange) with respect to the embryo (gray) and yolk (yellow). The amniotic cavity (white) is the region between the amnion and the ventral surface of the embryo. a: antenna, am: amnion, ab: abdominal region, gn: gnathal region, h: head, ser: serosa, t1–t3: thoracic segments/legs 1–3. (B) Blastokinesis in five steps. From left to right: immersion anatrepsis (1), extended germ band stage (2), early katatrepsis (3), mid katatrepsis (4), and late katatrepsis (5). Black arrows indicate the direction of motion. From Panfilio (2008), with permission.

ametabolan and hemimetabolan species, follow the short and intermediate germ band type, whereas more modified holometabolan groups follow the long germ band type (Fig. 2.5A) (Chipman, 2015; Davis and Patel, 2002; Liu and Kaufman, 2005). However, there are exceptions, notably that of the coleopteran *Tribolium castaneum*, an holometabolan that follows an intermediate germ band development (Lynch et al., 2012).

The formation of the blastoderm is followed by a process of cell invagination or gastrulation, which results in the formation of an embryo with two main layers. Cells remaining at the blastoderm periphery constitute the ectoderm, and those that invaginate below the ectoderm constitute the mesoderm. An important process that occurs around mid-embryogenesis is blastokinesis, a movement of the embryo into the yolk mass involving the amnion and serosa membranes, and that usually results in partial revolution of the embryonic body (Panfilio, 2008). Blastokinesis is divided into anatrepsis and katatrepsis. During the anatrepsis, the embryo, starting from the posterior end, is pulled into the yolk, while reversing its axes with respect to those of the egg. Later in development, through the reversal movement of katatrepsis, the embryo returns to its original position on the ventral side of the egg (Fig. 2.5B). Blastokinesis is typical of short and intermediate germ band, hemimetabolan insects. Similar movements in long germ band, holometabolan species, like the thoroughly studied *Drosophila melanogaster*, are oversimplified and lack formal names (Panfilio, 2008).

As the embryo develops, the two-layered, essentially flat germ band, transforms into a three-dimensional nymph or larva. Differentiated segments first become apparent in the anterior end, where the brain and eyes are developed from the ectoderm, and protrusions forming anterior to the mouth opening grow to form the labrum and the antennae. Each of the first three segments behind the mouth develops paired appendages that become the mandibles, maxillae, and labium, respectively. The next three segments constitute the thorax and form appendages that become legs. As organs grow and differentiate, the flanks of the germ band, both ectoderm and mesoderm, grow laterally wrapping the yolk, until the two edges meet and fuse along the dorsal midline in a process called dorsal closure.

The next steps of embryogenesis comprise the differentiation of the ectoderm and mesoderm into the organ systems of the nymph or the larva. Most of the morphology, conspicuously the "skin" of the nymphs or larvae, punctuated by bristles and sensory devices, comes from the ectoderm. Moreover, the nervous system forms from the ventral ectoderm, and the tracheal system develops from invaginations of the lateral ectoderm. Ocelli, prothoracic gland, corpora allata–corpora cardiaca, salivary glands, silk glands, and oenocytes, also form as ectodermal invaginations. Finally, two additional invaginations of the ectoderm occur that form the stomodeum and the proctodeum. The stomodeum occurs in a central position near the anterior of the germ band, and, once invaginated, their cells proliferate in a posterior direction, forming the foregut. The proctodeal invagination occurs in the terminal segment, and their cells grow anteriorly forming the hindgut. Malpighian tubules develop from outpocketings of the proctodeum. From a pair of transient coelomic sacs formed in each segment from the invaginated mesoderm develop the dorsal vessel, internal reproductive organs, muscles, fat body, subesophageal gland, and hemocytes. The midgut is formed from a third germ layer, the endoderm, which develops at the edge of the foregut and hindgut invaginations and fuses with them, thus completing the digestive system.

REFERENCES

Anderson, D.T., 1972a. The development of Holometabolous insects. In: Counce, S.J., Waddington, C.H. (Eds.), Developmental Systems: Insects. Academic Press, London, pp. 165–242.

Anderson, D.T., 1972b. The development of Hemimetabolous insects. In: Counce, S.J., Waddington, C.H. (Eds.), Developmental Systems: Insects. Academic Press, London, pp. 95–163.

Berlese, A., 1913. Intorno alle metamorfosi degli insetti. Redia 9, 121–136.

Chipman, A.D., 2015. Hexapoda: comparative aspects of early development. Evolutionary Developmental Biology of Invertebrates 5. Springer Vienna, Vienna, pp. 93–110. Available from: https://doi.org/10.1007/978-3-7091-1868-9_2.

Clark-Hachtel, C.M., Tomoyasu, Y., 2016. Exploring the origin of insect wings from an evo-devo perspective. Curr. Opin. Insect Sci. 13, 77–85. Available from: https://doi.org/10.1016/j.cois.2015.12.005.

Clark-Hachtel, C.M., Linz, D.M., Tomoyasu, Y., 2013. Insights into insect wing origin provided by functional analysis of vestigial in the red flour beetle, *Tribolium castaneum*. Proc. Natl. Acad. Sci. U.S.A. 110, 16951–16956. Available from: https://doi.org/10.1073/pnas.1304332110.

Davis, G.K., Patel, N.H., 2002. Short, long, and beyod: molecular and embryological approaches to insect segmentation. Annu. Rev. Entomol. 47, 669–699. Available from: https://doi.org/10.1146/annurev.ento.47.091201.145251.

Engel, M.S., Grimaldi, D.A., 2004. New light shed on the oldest insect. Nature 427, 627–630. Available from: https://doi.org/10.1038/nature02291.

Farrell, B.D., 1998. "Inordinate fondness" explained: why are there so many beetles? Science 281, 555–559.

Field, K.J., Cameron, D.D., Leake, J.R., Tille, S., Bidartondo, M.I., Beerling, D.J., 2012. Contrasting arbuscular mycorrhizal responses of vascular and non-vascular plants to a simulated Palaeozoic CO_2 decline. Nat. Commun. 3, 835. Available from: https://doi.org/10.1038/ncomms1831.

Grimaldi, D., Engel, M.S., 2005. Evolution of the Insects. Cambridge University Press, Cambridge Evolution Series.

Haug, C., Haug, J.T., 2017. The presumed oldest flying insect: more likely a myriapod? PeerJ 5, e3402. Available from: https://doi.org/10.7717/peerj.3402.

Johannsen, O.A., Butt, F.H., 1941. Embryology of insects and myriapods. McGraw-Hill Book Company, Inc, New York and London.

Jura, C., 1972. Development of Apterygote insects. In: Counce, S.J., Waddington, C.H. (Eds.), Developmental Systems: Insects. Academic Press, London, pp. 49–94.

Kukalová-Peck, J., 1978. Origin and evolution of insect wings and their relation to metamorphosis, as documented by the fossil record. J. Morphol. 156, 53–125. Available from: https://doi.org/10.1002/jmor.1051560104.

Liu, P.Z., Kaufman, T.C., 2005. Short and long germ segmentation: unanswered questions in the evolution of a developmental mode. Evol. Dev. 7, 629–646. Available from: https://doi.org/10.1111/j.1525-142X.2005.05066.x.

Lozano-Fernandez, J., Carton, R., Tanner, A.R., Puttick, M.N., Blaxter, M., Vinther, J., et al., 2016. A molecular palaeobiological exploration of arthropod terrestrialization. Philos. Trans. R. Soc. Lond. B: Biol. Sci. 371, 20150133. Available from: https://doi.org/10.1098/rstb.2015.0133.

Lynch, J.A., El-Sherif, E., Brown, S.J., 2012. Comparisons of the embryonic development of *Drosophila*, *Nasonia*, and *Tribolium*. Wiley Interdiscip. Rev. Dev. Biol. 1, 16–39. Available from: https://doi.org/10.1002/wdev.3.

Mayhew, P.J., 2002. Shifts in hexapod diversification and what Haldane could have said. Proc. Biol. Sci. 269, 969–974. Available from: https://doi.org/10.1098/rspb.2002.1957.

Misof, B., Liu, S., Meusemann, K., Peters, R.S., Donath, A., Mayer, C., et al., 2014. Phylogenomics resolves the timing and pattern of insect evolution. Science 346, 763–767.

Morris, J.L., Puttick, M.N., Clark, J.W., Edwards, D., Kenrick, P., Pressel, S., et al., 2018. The timescale of early land plant evolution. Proc. Natl. Acad. Sci. U.S.A. 115, E2274–E2283. Available from: https://doi.org/10.1073/pnas.1719588115.

Nüesch, O., 1987. Metamorphose bei Insekten: direkte und indirekte Entwicklung by Apterygoten and Exopterygoten. Zool. Jahrb. (Anat.) 115, 453–487.

Panfilio, K.A., 2008. Extraembryonic development in insects and the acrobatics of blastokinesis. Dev. Biol. 313, 471–491. Available from: https://doi.org/10.1016/j.ydbio.2007.11.004.

Prokop, J., Pecharová, M., Nel, A., Hörnschemeyer, T., Krzemińska, E., Krzemiński, W., et al., 2017. Paleozoic nymphal wing pads support dual model of insect wing origins. Curr. Biol. 27, 263–269. Available from: https://doi.org/10.1016/j.cub.2016.11.021.

Puttick, M.N., Morris, J.L., Williams, T.A., Cox, C.J., Edwards, D., Kenrick, P., et al., 2018. The interrelationships of land plants and the nature of the ancestral embryophyte. Curr. Biol. 28, 733–745. Available from: https://doi.org/10.1016/j.cub.2018.01.063.

Wang, Y.-H., Engel, M.S., Rafael, J.A., Wu, H.-Y., Rédei, D., Xie, Q., et al., 2016. Fossil record of stem groups employed in evaluating the chronogram of insects (Arthropoda: Hexapoda). Sci. Rep. 6, 38939. Available from: https://doi.org/10.1038/srep38939.

Ward, P., 2006. Out of Thin Air. Joseph Henry Press, Washington, D.C. Available from: https://doi.org/10.17226/11630.

Weber, H., 1949. Grundriss der Insektenkunde. Gustav Fischer, Jena.

Chapter 3

The ametabolan development

Ametabolan insects undergo a direct, gradual development from the first nymphal instar to the adult. It is the mode of development of apterygote insects, comprising the orders Archaeognatha, commonly known as bristletails, and Zygentoma, which include the so-called silverfish. A synonym of Archaeognatha is Microcoryphia, which is used by a significant number of authors. In classical entomological literature, Archaeognatha and Zygentoma are considered simple suborders of the old order Thysanura. However, in the late 20th century, each of those two suborders was raised to the status of an independent order; thus the order Thysanura became redundant and the use of this name is now discouraged (Mendes, 2002).

A characteristic of both bristletails and silverfish is that all have three filiform appendages at the distal part of the abdomen. The two laterals are cerci, and the medial one is the appendix dorsalis or the paracercus. In the Archaeognatha, the appendix dorsalis is considerably longer than the two cerci, whereas in the Zygentoma, the three filiform appendages are subequal in length (Fig. 3.1A and B). Another feature of Archaeognatha and Zygentoma is that the body becomes covered by scales in most of the species although there are some differences between both orders. Most of the Archaeognatha are scaled (except the first two nymphal instars), whereas a number of groups of Zygentoma are characteristically unscaled in all stages. The early-branching family Lepidotrichidae, for example, lacks scales, and other unscaled families are the Protrinemuridae and Maindroniidae. In the family Nicoletiidae, there are several subfamilies that lack scales, both euedaphic and cave species (Fig. 3.1C), like in the spectacular troglobite *Coletinia majorensis*, from lava caves in the Canary Islands (Fig. 3.1D). In Archeognatha, the scales are composed of upper and lower lamellae that form a lumen connected to the exterior through pores. In Zygentoma, the scales are composed of a continuous membrane with reinforcing longitudinal ribs running along the upper side of them, parallel to the longitudinal axis (Fig. 3.1E and F). Scales appear after the second or third molt; they play a protective role and can function as mechanoreceptors (Eisenbeis and Wichard, 1987).

Archaeognatha consists of about 350 described species. They live in grassy or wooded habitats where they usually are observed under bark,

Insect Metamorphosis. DOI: https://doi.org/10.1016/B978-0-12-813020-9.00003-X

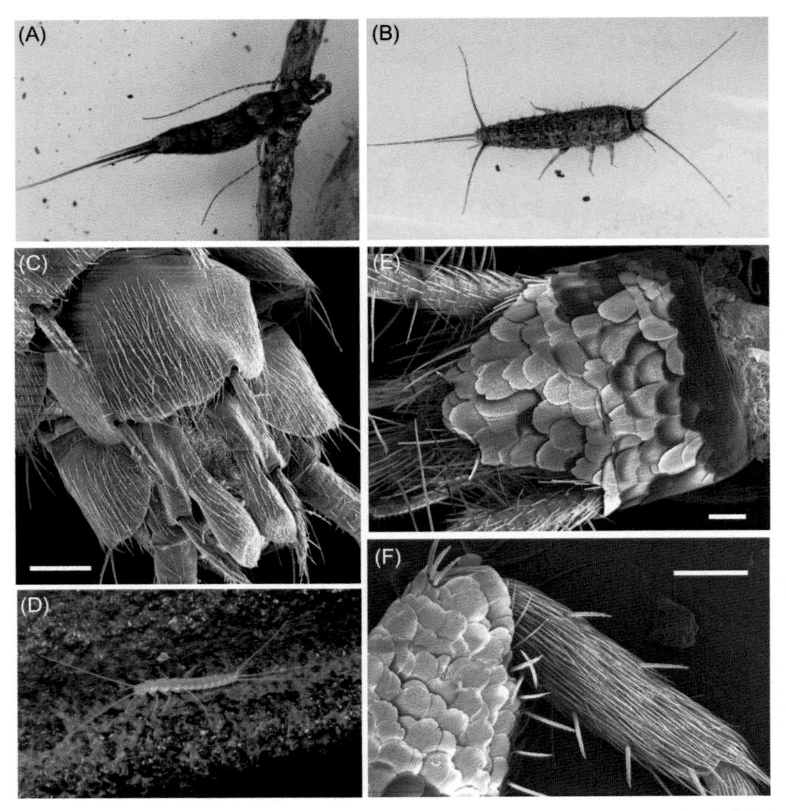

FIGURE 3.1 Habitus and details of scales of Archaeognatha and Zygentoma. (A) A bristletail of the genus *Catamachilis* sp. (B) The silverfish *Ctenolepisma ciliata*. (C) Ventral view of the apical region of the abdomen of the cave silverfish *Coletinia tinauti*, showing the absence of scales. (D) The spectacular troglobite *Coletinia majorensis* photographed in its natural habitat in Cueva del Llano (Fuerteventura island). (E) Scales in the uroterguite region of the bristletail *Allacrotelsa kraepelini*. (F) Femur and tibia of the silverfish *C. ciliata*. Scale bar in (C), (E), and (F): 200 μM. Photos (A) and (B) courtesy of Miquel Gaju-Ricart (Universidad de Córdoba, Spain). Photo (D) courtesy of Pedro Oromí (Universidad de La Laguna, Spain); electron micrographs (C), (E), and (F) courtesy of Rafael Molero-Baltanás (Universidad de Córdoba, Spain).

among stones, in leaf litter, or near the upper tidal line in coastal areas. Bristletails feed as herbivores or scavengers on algae, mosses, lichens, or decaying organic matter, being especially active at night. Zygentoma, with some 370 described species, may be found in dark, wet environments, like caves, or under dry conditions, as free-living organisms or associated to nests, always feeding on organic detritus. A few species of silverfish live in anthropic habitats, feeding mostly on paper, cereals, paste, starch in clothes, and even rayon fabrics. Characteristically, fertilization of Archaeognatha and Zygentoma is achieved by indirect transfer of sperm. Males produce a packet of sperm or spermatophore and leave it on the ground, and then the female picks up the spermatophore with her genital opening and sperm are released into her reproductive system. Elaborate courtship behaviors to ensure that the female finds the spermatophore have been described in several species, especially in silverfish (Eisenbeis and Wichard, 1987; Mendes, 2002).

EMBRYOGENESIS IN AMETABOLAN SPECIES

Embryo development in Archaeognatha and Zygentoma has been comprehensively reviewed by Jura (1972) and more recently by Larink (1983). In general, embryogenesis follows the developmental steps observed in insects, which have been described in Chapter 2 (A spectacular diversity of forms and developmental modes): formation of the blastoderm and the germ band; formation of the serosa and amnion membranes followed by the amniotic cavity; gastrulation, with the differentiation of the ectoderm, mesoderm, and endoderm; segmentation; blastokinesis; dorsal closure; and organogenesis. Bristletails and silverfish embryos follow a short germ band development and undertake a conspicuous blastokinesis. Embryogenesis finishes with the formation of a nymph that possesses the basic shape and features of the adult (Fig. 3.2).

Differences with respect to pterygote insects include the extremely short germ band anlage, the well-developed abdominal appendages, the early deep ventral flexure between the thorax and the abdomen, the formation of a tentorium with the anterior and posterior arms not fused, among others (Larink, 1983). Most authors consider that the amniotic cavity of Archaeognatha and Zygentoma is incompletely closed. This would be a distinctive feature of these ametabolan apterygote orders in opposition to hemimetabolan and holometabolan pterygotes that have the amniotic cavity closed. However, this does not seem a generality. For example, *Trigoniophthalmus alternatus* does not have an amniotic cavity at all, whereas *Petrobius brevistylis* and *Pedetontus unimaculatus*, all of them Archaeognatha, have a wide open cavity. In *Thermobia domestica* and *Lepisma saccharina*, the amniotic cavity is not completely closed but opens through a narrow canal or amniopore, whereas in *Ctenolepisma lineata*, all of them Zygentoma, the cavity is perfectly closed (Larink, 1983; Masumoto and Machida, 2006).

A major distinctive feature of embryos of ametabolan apterygote species concerns the deposition of only two embryonic cuticles (EC1 and EC2), as shown by the silverfish *T. domestica* (Konopová and Zrzavý, 2005). In contrast, the embryos of pterygotes deposit an additional embryonic cuticle, EC3. Among holometabolan pterygotes, the Diptera Brachycera secrete only EC1 and EC3, apparently due to the secondary loss of EC2. Differences of cuticulogenesis between silverfish and pterygote embryos suggest that the apterygote first nymphal instar was "embryonized," becoming a totally embryonic pronymph, in pterygotes (Konopová and Zrzavý, 2005).

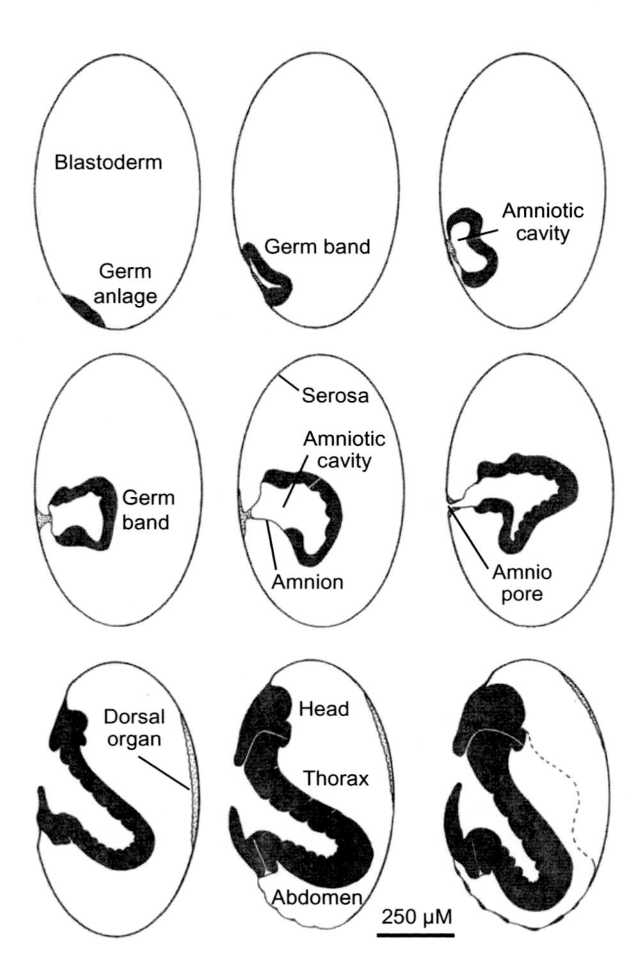

FIGURE 3.2 Embryonic development of *Lepisma saccharina*. From Larink (1983), modified.

POSTEMBRYONIC DEVELOPMENT

Archaeognatha and Zygentoma are ametabolan, following a direct, progressive development from the first nymphal instar to the adult. Thus the morphological features peculiar to the adult are gradually added to the still incomplete organization of the nymphs. Characteristically, they continue to molt throughout their life, with several sexually mature instars. Nevertheless, postembryonic development does not consist only of a progressive increase in size, but it includes some metamorphic changes, like the loss of hatching devices or the acquisition of scales and adult genital structures. A number of authors have compared this progressive development with minor metamorphic changes to the epimorphosis shown by some centipedes (Bitsch, 1964). In the case of Archaeognatha, Verhoeff (1910a,b) even came to categorize up to six characteristic phases within the postembryonic development of bristletails: nymph without scales, nymph with scales, immatures, praematures, pseudomatures, and matures, proposing the term "orthomorphosis" to design this type of development. These old classifications are not in use today, but they are useful to underline that the so-called ametabolan development includes a few but clear morphogenetic changes in diverse moments of the postembryonic life. More modern works simply distinguish unscaled stages, scaled juveniles (including subadults) and adults, as distinct morphofunctional types in the postembryonic development in bristletails and silverfish (Sturm and Machida, 2001).

Unscaled juveniles

Archeognatha are difficult to rear, and the first nymphal instar (N1) is of short duration, often about a day or two, thus it is difficult to find it in nature. Therefore descriptions are limited to a few species, as reviewed by Bach de Roca and Gaju-Ricart (1988). Specially detailed information has been reported for the species *Promesomachilis hispanica* (Bach de Roca and Gaju-Ricart, 1988), *T. alternatus*, and *Lepismachilis* cf. *y-sygnata* (Sturm and Machida, 2001). Conversely, a number of Zygentoma, especially species living in anthropic habitats, are easy to rear in the laboratory, and the freshly emerged N1 has been described by a number of authors. This is the case, for example, of Sweetman (1938) for *T. domestica*, Lindsay (1940) for *Ctenolepisma longicaudata*, Laibach (1952) for *L. saccharina*, and Sweetman (1952) for *C. lineata* (= *Ctenolepisma quadriseriata*).

In Archeognatha, the N1, in addition to being unscaled, presents a number of specific features. Among them are the conspicuous prognathy, the distinct forward projection of the laciniae and the maxillae (which play the role of oviruptors, for breaking the egg corion at the time of hatching), the fine cuticular teeth of distinct cephalic regions and the rod-like setae on the head and terga (Fig. 3.3A). Importantly, all these morphological features disappear after the first molt to the second nymphal instar (N2). The midgut is filled with yolk material, which allows the insect to remain hidden without eating during the short duration of the first instar. In the N2, the above N1 features disappear. N2 is orthognathous (Fig. 3.3B) and does not have the rod-like setae, which have been replaced by specialized macrochetae. In general, N2 already has the shape of scaled juveniles, but it is still unscaled, although it can present characteristic pigment patterns (Fig. 3.3C). In contrast to N1, the midgut is filled with food ingested after hatching (Sturm and Machida, 2001). In the studied Archaeognatha, scales are acquired in N3.

The first nymphal unscaled stages of Zygentoma are similar in shape to midstaged juveniles. The main difference between N1 and N2 is the apical part of the frons that forms a small ridge, which functions as an oviruptor (Fig. 3.3D) and disappears after the molt to N2. Laibach (1952) has vividly

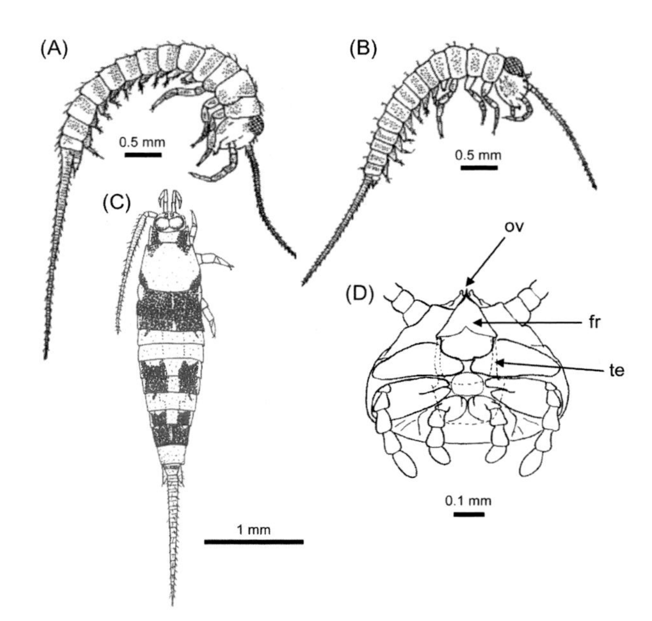

FIGURE 3.3 Morphology of Archeaeognatha and Zygentoma. (A and B) First and second nymphal instars of Archaeognatha. (C) Pigment of the first nymphal instar of *Promesomachilis hispanica*. (D) Ventral side of the head of the first nymphal instar of *Ctenolepisma longicaudata* showing the oviruptor (ov). Fr: frons; te: tentorium showing the anterior and posterior arms not fused. (A) and (B) from Sturm and Machida (2001); (C) from Bach de Roca and Gaju-Ricart (1988), and (D) from Lindsay (1940), with permission.

described the hatching behavior of *L. saccharina* using this structure. Another difference between the first nymphal instars has been recorded in *C. longicaudata*, where N1 has two tarsomeres in each leg, N2 shows a kind of septum on the second tarsomere of the metathoracic leg, and N3 has all tarsi with three tarsomeres, as in the adult (Lindsay, 1940). As observed in Archaeognatha, the N1 of Zygentoma survives without eating due to the reserves of yolk enclosed in the digestive tube. The anus can be even closed in N1, as observed by Lindsay (1940) in *C. longicaudata*. Only from N2 are silverfish able to eat external food, as demonstrated by Laibach (1952) after supplying *L. saccharina* N2 with starch colored with Rhodamine-red and observing that the midgut quickly became red colored. The stages where Zygentoma acquire the scales are N3 or N4.

Scaled juveniles

In the species studied of Archaeognatha, scales cover all terga, abdominal coxites, cerci, and appendix dorsalis of N3. In adult Machilidae (but not in Meinertellidae), the scales are additionally present on the antennae, head, maxillary and labial palps, legs, and abdominal styli. However, at least in some species, like *T. alternatus*, the scale cover of N3 and N4 is incomplete (Sturm and Machida, 2001). In *P. brevistylis*, it is even possible to distinguish the first scaled instars by means of the number of scale rings of the cerci (Fig. 3.4A) (Delany, 1957). Concerning Zygentoma, the body can

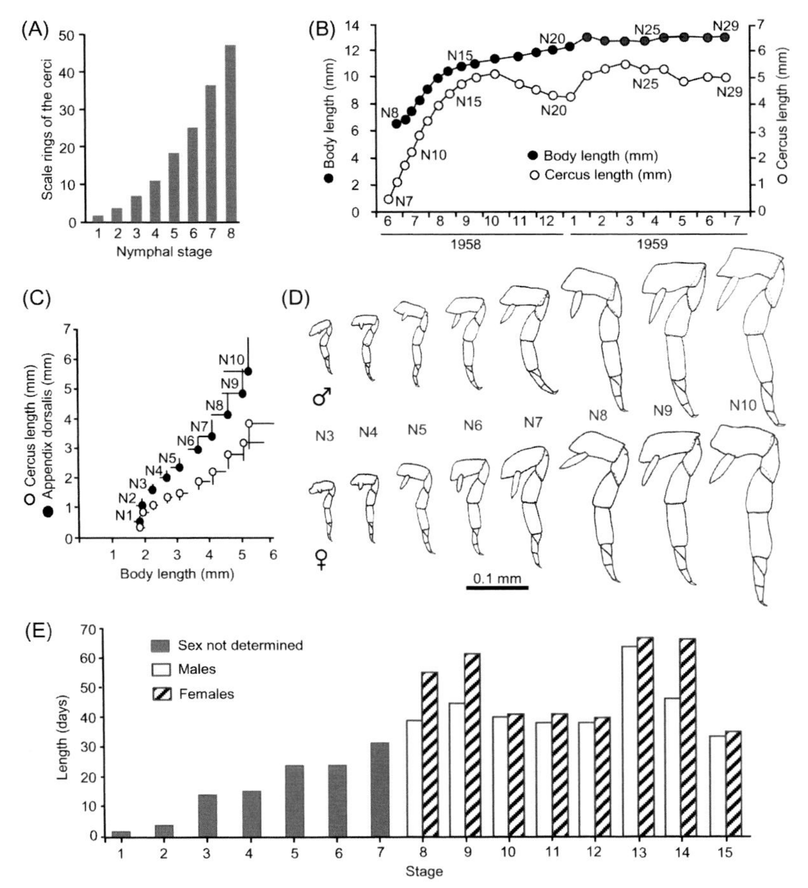

FIGURE 3.4 Progressive growth in Archeaeognatha and Zygentoma. (A) Number of scale rings of the cerci during the first nymphal instars of *Petrobius brevistylis.* (B) Total length and cercus length during development from N7 to N29 of *Machilis burgundiae* collected near Dijon (France) and then reared in the laboratory over 14 months of the years 1958 and 1959, as indicated in the abscissae; initially values refer to 22 individuals, but from N25 onwards only one individual was monitored. (C) Ratio between the body length and the length of the cerci or the appendix dorsalis during the first nymphal instars (N1–N10) of *Thermobia domestica*; standard deviations for each parameter are indicated by bars on each point. (D) Development of the metathoracic styles during the mid nymphal instars N3–N10 of *Promesomachilis hispanica.* (E) Length in days of the first 15 instars of *Lepisma saccharina*; adult morphology, including genitalia, is reached in N8; the bars represent the average number of days, based on 30 nymphs (between N1 and N7) and 8 adult males and 14 adult females (between N8 and N15). (A) Drawn with quantitative data reported by Delany (1957); (B) drawn with quantitative data reported by Bitsch (1964); (C) drawn with quantitative data reported by Kränzler and Larink (1980); (D) from Bach de Roca and Gaju-Ricart (1987), with permission; (E) drawn with quantitative data reported by Laibach (1952).

appear covered with scales from N3, as in *L. saccharina* (Laibach, 1952), or from N4, like in *T. domestica*, *C. lineata*, and *C. longicaudata* (Lindsay, 1940; Sweetman, 1952, 1938).

A number of morphological characters allow for the distinguishing of different instars, like the progressive apparition of ventral abdominal styli in *C. lineata* (Sweetman, 1952) or the increase in number and variation in patterning of the sensilla in the apical segment of the labial palp in *L. saccharina* (Larink, 1983). However, the most common approach to quantitatively monitoring the growth of bristletails and silverfish has been to measure different parameters along the successive molts. With respect to bristletails, Bitsch (1964) followed the development of *Machilis burgundiae* in the laboratory at room temperature by measuring the total length of the insect and those of the cerci. The results showed a rapid growth during the first 15 instars, and then the growth became slower until the death of the specimens under observation (Fig. 3.4B). A similar pattern of growth has been observed in silverfish, for example, in *L. saccharina*, where the increase in size and weight was monitored in detail by Laibach (1952), or in *T. domestica* when studying the ratio between the body length and the length of the cerci or the appendix dorsalis (Kränzler and Larink, 1980) (Fig. 3.4C). The development of most of the appendages and body segments is progressive, as shown in *C. longicaudata*, where the length of palps, tibiae, and cerci were measured (Lindsay, 1940), or in *P. hispanica*, where the labial palps, metathoracic styles, first urosternite, and genital structures were studied in detail (Bach de Roca and Gaju-Ricart, 1987) (Fig. 3.4D). In bristletails and in silverfish, the length of each instar is variable. However, the earlier (unscaled) stages are the shortest; throughout the juvenile scaled stages, the length increases progressively; and the imaginal stages show the longest length. This can be exemplified by *L. saccharina*, as reported by Laibach (1952) (Fig. 3.4E).

Adults

Genital structures develop progressively during scaled juvenile stages, as shown, for example, in the bristletails *P. brevistylis* (Delany, 1959) (Fig. 3.5A) and *P. hispanica* (Bach de Roca and Gaju-Ricart, 1987; Gaju-Ricart and Bach de Roca, 1989), and in the silverfish *C. longicaudata* (Lindsay, 1940). Scaled juveniles with the reproductive organs and structures practically differentiated as in the adult but not yet reproductively functional can be categorized as subadults (Sturm and Machida, 2001).

In general, the genitalia become completely formed in N8 or N9, both in Archaeognatha and in Zygentoma, and attainment of sexual maturity, fertilization, and oviposition can occur after one or two additional molts. This has been shown in the species mentioned above and in the silverfish *L. saccharina* (Laibach, 1952), among others. In the bristletail *P. hispanica*, Gaju-Ricart and Bach de Roca (1989) have reported that the N9 shows all the features

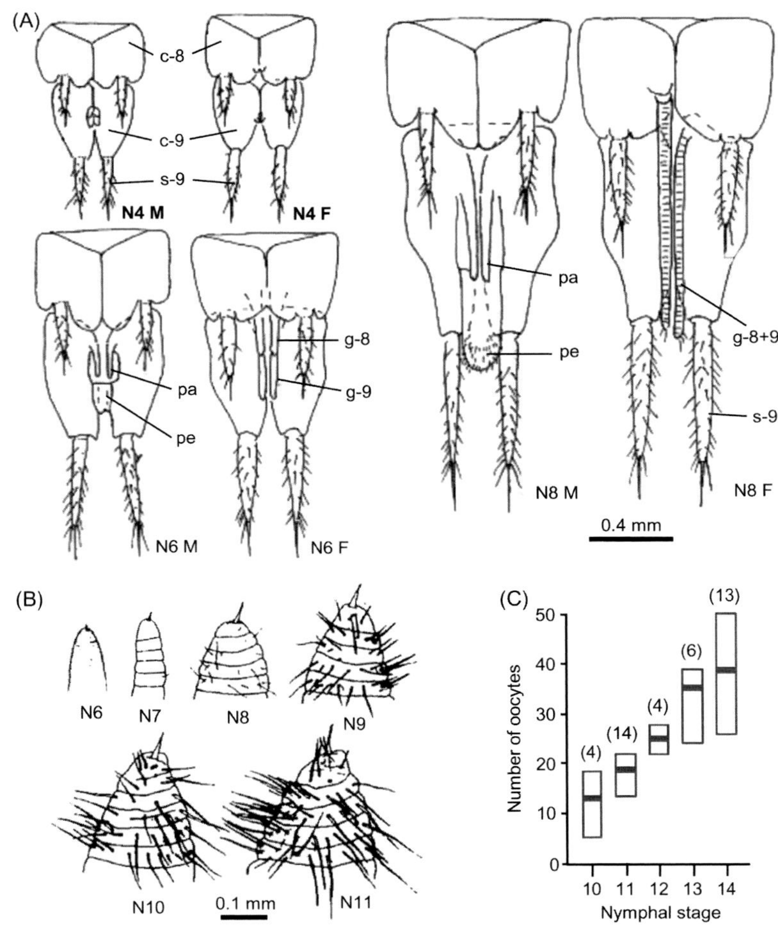

FIGURE 3.5 The development of adults in Archeaeognatha. (A) Development of genital structures in *Petrobius brevistylis* from N4 to N8, in males (M) and females (F); c-8, c-9: abdominal coxites VIII and IX; s-9: stylus abdominal segment IX; pa: parameres; pe: aedeagus of penis; 9-8, g-9: gonapophyses VIII and IX. (B) Development of the chaetotaxy of the ovipositor from N6 to N11 in *Promesomachilis hispanica*. (C) Increase in number of oocytes in the ovaries of *P. hispanica* from N10 to N14; the black segment indicates the average number and the total length of the white bar indicates the maximum and minimum values; the number in parenthesis at the top of each bar indicates the number of specimens used. (A) From Delany (1959), with permission; (B) from Gaju-Ricart and Bach de Roca (1989), with permission; and (C) drawn with quantitative data reported by Gaju-Ricart and Bach de Roca (1989).

characterizing the adult stage, including the complete chaetotaxy of the ovipositor (Fig. 3.5B). However, the onset of oocyte development occurs in N10, when an average of 13 oocytes is observed in the studied females; then oocyte number increases with age in successive stages (Fig. 3.5C).

After acquiring the adult features and being able to reproduce, both bristletails and silverfish continue to molt regularly, and alternate reproductive cycles, with the corresponding oviposition, with molting cycles (Bitsch, 1967). Moreover, they can live for a long time, up to 3 or 4 years. In bristletails, the longest period of monitoring in captivity was obtained in *M. burgundiae* (Bitsch, 1964). The longest-lived specimen was collected in nature as N7 and molted 22 times in 395 days. More precise data are available for silverfish as they are easier to rear in captivity. Laboratory strains of *T. domestica* can molt more than 60 times (Sweetman, 1938). According to Laibach (1952), the most long-lived specimen of his cultures of *L. saccharina* molted 35 times, reaching a weight of 16.50 mg and a size of 12 mm, in 1208 days. Possibly, the longest precisely recorded lifetime has been reported by Sweetman (1952) for *C. lineata*: two specimens of his laboratory colony died in N66 and lived 1417 days.

REFERENCES

Bach de Roca, C., Gaju-Ricart, M., 1987. An study on the postembrionic growth of *Promesomachilis hispanica* Silvestri, 1923 (Thysanura). In: Striganova, B.R. (Ed.), Soil Fauna and Soil Fertility. Proceedings of the 9th International Colloquium on Soil Zoology. Nauka, Moscow, pp. 611–618.

Bach de Roca, C., Gaju-Ricart, M., 1988. Descripción de los dos primeros estadíos de *Promesomachilis hispanica* Silvestri, 1923 (lnsecta: Apterygota, Microcoryphia). In: *Actas Del II Congreso Ibérico de Entomología. Asociación Española de Entomología*, Granada, pp. 121–132.

Bitsch, J., 1964. Observations sur le développement postembryonaire des Machilidés (Insecta Thysanura). Trav. Lab. Zool. la St. Aquic. Grimaldi 54, 1–17.

Bitsch, J., 1967. Le cycle de ponte chez les Machilides (Insecta Thysanura). C. R. Acad. *Sci.* Paris 264, 257–260.

Delany, M.J., 1957. Life histories in the Thysanura. Acta Zool. Cracoviensia 2, 61–90.

Delany, M.J., 1959. The life history and ecology of two species of *Petrobius* (Leach) *P. brevistylis* and *P. maritimus*. Trans. R. Soc. Edinbg. 63, 501–533.

Eisenbeis, G., Wichard, W., 1987. Atlas on the Biology of Soil Arthropods. Springer, Berlin Heidelberg.

Gaju-Ricart, M., Bach de Roca, C., 1989. Ovocyte number in *Promesomachilis hispanica* Silvestri, 1912 (Apterygota: Microcoryphia). In: Dallai, R. (Ed.), 3rd International Seminar on Apterygota. University of Siena, Siena, pp. 477–485.

Jura, C., 1972. Development of Apterygote insects. In: Counce, S.J., Waddington, C.H. (Eds.), Developmental Systems: Insects. Academic Press, London, pp. 49–94.

Konopová, B., Zrzavý, J., 2005. Ultrastructure, development, and homology of insect embryonic cuticles. J. Morphol. 264, 339–362. Available from: https://doi.org/10.1002/jmor.10338.

Kränzler, L., Larink, O., 1980. Postembryonale Veränderungen und Sensillenmuster der abdominalen Anhänge von *Thermobia domestica* (Packard) (Insecta: Zygentoma). Braunschw. Naturk. Schr. 1, 27–49.

Laibach, E., 1952. *Lepisma saccharina* L., das Silberfischchen. Z. Hyg. Zool. 40, 1–50.

Larink, O., 1983. Embryonic and postembryonic development of Machilidae and Lepismatidae (Insecta: Archaeognatha et Zygentoma). Entomol. Gen. 8, 119–133.

Lindsay, E., 1940. The biology of the silverfish, *Ctenolepisma longicaudata* Esch. with particular reference to its feeding habits. Proc. R. Soc. Victoria 52, 35–83.

Masumoto, M., Machida, R., 2006. Development of embryonic membranes in the silverfish *Lepisma saccharina* Linnaeus (Insecta: Zygentoma, Lepismatidae). Tissue Cell 38, 159–169. Available from: https://doi.org/10.1016/j.tice.2006.01.004.

Mendes, L.F., 2002. Taxonomy of Zygentoma and Microcoryphia: historical overview, present status and goals for the new millennium. Pedobiologia (Jena) 46, 225–233. Available from: https://doi.org/10.1078/0031-4056-00129.

Sturm, H., Machida, R., 2001. Archaeognatha. Handbook of Zoology. W. de Gruyter, Berlin, New York, 213 pp.

Sweetman, H.L., 1938. Physical ecology of the firebrat, *Thermobia domestica* (Packard). Ecol. Monogr. 8, 285–311. Available from: https://doi.org/10.2307/1943252.

Sweetman, H.L., 1952. The number of instars among the Thysanura as influenced by environment. In: 9th Int. Congr. Ent. 1, 411–415.

Verhoeff, K.W., 1910a. Über felsenspringer Machiloidea. 4. Aufsatz: Systematik und Orthomorphose. Zool. Anz. 36, 425–438.

Verhoeff, K.W., 1910b. Über Felsenspringer, Machiloidea 3. Aufsatz: Die Entwicklungsstufen. Zool. Anz. 36, 385–399.

Chapter 4

The hemimetabolan development

Hemimetaboly appeared with the emergence of wings and the pterygote insects, and it is the metamorphosis mode of three insect superorders: Palaeoptera, Polyneoptera, and Paraneoptera. Hemimetabolan metamorphosis is characterized by the morphology (and often the lifestyle) of the juvenile stages, which is similar to that of the adult, implying that embryogenesis gives rise to a nymph that has the basic adult-like features. Thus, nymphal development consists essentially of growing, during which the growth ratio of the exoskeleton linear dimensions tends to remain constant throughout all nymph-to-nymph molts. This empirically broadly supported pattern is known as Dyar's rule (Kivelä et al., 2016; Sehnal, 1985). The transition from the last nymphal instar to the adult is initiated when the nymph reaches a given critical weight, which is an important parameter that has been best studied in holometabolan insects (Nijhout et al., 2014; Nijhout and Callier, 2015). Then, in the transition from the final nymphal instar to the adult, significant qualitative changes occur, like the complete formation of the wings and functional genital structures.

EMBRYOGENESIS IN HEMIMETABOLAN SPECIES

Embryo development in Palaeoptera, Polyneoptera, and Paraneoptera fits the developmental steps generally observed in insects, from the formation of the blastoderm and the germ band to organogenesis (Anderson, 1972) (see Chapter 2: A spectacular diversity of forms and developmental modes). Embryo segmentation belongs to the short germ band type although with a notable variation depending on the species. Most species show the canonical short type, forming a blastoderm that comprises the head lobes, the most apical trunk segment, and the terminus (Davis and Patel, 2002). Others, however, like the cricket *Gryllus bimaculatus*, show an intermediate germ band type, as they reach gastrulation with some abdominal segments (Mito et al., 2005). Blastokinesis is characteristically complex in hemimetabolan species

Insect Metamorphosis. DOI: https://doi.org/10.1016/B978-0-12-813020-9.00004-1

(Panfilio, 2008), although there are species where it is reduced to discrete displacements of the embryo in the same egg side, like in the cockroach *Blattella germanica* (Tanaka, 1976).

The hemimetabolan nymphal morphology is shaped during the organogenesis, when cells differentiate into organ systems and epidermis, so that at the end of this process a functional nymph, morphologically similar to what the adult will be, hatches form the egg. The different organs are formed from the mesoderm, endoderm, and ectoderm. Essentially, the mesoderm gives rise to muscular systems and fat body tissues. The midgut and associated tissues are formed from the endoderm, whereas the nervous system and the cuticle structures develop from the ectoderm. Therefore most of the external morphology of the nymph, including, in particular, the appendages, wing primordia, and eyes, derive from the ectodermal germ layer.

The appendages (antennae, mouth pieces, legs) of a newly hatched hemimetabolan nymph arise from embryonic limb buds that start to be formed early in embryogenesis, and they complete the segmentation around organogenesis. Mouth pieces are especially diverse as they are usually adapted to specific alimentary regimes. The formation of the nymphal piercing and sucking mouth pieces of selected Thysanoptera, Hemiptera, and Phthiraptera has been studied in detail. The extremely specialized mouthparts of these insects differentiate in late embryogenesis through dramatic modifications of the mandibles and maxillae (Thysanoptera, Hemiptera) or the hypopharinge and labium (Phthiraptera) (Heming, 2003). The wing primordia also have an ectodermal origin, and their formation starts out as a cluster of cells growing underneath this germ layer in early embryogenesis. The wing primordia can be externally exposed from the first nymphal instar, forming wing pads, like in the locust *Locusta migratoria*, or from later instars, as in most species (Joly, 1968). In others, like the cockroach *B. germanica*, the wing primordia are encapsulated into a pteroteca, a sort of cuticle pocket located in the laterobasal part of the prothorax and metathorax.

The hemimetabolan nymphs have compound eyes similar to those of the adult although smaller in size. Compound eyes consist of a number of ommatidia; each ommatidium contains a dioptric apparatus, some pigment cells, and eight rhabdomeric-type photoreceptive cells. During embryogenesis, each ommatidium develops through cell proliferation, invagination, and differentiation of ommatidial precursors within an optic placode on either side of the embryonic head (Heming, 2003). Later, during the nymphal period, new ommatidia differentiate from the anterior growth zone, so that eyes continue to expand through successive molts (Paulus, 1989). Importantly, hemimetabolan embryos secrete and deposit three cuticles (EC1, EC2, and EC3) during embryogenesis, as observed in representatives of paleopteran, polyneopteran, and paraneopteran species (Konopová and Zrzavý, 2005).

THE HEMIMETABOLY IN PALAEOPTERA

The superorder Palaeoptera comprises two extant orders, Odonata and Ephemeroptera. They have in common the features of wing articulation and wing position in rest. Unlike in Neoptera, the wing muscles of Palaeoptera insert directly at the base of the wing. Therefore a movement downward of the wing base lifts the wing upward, as if the wings were paddling in the air. Moreover, and also unlike in Neoptera, Odonata and Ephemeroptera lack the ability to fold the wings back over the abdomen, which is an ancestral feature of winged insects. Odonata, commonly known as dragonflies, are divided into two suborders, the Zygoptera or damselflies and the Anisoptera or true dragonflies. Over 6000 species of Odonata are known worldwide. The Ephemeroptera are most commonly known as mayflies and comprise approximately 3000 described species. With the exception of a few cases, in both orders, the nymphs are aquatic and adults are aerial, with quite different and noncompeting styles of life (Thorp and Rogers, 2015).

Odonata

Dragonflies are generally large-sized insects, the adult of the biggest species reaching a size of 19 cm across the wings and a body length of over 12 cm. They are predatory in the aquatic nymphal stages and the aerial adult period (Fig. 4.1). Nymphs feed on a range of freshwater invertebrates, the larger ones being even able to prey on tadpoles and small fishes. Adults capture insect prey in the air, taking advantage of their acute vision and precisely controlled flight. The mating system is complex because dragonflies are among the few insect groups that use indirect sperm transfer along with sperm storage, delayed fertilization, and sperm competition (Corbet, 1999; Stoks and Córdoba-Aguilar, 2012; Tillyard, 1917).

FIGURE 4.1 The Odonata *Somatochlora viridiaenea*. From left to right, mature nymph and adult female and male just after emerging from the nymphal exuvia. Photos courtesy of Akira Ozono (nymph) and Ryo Futahashi (adults).

The female lays the eggs directly into the water or on a suitable aquatic substrate, usually over emerged or immersed vegetation. The eggs take from several days to several months (some dragonflies overwinter as embryos) to hatch, giving aquatic nymphs that, in different species, will molt between 9 and 15 times before metamorphosing into adults (Corbet, 1999; Suhling et al., 2015). The duration of the nymphal period ranges between 1 month and several years, depending on the species, although it is generally associated with the habitat rather than with the species size. Species living in running water have longer nymphal periods than those inhabiting still water. When the nymph prepares for metamorphosis, it stops eating, goes to the surface of the water, climbs up a semisubmerged plant, begins to habituate to air breathing, and culminates the molt. Nymphs have an oval slender or flattened body, with three pairs of long segmented legs extending from the thorax and generally wing pads in the mesothorax and metathorax (Fig. 4.1). Dragonfly nymphs breathe underwater through gills, which are internal, lining in a specially modified chamber in the rectum (Fig. 4.2). Damselflies also have external gills that are placed at the distal end of the abdomen. A particular feature of the dragonfly nymphs is the hinged labium, also known as "mask," which can project forward to capture prey and then rapidly retract (Fig. 4.3A).

The adult shows a slender body (Fig. 4.1). The head is dominated by the two compound eyes that can contain more than 20,000 ommatidia, which provide complete vision in the frontal hemisphere of the insect (Futahashi, 2016). The mouthparts are adapted for biting with a toothed jaw, and the

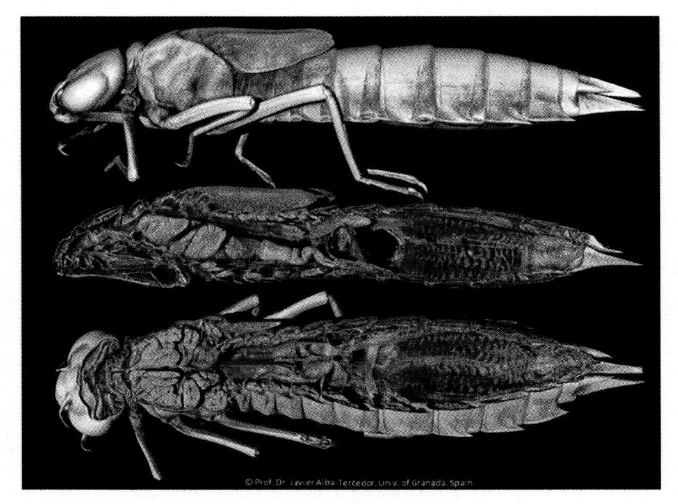

FIGURE 4.2 Microtomographic images of a mature nymph of the Odonata *Anax imperator*. Note the gills lining the modified chamber in the rectum. Colors represent the opacity to X-ray and are not real. Image courtesy of Javier Alba Tercedor (Universidad de Granada, Spain).

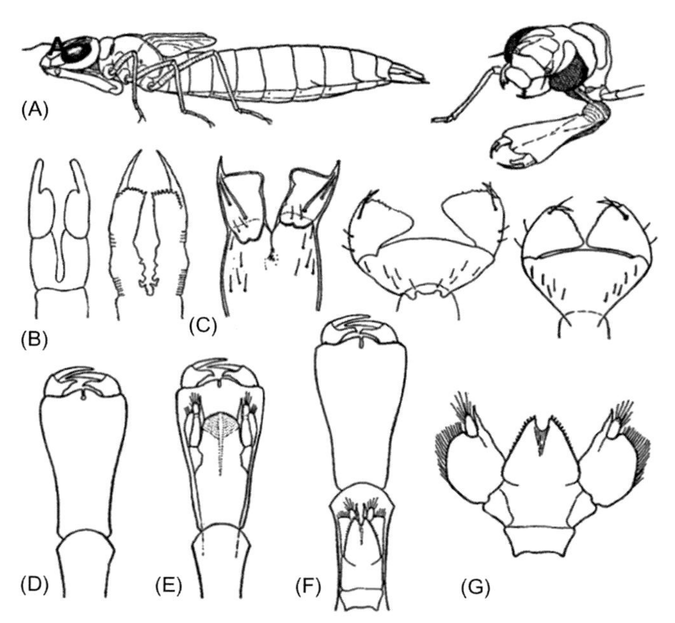

FIGURE 4.3 The labium of Odonata. (A) Nymph of *Anax* sp. with the labium folded (left) and detail of the labium being protracted (right). (B) Labium of a 17-day-old embryo (left) and 20-day-old embryo of *Anax* sp. (right). (C) Labium development in *Sympetrum* sp., from left to right: labium from a first instar nymph (containing that of the second instar), from a second instar nymph partly expanded, and the same fully expanded. (D–G) Labium development in *Anax* sp.: a mature nymph (D); an older nymph in an early stage of the formation of the imaginal labium within the prementum (E); a later nymphal instar, when the imaginal labium has retracted into the postmentum and has adopted the adult structure (F); and adult labium dissected out from the stage shown in (F), and extended on a slide (G). From Snodgrass (1954), with permission.

flap-like labrum can be shot rapidly forward for prey capturing. The mesothorax and metathorax are fused into a hard structure that provides a robust attachment for the wing muscles inside it (Suhling et al., 2015). The wings are long, narrower at the tip and wider near the base, and generally hold horizontally both in flight and at rest, although damselflies hold the wings upright above the thorax and orthogonal to the main body axis. The abdomen is long and slender, with 10 segments and a terminal appendage-bearing segment. In males, the ventral side of the second and third abdominal segments contains a pair of claspers and the penis, whereas the spermaries open on the ninth segment. In females, the genital opening is on the underside of the eighth segment and is covered by a simple flap or an ovipositor (Suhling et al., 2015).

The metamorphosis in Odonata involves extensive morphological remodeling, especially regarding the labium. The embryonic labium has a primitive shape, with the lobes of the prementum clearly separated and the unsegmented palpi, which show fingerings that will later become apical hooks (Fig. 4.3B) (Butler, 1904). Then, in the first nymphal instar, the prementum is undivided, and the labial palpi are placed at the distal part, close together (Fig. 4.3C). After the molt to the second instar, the prementum stretches transversally and later becomes more elongated (Fig. 4.3C) (Corbet, 1951). We can see thus that the labium of the embryo shows a shape that directly transforms into the specialized structure of the nymph (Fig. 4.3C and D). Later the labium of the adult is first formed in the prementum of the nymphal labium, approximately 5 days after feeding has ceased (Fig. 4.3E). Subsequently, the labium retracts into the postmentum and acquires a more transverse form, adopting the approximate adult shape (Fig. 4.3F and G). The total transformation of the nymphal to the adult labium in *Aeshna cyanea* takes 12 days (Munscheid, 1933). During this time, a total histolysis of the nymphal labial muscles followed by the formation of new imaginal muscles, and the corresponding tonofibrillar muscle attachments on the imaginal cuticle, takes place. In this process, two pairs of nymphal muscles disintegrate and are not replaced (Munscheid, 1933). A number of authors (e.g., Munscheid, 1933; Snodgrass, 1954) consider that the changes in the labium of dragonflies are equivalent to the changes in the pupa of the holometabolan insects. The long quiescent transformation period apparently allows the regenerated muscles to attach directly on the new imaginal cuticle without the bridge of another molt.

The general changes undertaken by dragonfly nymphs have been described in detail by Tillyard (1917). These changes, besides those of the labium, include the growth of compound eyes, development of the ocelli, increase of the segment number in the antennae and tarsi, increase of the Malpighian tubule number, changes in the structure of the thorax associated with the development of the wings, and progressive development of the nervous system (Tillyard, 1917). Among the above changes, perhaps the more apparent is the transformation of the compound eyes, which require new capacities in the aerial life. Indeed, the metamorphosis of the eyes includes a new program of spatial–temporal expression of opsins that is more suitable for prey capture in the air (Futahashi et al., 2015). In any case, the above progressive changes fit into the general definition of hemimetabolan metamorphosis, although Berlese (1913) categorized this mode of metamorphosis as paurometaboly, because of the partial similarity between nymphs and adult, and Weber (1949) as archimetaboly essentially because the nymphs are aquatic.

Ephemeroptera

Mayflies are minute insects: in the adult, the body length is between 2 and 40 mm. They are unique among insects in that they molt one more time after forming functional wings. Thus the last molt of the nymphal period does not give the definitive adult form, but a winged stage called subimago that morphologically resembles the adult. This peculiar mode of metamorphosis was classified as prometaboly by Weber (1949), a category that is still used by a number of modern authors. Berlese (1913) placed mayflies among their paurometabolan, considering the relative similarity of nymphs and adults. Nymphal stages of Ephemeroptera are aquatic, showing herbivore or detritivore regimes (only a few species are predators of other small insects). Conversely, subimagos and adults practically do not feed, which is consistent with the short duration of these stages. Generally, subimagos and adults exhibit either of two basic longevity patterns. The longer and more predominant includes a subimaginal period that lasts between 8 hours and 3 days, and an adult period that can last between 1 day and 2 weeks or more (but rarely beyond 3 weeks, like in the abundant and largely distributed species *Cloeon dipterum*). The second and shorter longevity pattern includes a subimaginal period of a few minutes and an adult period of a few hours at most. Within these short times, adult males perform a courtship dance, flying a few meters above the water usually in large swarms. Females fly into these swarms, and mating takes place in the air (Alba-Tercedor, 2015; Edmunds and McCafferty, 1988; Lancaster and Downes, 2013).

Females typically lay the eggs dropped onto the surface of the water, although the species of the genus *Baetis* actively walk until they submerge and lay the eggs attached to submerged substrates, usually stones. The period of embryo development until hatching is quite variable and may vary from a few days to nearly a year. Some species can present egg diapause (Clifford, 1982). From the egg hatches an aquatic nymph that, depending on the species, may live between 2 weeks and 2 years in the water, undertaking between 12 and 50 molts. During this period, nymphs undergo significant morphological changes that affect mainly the formation of gills and wing pads.

Nymphs have an elongated, cylindrical, or somewhat flattened body. The mouthparts are specialized in chewing and include a flap-like labrum and a pair of strong mandibles. Each thoracic segment bears a pair of legs and, in mature nymphs, wing pads are visible on the mesothorax, and in some species, also on the metathorax. The abdomen has 10 segments, some of which with a large pair of operculate gills. In many species, up to seven pairs of gills arise from the top or sides of the abdomen (Fig. 4.4A–C), but in some species they are under the abdomen, and in a very few species, the gills are instead located on the coxae or the base of the maxillae. The tip of the abdomen presents a pair or three slender thread-like projections, called caudal filaments (Lancaster and Downes, 2013) (Fig. 4.4A and C).

The last nymphal instar molts into a winged subimago and then to the adult stage. Although the subimago resembles the adult, both winged stages are easily distinguished from each other in practically all species. Subimagos

FIGURE 4.4 Nymphs, subimagos, and adults of mayflies. (A) Mature nymph of *Baetis maurus*, typical of water courses with strong current, showing the central terminal filament markedly reduced, which helps to avoid the force of the current. (B) Mature nymph of *Baetis rhodani* browsing among the immersed vegetation. (C) Microtomographic images of a mature nymph of *Ephemera danica* showing the gills lining in the abdomen. Colors represent the opacity to X-ray and are not real. (D) Subimago female of *Ecdyonurus venosus*. (E) Adult male of the same species. (A), (B), (D), and (E), From Alba-Tercedor (2015), with permission; (C) image courtesy of Javier Alba Tercedor (Universidad de Granada, Spain).

generally have dull, opaque, or translucent wings, whereas those of the adult are shiny and transparent (Fig. 4.4D and E). The outer and hind edges of the subimago wings are fringed with a row of fine cilia, and their surface is covered with microtrichia. Conversely, the wings of the adult of most species lack these hairy structures. The body surface of the subimago is more or less covered with microtrichia or microspines, whereas almost all adults lack coverings of the body surface and appear glossy. The forelegs and caudal filaments of subimagos are shorter than in adults, especially in males, and the male genitalia and sometimes the eyes are not yet full sized. Subimagos tend to be slow fliers with little agility in comparison with adults of the same species.

All male mayflies and most females molt from subimago to adult, but in females of at least two species of Leptophlebiidae, the apolysis of the subimago to the adult occurs but the ecdysis does not. Oligoneuriinae mayflies are also peculiar because when molting to the adult stage they shed the exuvia from the body but retain the subimaginal cuticle on the wings. Finally, females of a few specialized species of Palingeniinae seem to have lost the adult stage, and the subimago is the last one (Edmunds and McCafferty, 1988). Interestingly, this feature had already been reported by Swammerdam in the 17th century, when he discovered that "adult" males of *Palingenia longicauda* molt once again but the females do not (Swammerdam, 1675).

The evolution and functional sense of the subimago stage has been extensively discussed (see Edmunds and McCafferty, 1988). Paleontological information suggests that the subimago is a primitive feature. Kukalová-Peck (1978, 1983) has reported that early immature stages of fossil Ephemeroptera possess freely articulated developing wings, and that wing development proceeded gradually through successive molts. In later stages, wings appear incompletely developed but perfectly articulated, whereas the body shape is similar to that of an adult, with fully formed wings. Considering paleontological data, Snodgrass (1954) and Schaefer (1975) suggested that the subimago stage of extant mayflies do not have any special functional sense, but it is a kind of relic of one or more preadult winged instars of ancestral mayflies.

Ide (1937) proposed that the hydrofuge properties of the hairy surface of the body, legs, and wings of the subimago would have a functional sense. They would allow the insect to overcome the hazards of a metamorphic transition from an aquatic nymph to a winged terrestrial form at the water–air interface. Therefore, the subimago stage, with their peculiar hydrofuge properties, would have the functional sense of facilitating the habitat transition from water to air. Hydrofuge properties would be especially important for those species that complete the emergence underwater, although there are others in which the nymph exits from the water and then starts the transformation into the (also hairy) subimago in the aerial environment (Edmunds and McCafferty, 1988). These authors speculated that the hydrofuge structures may have been selected to prevent the membrane of the wing from sticking to itself in the folded, furled, or convoluted position and may thus facilitate unfolding at emergence (Edmunds and McCafferty, 1988). It is also plausible that the hairy surface of the subimago, particularly on wings, have a function during the delicate ecdysis and exuvia removal in the wings, within the molting process of subimago to adult.

Maiorana (1979) proposed that the function of the subimago is to allow necessary growth from the nymphal to the adult morphology, which could not otherwise be accomplished in a single molt. The mayfly subimago was considered thus as experiencing a transformation function similar to that of the holometabolan pupa (Maiorana, 1979). In support of this notion are data indicating that almost all adult structures are formed before the subimaginal molt, but that their full expansion or unfolding, for example, of forelegs and caudal filaments, may not be completed until the adult molt (Edmunds and McCafferty, 1988). However, the possible homology of the ephemeropteran subimago stage with the holometabolan pupa is another question. Recent transcriptomic analyses comparing young nymphs, mature nymphs, subimago, and imago stages of the mayfly *Cloeon viridulum* indicate that the mayflies' life cycle is typically hemimetabolan, and that the closest similarity of the subimago stage is with the adult (Si et al., 2017).

THE POLYNEOPTERA

Polyneopterans include the orders Plecoptera, or stoneflies; Dermaptera, or earwigs; Orthoptera (grasshoppers, katydids, bush crickets, crickets, and locusts); Zoraptera, or angel insects; Mantodea, or mantises; Blattodea, or cockroaches and termites; Mantophasmatodea, most commonly known as gladiators; Grylloblattodea, commonly called rock crawlers, ice crawlers, or ice bugs; Phasmatodea, or stick insects; and Embiodea, commonly known as webspinners (Misof et al., 2014). Some 40,000 species of polyneopterans have been described. The most diverse order is Orthoptera (25,000 species) and the least Zoraptera (30 species), Mantophasmatodea (16 species), Grylloblattodea (34 species), and Embiodea (360 species). The other orders have between 2000 and 4000 species.

All polyneopteran species exhibit the typical hemimetabolan mode of metamorphosis; thus the embryo development produces a nymph that shows a morphology and lifestyle close to those of the adult. The best known groups are Dermaptera, Orthoptera, Mantodea, Blattodea, and Phasmatodea, which show a postembryonic development that is moderately long, from weeks to more than a year, divided into a number of molts that range from 5 to 12. The wing primordia can be externally visible from the first nymphal instar, as in the locust *L. migratoria* (Fig. 4.5A) or from more advanced instars, like in most of the species. In some groups, like cockroaches, wing primordia are hidden in cuticle pockets, or pterotecae, located in the basal angles of the mesothorax and metathorax (Fig. 4.5B). Although being hemimetabolan, termites are special because they are eusocial insects that divide labor among morphologically differentiated castes consisting of sterile male and female "workers" and "soldiers." Colonies have fertile males called "kings" and one or more fertile females called "queens." Termites were once classified in a separate order from cockroaches, but recent phylogenetic studies indicate that they evolved from

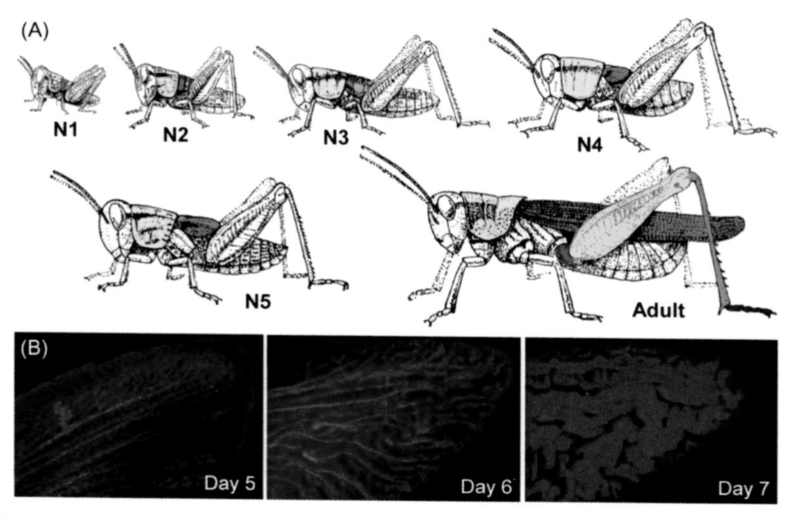

FIGURE 4.5 Wing development in Polyneoptera. (A) Life cycle of the orthopteran *Locusta migratoria*, showing the five nymphal instars (N1–N5) and the adult. Wing pads are shown in brown and the thorax in yellow; in the adult, the wings are completely formed, the thorax becomes more elongated, and all main segments of the hind, jumping legs (coxa in red, femur in orange, tibia in green, and tarsus in blue) are more robust and longer (the tarsus, in addition, adds a new tarsomere). (B) Development of wing buds in the blattarian *Blattella germanica*. Development of the wing buds within the pterotecae is shown on the last 3 days (5, 6, and 7) of the last nymphal instar; labeling corresponds to 5-ethynyl-2′-deoxyuridine (EdU) (discrete red spots labeling DNA synthesis), and 4′,6-diamidine-2′-phenylindole dihydrochloride (DAPI) (blue color labeling live cells). (A) Modified from Folsom (1914); (B) From Huang et al. (2013), with permission.

close ancestors of cockroaches during the Jurassic or Triassic (Harrison et al., 2018; Inward et al., 2007).

Considering metamorphosis, the order Plecoptera is the most peculiar. Stonefly life cycle is considerably longer than that of other polyneopterans, from months to few years, reaching more than 30 molts depending on the species and conditions (Grimaldi and Engel, 2005; Hoell et al., 1998; Sehnal, 1985). Moreover, stoneflies are exceptional among polyneopterans especially because adults are terrestrial (Fig. 4.6A), whereas the juvenile stages are aquatic (Fig. 4.6B–D). There are only a few interesting exceptions to this general lifestyle. For example, the adult of the species *Capnia lacustra* is not aerial but has been captured only in deep waters of Lake Tahoe, in the United States (Jewett, 1963). Also interesting is the case of some Gripopterigydae species from Patagonia and New Zealand, whose nymphs live in terrestrial habitats in cold and humid mountain areas (Hynes, 1976). However, the nymphs of the vast majority of stonefly species are aquatic although the adaptations of nymphs to the aquatic life are not very conspicuous. Indeed, in addition to the mild morphological differences between nymphs and adults that are due to immaturity, the only remarkable nymphal

FIGURE 4.6 Nymphs and adults of Plecoptera. Adult (A) and mature nymph (B) of Nemouridae. (C) Mature nymph of *Perla* sp. (D) Detail of the thorax showing the gills with a structure of tufts of delicate filaments. Images courtesy of J. Manuel Tierno de Figueroa and Manuel J. López-Rodríguez (Universidad de Granada, Spain).

feature is the occurrence of the gills needed for respiration in the water. In general, stonefly nymphs have an elongated shape with cylindrical or flattened section, with well-developed compound eyes and frontal ocelli; antennae multisegmented, long and slender; legs relatively robust, with three-segmented tarsi and two pretarsal claws; the three thoracic segments well differentiated, showing the characteristic wing pads well apparent at the base of the mesothorax and metathorax, which develop progressively through the numerous nymphal molts; and the abdomen, showing two symmetric caudal filaments in the distal segment (Fig. 4.6B and C). Although there are species of stoneflies whose nymphs do not have gills (in the genera *Leuctra*, *Nemoura*, *Capnioneura*, and *Capnia*, for example), many others present gills as a special adaptation to the aquatic life. The gills often consist of tufts of delicate filaments penetrated by tracheae than can be placed in the head (in the submentum) or, more frequently, in the thorax (in the prosternum, in the pleural area, or in the coxae) (Fig. 4.6D) or the abdomen (in general in the anal region) (Tierno de Figueroa and López-Rodríguez, 2015). It is important to stress that all nymphal features are essentially the same as in the adult, and the gills, when present, show in most of the cases a very simple structure. All these features denote that juvenile Plecoptera have secondarily adapted to the aquatic life but conserving most of the characteristics of a current polyneopteran terrestrial nymph. Weber (1949) categorized the plecopteran mode of metamorphosis as "archimetaboly" because, like in Odonata, the nymphs are aquatic.

In general, nymphs and adults have different feeding habits. This is especially pronounced in the Perloidea and Eusthenoidea superfamilies, where nymphs are mainly predators while adults feed mainly on pollen (in small and medium-sized species) or practically do not feed (in large species). Even in the phytophagous and detritivore species, food resources can be different in nymphs and adults, in part as a consequence of the different habitats where they live (e.g., diatoms is a usual nymph resource, whereas pollen is a much more frequent resource in adults). There is also a certain variability in the predatory habits of Perloidea and Eusthenoidea nymphs, which show a tendency to be detritivorous at the beginning of the development, becoming more predators when increasing in size (Tierno de Figueroa et al., 2003).

The number of molts to complete the nymphal development varies from species to species, but they are in the range of 12–33, and generally differ between males and females. The duration of the nymphal period is usually 1 year or less although some species may take up to 3 years or more to do so (Hynes, 1976). Just before metamorphosis, the mature nymph climbs plants or stones in order to complete the imaginal molt. The nymph may reach some distance from the shore before molting, which suggests that it still keeps some ability to manage on land. Adults possess two pairs of wings, which at rest are usually arranged on the abdomen, the hind wing folded longitudinally, forming in most cases a flat sheet (Fig. 4.6A) (Zwick, 2000). Adults, however, are generally poor fliers. As observed in species of the Northern hemisphere, they rather rely on communication through vibrations to find a mate (Stewart, 2001). After copulation, the adult female return to the water to lay the eggs, which are deposited in groups (or, more rarely, isolated) on the water surface. The eggs fall on the stones of the fluvial bed where they are fixed by means of specialized egg structures or gelatinous membranes, although in some cases they are partially or totally buried in the substrate. The longevity of the adult ranges from a few days to several weeks (Hynes, 1976; Tierno de Figueroa et al., 2003).

In spite of the peculiarity of the aquatic life of the juvenile phases, the stoneflies represent a very simple hemimetabolan metamorphosis, with a markedly ancestral character, both for the large number of nymph molts and for the morphological similarity between the nymphs and the adult.

THE PARANEOPTERA

Paraneopterans include the orders Psocoptera (booklice, barklice, or barkflies, with 5500 species described), Phthiraptera (lice, with 5000 species), Thysanoptera (thrips, with 6000 species), and Hemiptera. The Hemiptera are the most diverse and are divided into four suborders: Cicadomorpha (cicadas, leafhoppers, treehoppers, and spittlebugs, with 35,000 described species); Fulgoromorpha (planthoppers, with 12,500 species); Sternorrhyncha (psyllids or jumping plant louse, whiteflies, scale insects, aphids, phylloxera bugs, with 18,200 species); and Heteroptera (true or typical bugs, with 40,000 species).

In general, Psocoptera, Phthiraptera, and Hemiptera Heteroptera are small (the smallest Psocoptera are 1 mm in length) to medium sized (the biggest Heteroptera Belostomatidae can reach 12 cm in length). They exhibit the typical hemimetabolan mode of metamorphosis. Embryogenesis produces a nymph morphologically close to the adult, which develops through a nymphal period that can generally extend from weeks to years (like in some cicadas that can live some 15 years as underground nymphs), undergoing a moderate number of molts (from 3 to 4 in Phthiraptera to 4 to 9 in Heteroptera). This allows progressive growth, generally including allometric growth of the wing pads, as exemplified by Psocoptera (Alexander et al., 2015) (Fig. 4.7). At the end of the nymphal period, wings and genitalia complete development with the imaginal molt.

Most paraneopterans can reproduce parthenogenetically (Psocoptera, Thysanoptera, Hemiptera Sternorrhyncha), but some Aphidoidea and Phylloxeroidea have complex life cycles and modes of reproduction, with cyclical parthenogenesis and host alternation. In aphids, the simplest life cycle involves a single host during all the year and may alternate between sexual and asexual generations (holocyclic). Alternatively, all offspring can be produced parthenogenetically, without egg laying (anholocyclic). A number of aphid species can show both holocyclic and anholocyclic populations under different environmental contexts. Frequently, however, aphid reproduction is more complex, involving different host plants. In a few species, aphids alternate between primary hosts (usually woody plants), on which

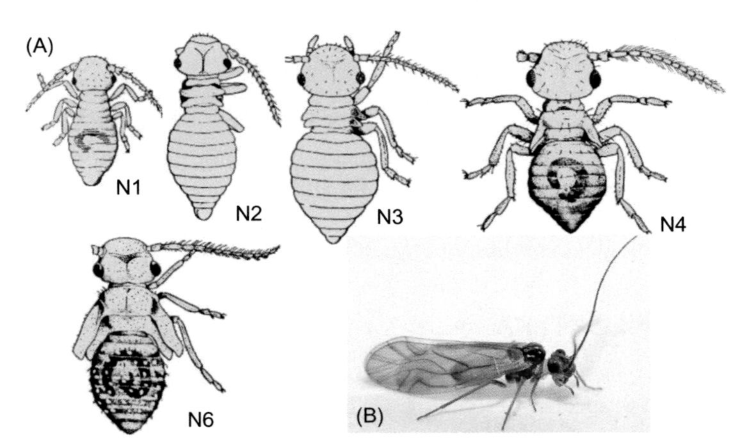

FIGURE 4.7 Nymphs and adults of Psocoptera. (A) Nymphal stages (N1–N4 and N6) of *Stenopsocus meridionalis* (Stenopsocidae). (B) Adult of *Valenzuela* sp. (Caeciliusidae). From Alexander et al. (2015), with permission.

they overwinter, and secondary hosts (usually herbaceous), where they reproduce massively in summer. Moreover, a few aphid species can produce a soldier caste, other species show a varied polyphenism under different environmental contexts, and some others are able to control the sex ratio of their progeny, also depending on the environmental context (Moran, 1992).

Despite all these life cycle peculiarities linked to reproductive strategies, the nymphs of Aphidoidea and Phylloxeroidea are morphologically similar to the adult, thus fitting in the hemimetabolan mode of metamorphosis. Weber (1949) categorized the Phylloxeroidea as homometabola because the wing pads appear in the last nymphal instar. The other groups of Stenorrhyncha (Psylloidea, Aleurodoidea, and Coccomorpha), as well as the order Thysanoptera (thrips), deserve a special attention as they present a peculiar postembryonic development.

Psylloidea, Aleyrodoidea, and Coccomorpha

These Stenorrhyncha groups are composed of minute species (the adult hardly reaches 5 mm) and include notorious agricultural pests. The postembryonic development of Psylloidea, or psyllids, lasts a few weeks and comprises five nymphal instars that, in general, morphologically resemble the adult. The wing pads appear during the third nymphal instar and grow progressively until the fifth instar. What makes the psyllid development peculiar is the deep metamorphosis of the legs, which has been described in detail by Weber (1930) for the species *Psylla mali*, and summarized by Snodgrass (1954). The first nymphal instar is mobile, and the legs are significantly separated from each other on the ventral side of the body. After the first molt, the psyllid starts a sessile life, and the legs become closer to each other and flexed transversally beneath the thorax, making them clasping organs. They also have reduced segmentation, as the joints between the femur and trochanter and those between the tibia and the tarsi disappear. With the molt to the fifth instar, the nymph acquires adult-like features: the body elongates, the antennae become longer, and rudiments of new leg segmentation appear. At the imaginal molt, the legs become fully segmented, and the rudimentary hind legs of the nymph transform into jumping legs in the adult, due to the enlargement of the coxae, the muscles of the trochanter, and those associated in the thorax (Weber, 1930). From this dramatic remodeling, adult psyllids can jump by rapid movements of the hind legs, whereas power is provided by the large thoracic muscles that depress the trochanters so that the two hind legs move in parallel planes on either side of the body. These movements accelerate the body to take-off in 1–1.5 m distances (Burrows and Wang, 2012).

The postembryonic development of Aleyrodoidea, or whiteflies, generally lasts a few weeks and includes four nymphal instars. Similarly to psyllids, the first dispersal stage (or "crawler") possesses functional legs with which they search a place to settle down, where it will feed on the plant phloem by inserting their buccal stylets in the leaf tissue. The following nymphal instars are then sessile, having reduced legs and the shape of a flattened ellipse, scale-like, fringed with bristles and waxy filaments. The last nymphal instar ceases to feed and undergoes dramatic metamorphic changes to become an adult (Fig. 4.8A). One of the most detailed description of the life cycle and metamorphosis of a whitefly has been reported by Weber (1931) for the species *Trialeurodes vaporariorum*. Briefly, the mobile first nymphal instar has slender antennae, relatively long legs, each showing three segments and an adhesive disc at the apex. The second and third nymphal instars have much shorter antennae, and the legs consist of nonfunctional, small, unsegmented stumps, although they retain the apical adhesive disc. In the fourth nymphal instar, the antennae and the legs become longer, the legs being two-segmented. The adult is formed through a metamorphic process that involves extensive tissue destruction and regeneration within the central part of the body of the fourth nymphal instar, which appears like a quiescent carcass. Finally, at ecdysis, the peripheral, superfluous parts of the nymphal body are cast off (Weber, 1931). The peculiarity of this metamorphosis had been categorized by Berlese (1913) as heterometaboly and by Weber (1949) as

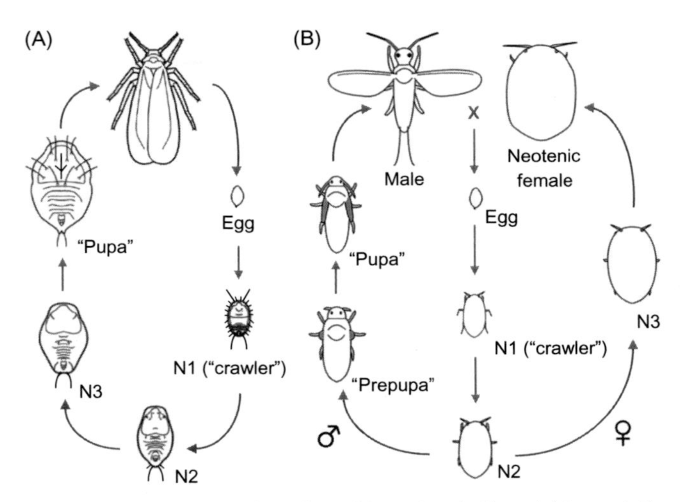

FIGURE 4.8 Schematic representation of the life cycle of Aleyrodoidea and Coccomorpha. (A) *Trialeurodes vaporariorum* (Aleyrodoidea), where the first nymphal instar (N1, "crawler") is mobile, the two following nymphal instars (N2 and N3) are sessile and feed on the plant phloem; the third instar "pupa" is also sessile, but it does not feed. (B) *Planococcus krauhniae* (Coccomorpha), where the N1 "crawler" is undifferentiated in males and females; during the first few days of N2, males and females are undifferentiated, whereas in mature nymphs, males are identifiable by filamentous secretions, which are absent in females; in females, N2 molts into N3 and then into a neotenic female, whereas in males, N2 molts into a "prepupa," a "pupa," and an adult.

allometaboly. A number of authors use the term "pupa" for the terminal nymphal instar of Aleyrodoidea because of its similarities with the holometabolan pupal stage.

The postembryonic development of Coccomorpha (or scale insects, from the wax coating that characterizes the group) is even more modified. The life cycle generally lasts a few weeks, with some exceptions where the complete development spans a few years, like in *Xylococcus filiferus* (Kosztarab and Kozár, 1987). Scale insects are characterized by extreme sexual dimorphism. Males generally undergo two crawling nymphal stages after which they develop into one or two quiescent stages called "prepupa" and "pupa." The male then emerges as a winged adult. In contrast, the female nymphs undergo three or four successive molts, but each stage remains wingless and the adult female emerges as a sexually mature individual that retains juvenile features (Fig. 4.8B).

The morphology of the different stages was described in detail by classical authors, like Suter (1932) and Snodgrass (1954), and, more recently, by Morales (1991) and Gavrilov-Zimin (2018). The first nymphal instar has well-developed eyes, antennae, and legs. It immediately searches for a favorable place on the plant to settle down and feed. Once settled, it molts into a second nymphal stage in which the legs and antennae can be well developed or reduced, depending on the species. Generally, and exemplified by the Pseudococcidae *Planococcus krauhniae* (Vea et al., 2016), the sexes start to differentiate during the second nymphal instar. The female undergoes a new molt to the third nymphal instar and then a final molt that gives a neotenic and sexually functional individual (Figs. 4.8B and 4.9). Her morphology is similar to that of a nymphal instar, lacking wings, with the legs reduced or absent in some cases. Nevertheless, the ovaries become functional and produce a high quantity of eggs in a short time. In the male, the second nymphal instar can already be differentiated because it wanders, settles in dark parts of the plant, stops feeding, and starts secreting filamentous protections. Females also have secretive protections, which can be composed of filament, wax, or even scale cover. Usually after the second instar, male nymphs go through two quiescent stages. The first stage, called "prepupa" (or "propupa," according to some authors), possesses well-developed antennae, legs, eyes, and wing pads and is covered by a secretion forming a light cocoon. In the following stage, usually called "pupa," the insect acquires a morphological shape closer to the adult, with the head, thorax, and abdomen clearly separated and showing elongated antennae and wing pads (Figs. 4.8B and 4.9). It is worth noting that most of the information available refers to the Pseudococcidae (mealybugs), which represent a small percentage of the total Coccomorpha. If we consider also other families, we found a notable diversity of life cycles. For example, the male of the apple mussel scale, *Lepidosaphes ulmi* (Diaspididae) pass through a single "pupa" stage (Suter, 1932), and there are species of Margarodidae that have apodous, cist-like nymphal instars (Gavrilov-Zimin, 2018).

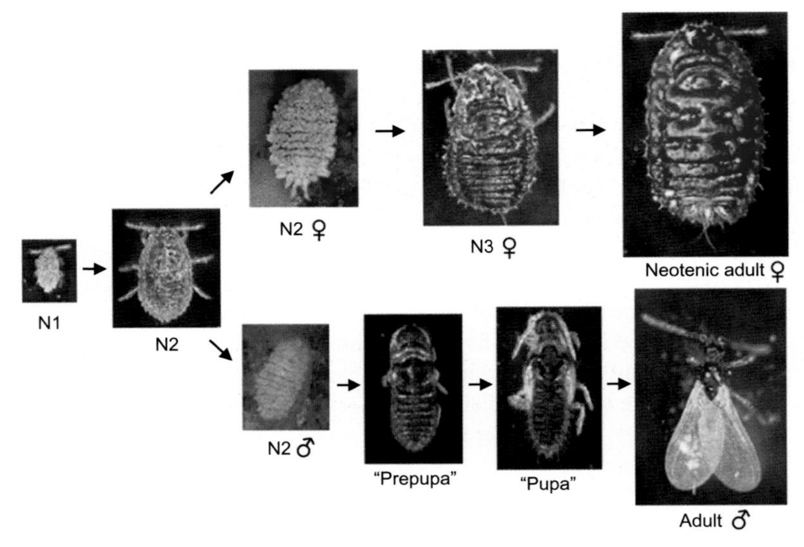

FIGURE 4.9 Life cycle stages of the Coccomorpha *Planococcus krauhniae*. N1-N3, first to third nymphal instars. From Vea et al. (2016), with permission.

External metamorphosis is accompanied by internal transformations, as exemplified by the dramatic changes in musculature undergone by species of the genus *Pseudococcus* reported in minute detail by Mäkel (1942). In her work, Mäkel distinguishes five groups of muscles. Nymphal muscles that do not change significantly in the transition to adult; nymphal muscles that undergo significant changes, such as splitting, merging, or changing in location; nymphal muscles that are eliminated and not regenerated; remodeled muscles formed by the addition of adult structures to the nymphal muscles; and adult muscles formed de novo during the "prepupa" stage. The dramatic morphological and anatomical transformations experienced by Coccomorpha during the postembryonic development led to the suggestion by Snodgrass (1954) that they should be considered as true holometabolan insects. Moreover, Berlese (1913) had already distinguished the Coccomorpha in a special category, neometaboly, and Weber (1949) categorized them as parametabolan, defined by a life cycle with two or three feeding wingless stages followed by one or two nonfeeding stages with external wing primordia. However, the nymphs of Coccomorpha are not very different from the adult stage, and the specialized morphology of the intermediate stages only represents an extreme adaptation to parasitism on plants and phloem feeding.

Thysanoptera

Thrips are very small insects (most species are 1–3 mm long or less, and only a few Australian endemics can reach 12–15 mm), with a slender and cylindrical body. Adults can be winged or apterous. In the first case, the four wings are elongated, narrow, and with fringes on the edges, which increase the wing surface when they are in flight. However, thrips fly only weakly, as their peculiar wings are unsuitable for conventional flight. Instead, they exploit the unusual mechanism of clap and fling, with which they generate lift forming transient vortices near the wings. Thrips have unique asymmetrical mouthparts, as only the left mandible is developed, forming a characteristic cone in the apical part. The mouth apparatus is diversely formed according to the type of alimentary regime, since there are phytophagous, carnivorous, ectoparasitic, and mycophagous species. The Thysanoptera are divided into two suborders, Terebrantia and Tubulifera. Females of Terebrantia species have a sickle-shaped ovipositor, which allows embedding eggs in the plant tissue. In contrast, Tubulifera females do not have ovipositors, so they simply deposit the eggs on the soil or on plants. The life cycle, from egg to adult, can be completed in around 5–20 days, and adult life span can reach 1 or 2 months (Moritz, 2006). Thrips are haplodiploid, with haploid males and diploid females able to reproduce by parthenogenesis, either by arrhenotoky, which is the most common mode, or by thelytoky (van der Kooi and Schwander, 2014).

The postembryonic development of thrips is peculiar because it includes two or three quiescent stages. The first two nymphal instars resemble small adults without wings and genitalia, and they feed on plant tissue. In the Terebrantia (Thripidae), the third and fourth instars are nonfeeding and quiescent, being reminiscent of the holometabolan pupa, which are called "prepupa" (or "propupa", according to some authors) and "pupa," respectively (Fig. 4.10). The Tubulifera (Phlaeothripidae) have three quiescent instars

FIGURE 4.10 Life cycle of the thrips *Frankliniella occidentalis*. Scanning electron microscopy images of the first and second nymphal instars (N1 and N2), the "prepupa," the "pupa," and the male and female adults. The magnification is the same for all images (adult length: 1.6 mm). From Moritz et al. (2004), with permission.

after the two first nymphal instars: the "prepupa," the first "pupa," and the second "pupa" (Moritz, 2006). In these quiescent instars not only do external features develop (including wing pads and genitalia), but there is also a considerable internal remodeling.

Externally, and compared with the two previous mobile nymphal instars, the bristles of the "prepupa" and "pupa" are much longer, and the cuticle is smoother and membranous in most body regions. The hind coxae shift from the pleural region to a more sternal position. The ovipositor valves covered by a cuticular sheath can be observed in the eighth and ninth sternites. The compound eyes develop in association with deep changes that occur in the first neuropil of the optic lobes (Moritz, 1989). Most dramatic is the reorganization of the muscular system, studied in detail in *Liothrips oleae* by Melis (1935) and in *Hercinothrips femoralis* by Moritz (1989). During the "prepupal" and "pupal" stages, processes of myolysis and muscle formation de novo especially affect the head and the three thoracic segments. Especially dramatic are those of the pterothorax associated with wing functionality and flight. In the abdomen, muscular reorganization is less dramatic, in general, but in *H. femoralis*, Moritz (1989) reports that 21 muscles associated with the ovipositor and gonoducts are formed de novo.

The detailed study of the juvenile development of thrips has been of paramount importance to understand the mechanisms of tospoviruses transmission. As shown in *Frankliniella occidentalis*, the acquisition of tospoviruses is restricted to a narrow temporal window during the first and early second nymphal instars. In this period, and resulting from a displacement of the brain into the prothoracic region, there is a temporary association between the midgut, the visceral muscles, and the salivary glands. When this association is subsequently lost, tospovirus particles massively enter into the Malpighian tubules via the hemocoele (Moritz et al., 2004).

The peculiar metamorphosis of thrips has been considered as an intermediate between the hemimetabolan and holometabolan modes (Moritz, 1989; Snodgrass, 1954). Berlese (1913) classified the Thysanoptera as paurometabola, because of the differences between the nymphs and the adults, whereas Weber (1949) categorized them as remetabola, because two wingless nymphal instars are followed by one mobile and then by one, two, or three quiescent instars showing wing primordia. The morphological similarity of the nymphal instars (including the "prepupa" and "pupa") with the adult, suggest that Thysanoptera should be considered hemimetabolan insects.

REFERENCES

Alba-Tercedor, J., 2015. Clase Insecta. Orden Ephemeroptera. Rev IDE@ – SEA Ibero Divers Entomológica @ccesible 40, 1–17.

Alexander, K.N., Ribera, I., Melic, A., 2015. Clase Insecta. Orden Psocoptera. Rev IDE@ – SEA Ibero Divers Entomológica @ccesible 50, 1–13.

Anderson, D.T., 1972. The development of Hemimetabolous insects. In: Counce, S.J., Waddington, C.H. (Eds.), Developmental Systems: Insects. Academic Press, London, pp. 95–163.

Berlese, A., 1913. Intorno alle metamorfosi degli insetti. Redia 9, 121–136.

Burrows, M., Wang, R., 2012. Jumping mechanisms in jumping plant lice (Hemiptera, Sternorrhyncha, Psyllidae). J. Exp. Biol. 215, 3612–3621. Available from: https://doi.org/10.1242/jeb.074682.

Butler, H., 1904. The labium of the Odonata. Trans. Am. Ent. Soc. 30, 111–133.

Clifford, H.F., 1982. Life cycles of mayflies (Ephemeroptera), with special reference to voltinism. Quaest. Entomol. 18, 15–90.

Corbet, P.S., 1951. The development of the labium of *Sympetrum striolatum* (Charp.) (Odonata). Ent. Mon. Mag 87, 289–296.

Corbet, P.S., 1999. Dragonflies, Behavior and Ecology of Odonata. Cornell Unversity Press, Ithaca.

Davis, G.K., Patel, N.H., 2002. Short, long, and beyod: molecular and embryological approaches to insect segmentation. Annu. Rev. Entomol. 47, 669–699. Available from: https://doi.org/10.1146/annurev.ento.47.091201.145251.

Edmunds, G.F., McCafferty, W.P., 1988. The mayfly subimago. Annu. Rev. Entomol. 33, 509–527. Available from: https://doi.org/10.1146/annurev.en.33.010188.002453.

Folsom, J.W., 1914. Entomology, with special reference to its biological and economic aspects, P. Blakiston's son & co., Philadelphia.

Futahashi, R., 2016. Color vision and color formation in dragonflies. Curr. Opin. Insect Sci. 17, 32–39. Available from: https://doi.org/10.1016/j.cois.2016.05.014.

Futahashi, R., Kawahara-Miki, R., Kinoshita, M., Yoshitake, K., Yajima, S., Arikawa, K., et al., 2015. Extraordinary diversity of visual opsin genes in dragonflies. Proc. Natl. Acad. Sci. U.S.A. 112, E1247–E1256. Available from: https://doi.org/10.1073/pnas.1424670112.

Gavrilov-Zimin, I.A., 2018. Ontogenesis, morphology and higher classification of archaeococcids (Homoptera: Coccinea: Orthezioidea). Zoosyst. Ross. Suppl. 2, 1–260.

Grimaldi, D., Engel, M.S., 2005. Evolution of the Insects. Cambridge University Press, Cambridge Evolution Series.

Harrison, M.C., Jongepier, E., Robertson, H.M., Arning, N., Bitard-Feildel, T., Chao, H., et al., 2018. Hemimetabolous genomes reveal molecular basis of termite eusociality. Nat. Ecol. Evol. 2, 557–566. Available from: https://doi.org/10.1038/s41559-017-0459-1.

Heming, B.S., 2003. Insect Development and Evolution. Comstock Pub. Associates.

Hoell, H.V., Doyen, J.T., Purcell, A.H., 1998. Introduction to Insect Biology and Diversity, second ed. Oxford University Press, Oxford.

Huang, J.-H., Lozano, J., Belles, X., 2013. Broad-complex functions in postembryonic development of the cockroach *Blattella germanica* shed new light on the evolution of insect metamorphosis. Biochim. Biophys. Acta - Gen. Subj. 1830, 2178–2187. Available from: https://doi.org/10.1016/j.bbagen.2012.09.025.

Hynes, H.B.N., 1976. Biology of Plecoptera. Annu. Rev. Entomol. 21, 135–153. Available from: https://doi.org/10.1146/annurev.en.21.010176.001031.

Ide, F.P., 1937. The subimago of *Ephoron leukon* Will., and a discussion of the imago instar (Ephem.). Can. Entomol. 69, 25–29. Available from: https://doi.org/10.4039/Ent6925-2.

Inward, D., Beccaloni, G., Eggleton, P., 2007. Death of an order: a comprehensive molecular phylogenetic study confirms that termites are eusocial cockroaches. Biol. Lett. 3, 331–335. Available from: https://doi.org/10.1098/rsbl.2007.0102.

Jewett, S.G., 1963. A stonefly aquatic in the adult stage. Science 139, 484–485. Available from: https://doi.org/10.1126/science.139.3554.484.

Joly, P., 1968. Endocrinologie des Insectes. Masson et Cie, Paris.

Kivelä, S.M., Friberg, M., Wiklund, C., Leimar, O., Gotthard, K., 2016. Towards a mechanistic understanding of insect life history evolution: oxygen-dependent induction of moulting explains moulting sizes. Biol. J. Linn. Soc. 117, 586–600. Available from: https://doi.org/10.1111/bij.12689.

Konopová, B., Zrzavý, J., 2005. Ultrastructure, development, and homology of insect embryonic cuticles. J. Morphol. 264, 339–362. Available from: https://doi.org/10.1002/jmor.10338.

Kosztarab, M., Kozár, F., 1987. Scale Insects of Central Europe. Springer, Netherlands, Dordrecht. Available from: https://doi.org/10.1007/978-94-009-4045-1.

Kukalová-Peck, J., 1978. Origin and evolution of insect wings and their relation to metamorphosis, as documented by the fossil record. J. Morphol. 156, 53–125. Available from: https://doi.org/10.1002/jmor.1051560104.

Kukalová-Peck, J., 1983. Origin of the insect wing and wing articulation from the arthropodan leg. Can. J. Zool. 61, 1618–1669. Available from: https://doi.org/10.1139/z83-217.

Lancaster, J., Downes, B.J., 2013. Aquatic Entomology. Oxford University Press, Oxford.

Maiorana, V.C., 1979. Why do adult insects not moult? Biol. J. Linn. Soc. 11, 253–258.

Mäkel, M., 1942. Metamorphose und Morphologie des *Pseudococcus*-Männchens mit besonderer Berüchsichtigung des Skelettmuskelsystems. Zool. Jahrb. Anat. 67, 461–588.

Melis, A., 1935. Tisanotteri italiana. Studio anatomo-morphologico e biologico del Liothripide dell'olivo ("*Liothrips oleae*" Costa). Redia 21, 1–188.

Misof, B., Liu, S., Meusemann, K., Peters, R.S., Donath, A., Mayer, C., et al., 2014. Phylogenomics resolves the timing and pattern of insect evolution. Science 346, 763–767.

Mito, T., Sarashina, I., Zhang, H., Iwahashi, A., Okamoto, H., Miyawaki, K., et al., 2005. Non-canonical functions of hunchback in segment patterning of the intermediate germ cricket *Gryllus bimaculatus*. Development 132, 2069–2079. Available from: https://doi.org/10.1242/dev.01784.

Morales, C.F., 1991. Margarodidae (Insecta: Hemiptera). Fauna New Zeal. 21, 1–124.

Moran, N.A., 1992. The evolution of aphid life cycles. Annu. Rev. Entomol. 37, 321–348. Available from: https://doi.org/10.1146/annurev.en.37.010192.001541.

Moritz, G., 1989. Die Ontogenese der Thysanoptera (lnsecta) unter besonderer Berücksichtigung des Fransenflüglers *Hercinothrips femoralis* (0. M. Reuter, 1891) (Thysanoptera, Thripidae, Panchaetothripinae) III. Mitteilung: Praepupa und Pupa. Zool. Jb. Anat. 118, 15–54.

Moritz, G., 2006. Thripse: Fransenflügler, Thysanoptera. Die Neue Brehm-Bücherei. Westarp-Verlag, Hohenwarsleben.

Moritz, G., Kumm, S., Mound, L., 2004. Tospovirus transmission depends on thrips ontogeny. Virus Res. 100, 143–149. Available from: https://doi.org/10.1016/j.virusres.2003.12.022.

Munscheid, L., 1933. Die metamorphose des Labiums der Odonaten. Zeitschr. Wiss. Zool. 143, 201–240.

Nijhout, H.F., Callier, V., 2015. Developmental mechanisms of body size and wing-body scaling in insects. Annu. Rev. Entomol. 60, 141–156. Available from: https://doi.org/10.1146/annurev-ento-010814-020841.

Nijhout, H.F., Riddiford, L.M., Mirth, C., Shingleton, A.W., Suzuki, Y., Callier, V., 2014. The developmental control of size in insects. Wiley Interdiscip. Rev. Dev. Biol. 3, 113–134. Available from: https://doi.org/10.1002/wdev.124.

Panfilio, K.A., 2008. Extraembryonic development in insects and the acrobatics of blastokinesis. Dev. Biol. 313, 471–491. Available from: https://doi.org/10.1016/j.ydbio.2007.11.004.

Paulus, H.F., 1989. Das Homologisieren in der Feinstrukturforschung: Das Bolwig-Organ der hoeheren Dipteren und seine Homologisierung mit Stemmata und Ommatidien eines urspruenglichen Facettenauges der Mandibulata. Zool. Beitr. N.F. 32, 437–478.

Schaefer, C.W., 1975. The mayfly subimago: a possible explanation. Ann. Entomol. Soc. Am. 68, 183.

Sehnal, F., 1985. Growth and life cycles. In: Kerkut, G.A., Gilbert, L. (Eds.), Comprehensive Insect Physiology, Biochemistry and Pharmacology. Pergamon Press, Oxford, pp. 1–86.

Si, Q., Luo, J.-Y., Hu, Z., Zhang, W., Zhou, C.-F., 2017. De novo transcriptome of the mayfly *Cloeon viridulum* and transcriptional signatures of Prometabola. PLoS One 12, e0179083. Available from: https://doi.org/10.1371/journal.pone.0179083.

Snodgrass, R.E., 1954. Insect metamorphosis. Smithson. Misc. Collect. 122, 1–124.

Stewart, K.W., 2001. Vibrational communication (drumming) and mate-searching behavior of stoneflies (Plecoptera); evolutionary considerations. In: Domínguez, E. (Ed.), Trends in Research in Ephemeroptera and Plecoptera. Kluwer Academic & Plenum Publishers, New York, pp. 217–225.

Stoks, R., Córdoba-Aguilar, A., 2012. Evolutionary ecology of Odonata: a complex life cycle perspective. Annu. Rev. Entomol. 57, 249–265. Available from: https://doi.org/10.1146/annurev-ento-120710-100557.

Suhling, F., Sahlén, G., Gorb, S., Kalkman, V.J., Dijkstra, K.-D.B., van Tol, J., 2015. Order Odonata. In: Thorp, J., Rogers, J.D. (Eds.), Ecology and General Biology. Thorp and Covich's Freshwater Invertebrates, fourth ed. Academic Press, Cambridge, pp. 893–932.

Suter, P., 1932. Untersuchungen über Korperbau und Rassendifferenzierung der Kommashildlaus, *Lepidosaphes ulmi* L. Mitt. Schweiz. Ent. Ges. 15, 347–420.

Swammerdam, J., 1675. Ephemeri Vita of Afbeeldingh van 's Menschen Leven, vertoont in de wonderbaarelijcke en nooyt gehoorde historie van het vliegent ende een-daghlevent haft of oever-aas. Abraham Wolfgang, Amsterdam.

Tanaka, A., 1976. Stages in the embryonic development of the German cockroach, *Blattella germanica* Linné (Blattaria, Blattellidae). Kontyû Tokyo 44, 1703–1714.

Thorp, J., Rogers, D.C., 2015. Ecology and general biology, Thorp and Covich's Freshwater Invertebrates, fourth ed. Academic Press, Cambridge.

Tierno de Figueroa, J.M., López-Rodríguez, M., 2015. Clase Insecta. Orden Plecoptera. Rev IDE@ – SEA Ibero Divers Entomológica @ccesible 43, 1–15.

Tierno de Figueroa, J.M., Sánchez-Ortega, A., Membiela-Iglesia, P., Luzón-Ortega, J.M., 2003. Plecoptera. Fauna ibérica, 22. Museo Nacional de Ciencias Naturales, CSIC, Madrid.

Tillyard, R.J., 1917. The Biology of Dragonflies (Odonata or Paraneuroptera). Cambridge University Press, Cambridge.

van der Kooi, C.J., Schwander, T., 2014. Evolution of asexuality via different mechanisms in grass thrips (Thysanoptera: Aptinothrips). Evolution (N.Y.) 68, 1883–1893. Available from: https://doi.org/10.1111/evo.12402.

Vea, I.M., Tanaka, S., Shiotsuki, T., Jouraku, A., Tanaka, T., Minakuchi, C., 2016. Differential juvenile hormone variations in scale insect extreme sexual dimorphism. PLoS One 11, e0149459. Available from: https://doi.org/10.1371/journal.pone.0149459.

Weber, H., 1930. Biologie der Hemipteren. Springer, Berlin.

Weber, H., 1931. Lebensweise und Umweltbeziehungen von *Trialeurodes vaporariorum* (Westwood) (Homoptera-Aleurodina). Zeitschr. Morph. Ökol. Tiere 23, 575–735.

Weber, H., 1949. Grundriss der Insektenkunde. Gustav Fischer, Jena.

Zwick, P., 2000. Phylogenetic system and zoogeography of the Plecoptera. Annu. Rev. Entomol. 45, 709–746. Available from: https://doi.org/10.1146/annurev.ento.45.1.709.

Chapter 5

The holometabolan development

Holometabolan species form a superorder of insects within the infraclass Neoptera, which has received two names: Holometabola and Endopterygota. In this book, the superorder name Endopterygota is used, in order to avoid confusion with the word holometabola, used here to refer to a particular type of metamorphosis. The Endopterygota is a monophyletic group (Fig. 2.1 in Chapter 2: A spectacular diversity of forms and developmental modes) that differs from other winged insects by the different development of wings. Endopterygota develop wings inside the body and undergo an elaborate metamorphosis that requires a pupal stage. All other winged insects (Palaeoptera, Polyneoptera, and Paraneoptera) do not develop wings inside of the body and do not undergo a true pupal stage. Most importantly, the larvae of Endopterygota are different from the adult in terms of morphology and lifestyle. Other features exclusive of Endopterygota have been defined by Beutel and Pohl (2006).

The Endopterygota is the most diverse insect superorder, with c.850,000 living species distributed among the following 11 orders: Hymenoptera (which comprises the sawflies, wasps, bees, and ants), Megaloptera (alderflies, dobsonflies, fishflies, and relatives), Neuroptera (which includes lacewings, mantidflies, and antlions), Raphidioptera (snakeflies), Coleoptera (beetles), Strepsiptera (twisted-wing parasites), Trichoptera (caddisflies), Lepidoptera (butterflies and moths), Diptera (which includes flies, mosquitoes, and midges), Mecoptera (scorpionflies), and Siphonaptera (fleas). Most of the biodiversity concentrates in the following four orders, the "big four": Hymenoptera (with approximately 125,000 species described); Coleoptera (360,000 species); Lepidoptera (150,000 species); and Diptera (150,000 species). The other 65,000 species are distributed among the other, less diverse, seven endopterygote orders (Beutel and Pohl, 2006).

In holometabolan species, the embryo gives rise to a larva whose morphology and often the mode of life are different from those of the adult. Then, postembryonic development consists of a number of larval instars during which the insect essentially grows, followed by a pupal stage that bridges the gap between the larval and the adult morphology and finally the adult.

Insect Metamorphosis. DOI: https://doi.org/10.1016/B978-0-12-813020-9.00005-3

EMBRYOGENESIS IN HOLOMETABOLAN SPECIES

The best-known species with respect to the developmental genetic network that underlies embryogenesis is the dipteran *Drosophila melanogaster* (Lawrence, 1995). Moreover, the red flour beetle *Tribolium castaneum* and the parasitic wasp *Nasonia vitripennis* are rapidly consolidating as additional reference species (Peel, 2008). In general, the basic developmental transitions of a typical insect embryogenesis are also followed by the holometabolans (Anderson, 1972) (see Chapter 2: A spectacular diversity of forms and developmental modes), but showing a number of peculiarities, often depending on the species.

Most holometabolans follow the derived long germ band embryo development (Peel, 2008). The "big four" orders (Coleoptera, Lepidoptera, Hymenoptera, and Diptera) contain species following the long germ band embryogenesis, for example, the dipteran *D. melanogaster*, the lepidopteran *Manduca sexta*, the coleopteran *Callosobruchus maculatus*, and the hymenopterans, *Apis mellifera*, *N. vitripennis*, and *Bracon hebetor* (Davis and Patel, 2002; Grbić and Strand, 1998; Peel, 2008). Nevertheless, within the Lepidoptera, Coleoptera, and Hymenoptera, there are species that show the short or intermediate germ band development, like the lepidopteran *Bombyx mori*, the coleopteran *T. castaneum* (Davis and Patel, 2002; Peel, 2008), and the hymenopterans *Aphidius ervi* and *Macrocentrus cingulum*. The available data suggest that the evolutionary direction and sense were from short to long germ band, and that this transition occurred more than once independently in different groups during the endopterygote radiation (Davis and Patel, 2002; Peel, 2008). However, this evolution can be reversible, as in the case of some parasitic hymenopterans (Grbić and Strand, 1998; Sucena et al., 2014). Blastokinesis, although with species-specific variations, tends to be much less complex than in hemimetabolan species. The most extreme case is represented by *D. melanogaster*, which possesses an extremely reduced extraembryonic component, the amnioserosa, and blastokinesis practically does not occur (Panfilio, 2008).

As in hemimetabolans, larval appendages are formed from limb buds that start to appear early in embryogenesis. The antennae formed during the embryo development are much shorter, with fewer segments than in the adult, whereas the legs are simpler. For example, in the lepidopteran *M. sexta*, a thoracic leg has all the segments typical of any insect leg (coxa, trochanter, femur, tibia, and tarsus), but they are reduced, their functionality being practically limited to grasping movements (Tanaka and Truman, 2007). The larva of the most modified *D. melanogaster* is apodous, and the adult legs develop from imaginal discs (see next section), which form during the embryogenesis. Like in the hemimetabolan nymphs, the holometabolan larvae show a considerable diversity of mouth apparatus according to adaptations to different alimentary regimes. One of the most dramatic specializations is the mouth hooks of the *D. melanogaster* larva, a paired structure that brings food to the mouth. The head of these larvae is considerably reduced and withdrawn into the thoracic region, forming a modified cephalopharyngeal skeleton that includes the mouth hooks. In *D. melanogaster* and in other less modified species, mouth structures differentiate during embryogenesis from cephalic segments, in stage 10 (21%–25% embryo development) although the mandibles, maxillae, and labium are reduced to mere small lobes. Then, during organogenesis, practically all cells of the cephalic segments roll inward to the thoracic region undertaking a complex morphogenetic movement known as head involution. The last steps of head involution include the formation of the mouth hooks within the atriopharyngeal cavity from maxillary structures (Campos-Ortega and Hartenstein, 1997). In holometabolan embryogenesis, a cluster of primordial wing cells differentiates with those of other appendages early in embryogenesis. These cells remain internalized under the ectoderm, generally being encapsulated in wing discs (see below) in later stages.

Regarding the eyes, the holometabolan larva has one to several pairs of stemmata, which compose the specialized larval visual system. In *D. melanogaster*, coinciding with gastrulation, two small stemmata and associated optic lobes originate from a few cells in what is called the center of mitotic domain 20 (Heming, 2003). The stemmata move toward the anterior part of the embryo during the head involution, thus becoming nested against the sides of the cephalopharyngeal skeleton in the larva. The process is relatively similar in the blister beetle *Lytta viridana* (Heming, 2003). Imaginal precursor cells, or eye-antennal imaginal discs (in *D. melanogaster*) (see next section), are specified just after blastoderm formation, from which the compound eyes of the adult will be formed during metamorphosis. Larvae of Mecoptera (scorpion flies) and of *Chaoborus* dipterans (glassworm midges) are an exception, as they possess compound eyes instead of stemmata. In both cases, the larval compound eye differs in a number of features from the adult eye, and it is dramatically remodeled at during metamorphosis (Paulus, 1989).

Importantly, holometabolan embryos secrete and deposit three cuticles (EC1, EC2, and EC3) during embryogenesis. However, EC2 may be simplified, with only a reduced epicuticle, in the embryos of more derived holometabolans, or may even have been lost secondarily in the most evolutionarily modified Diptera Cyclorrhapha, like *D. melanogaster* (Konopová and Zrzavý, 2005).

IMAGINAL PRECURSOR CELLS, IMAGINAL DISCS, AND HISTOBLASTS

Morphologically, the holometabolan larva diverges with respect to the adult. There are modified larval structures that are different from the corresponding adult ones (antennae, legs, and eyes, for example), whereas other typically adult organs are simply absent externally (wings and genitalia are the most general). The transformation of larval antennae and legs into adult structures, as well as the emergence of adult wings and genitalia takes place during metamorphosis.

The more or less divergent larval morphology responds to ancestral specific adaptations although a general trend is to conceal the wing primordia (hence the name Endopterygota). Then, depending on the groups, other typically adult structures are also concealed during the larval life. The most modified and best-studied species is *D. melanogaster*, where the vermiform, larval body hardly shows any sign of what the adult animal will be. In less modified holometabolan groups, the development of typically adult structures has been studied in detail in few cases, whereas there is no information in some orders. In the studied species, the adult structures are formed from clusters of imaginal precursor cells that are more or less compacted, sometimes adopting a disc-like shape that lie in the epidermis, which are more easily observed in the penultimate or in the last larval instar. In the case of organs that are present in the larva as a simpler structure, like the antennae, legs, or eyes, these clusters are close or associated to the respective organ. Those of the wings are located in the sides of the mesothorax and metathorax and those of the genitalia in the corresponding abdominal segment (Švácha, 1992).

Regarding the main appendages, the leg has been studied in detail in the coleopteran *Tenebrio molitor* (Huet and Lenoir-Rousseaux, 1976; Tikhomirova, 1983) and the lepidopterans *Pieris brassicae* (Kim, 1959) and *M. sexta* (Tanaka and Truman, 2005) and the antennae in the lepidopteran *B. mori* (Švácha, 1992). Results show that the larval leg and antennae are remodeled into the more complex adult corresponding appendages. For example, in the leg of *M. sexta,* most of the adult leg epidermis derives from clusters of imaginal precursor cells located in specific regions of the larval appendage (Fig. 5.1), which rapidly proliferate in the final larval instar. The epidermal cells outside the imaginal precursor cells are composed by two different populations. The most numerous one is larval specific and disappears early in metamorphosis through apoptosis, whereas the second consists of polypotent cells that contribute to the larval, pupal, and adult leg epidermis (Tanaka and Truman, 2005, 2007).

M. sexta has been also a favorite model for the study of eye development (Allee et al., 2006; Champlin and Truman, 1998; Monsma and Booker, 1996). The adult eye is formed from an unpatterned monolayer epithelium located in the larval head capsule, between the larval ocelli and the antenna. During the final larval instar, a wave of cell proliferation and patterning progresses across this epithelium, leaving evenly spaced neural cell clusters that will become the ommatidia of the adult eye. The anterior edge of this wave, or the morphogenetic furrow, progressively sweeps across the primordium,

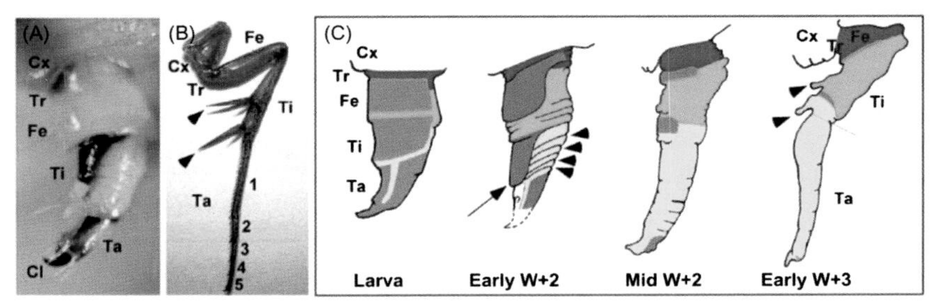

FIGURE 5.1 The development of the leg in the lepidopteran *Manduca sexta*. (A) Leg of a fifth (last) instar larva. (B) Metathoracic leg of an adult; the tibia presents two pairs of spines (arrowheads), and the tarsus (Ta) is divided into five tarsomeres (1–5). (C) Morphogenesis of the metathoracic adult leg during the prepupal stage; the adult primordia and the areas they give rise to are highlighted in colors, whereas the larval regions are shown in gray. Just 2 days after the wandering phase (early W + 2), the distal region (dashed line) is an estimate as this part is covered by the larval cuticle at this stage; the larval region in the tibia (arrow) extends over the ventral disc in the tibia, which shows signs of segmentation and will produce the adult tarsomeres (four arrowheads). Just 3 days after the wandering phase (early W + 3), the line indicates presumptive Ti/Ta border; two arrowheads point to the tibial spines. Cl, pretarsal claw; Cx, coxa; Fe, femur; Ta, tarsus; Ti, tibia; Tr, trochanter. From Tanaka and Truman (2007), with permission.

whereas cells behind it become organized into immature ommatidial clusters. Then they undergo maturation of adult eye structures, including rhabdomere formation, pigment synthesis, and lens cuticle secretion from cone cells. Similar development operates in the coleopteran *T. castaneum* and other early-branching holometabolan species (Friedrich, 2003, 2006).

As commented above, larvae of Mecoptera and *Chaoborus* dipterans possess compound eyes instead of stemmata. Mecopterans of the families Panorpidae and Bittacidae have been particularly well studied. The larva emerges with compound eyes, which are reminiscent of those of ametabolan and hemimetabolan nymphs. However, while nymphal eyes continue to expand during postembryogenesis by differentiation of new ommatidia from the anterior growth zone, the number of ommatidia in the larval eyes of Mecoptera does not increase after embryogenesis (Paulus, 1989). Moreover, Mecopteran larval eyes are different with respect to the adult eyes. Those of the larva possess a tiered retinula and a funnel-shaped rhabdom. In contrast, the adult compound eye has eight retinula cells arranged only in one tier, and the rhabdom is vertically in contact with the crystalline cone in the ommatidium (Bittacidae), instead of being arranged in a funnel shape. Alternatively (Panorpidae), the rhabdom is enveloped by the proximal end of the crystalline cone (Chen et al., 2012). As studied in Panorpidae (Paulus, 1989; Rottmar, 1966), the larval compound eyes of mecopteran insects degenerate almost totally during the pupal stage, whereas the adult compound eyes develop from imaginal precursor cells. The compound eyes of the mecopteran larvae suggest the evolutionary origin of holometabolan insect larval eyes from compound eyes (Friedrich, 2006). However, the phylogenetic position of Mecoptera as a sister group of Diptera (see Fig. 2.1 in Chapter 2: A spectacular diversity of forms and developmental modes) and isolated from other early-branching Endopterygota clades, like Lepidoptera, Coleoptera, and Hymenoptera, suggests that the larval compound eyes in Mecoptera evolved from a stemmata condition, which, arguably, would be the one presented by the last common ancestor of all Endopterygota. The alternative explanation presuming the retention of larval compound eyes in mecopterans, besides an independent evolution toward larval stemmata in all other endopterygote clades, including Hymenoptera, Coleoptera, Lepidoptera, and Diptera, is much less parsimonious. The same reasons suggest that larval compound eyes of the dipterans *Chaoborus* derive from ancestral stemmata.

Genitalia and wings are externally absent in the larva and appear during metamorphosis. Genitalia precursor cells are specified during embryogenesis (Horsfall and Ronquillo, 1970; Ronquillo and Horsfall, 1969), but only in late larval instars, genitalia precursor cells form a clearly visible disc-like structure in the epidermis. During the transition from the final larval instar to the pupa, the external genital appendages are formed, as shown in Lepidoptera (Mehta, 1933; Reinecke et al., 1983), Coleoptera (Aspiras et al., 2011), and Diptera Nematocera (Horsfall and Ronquillo, 1970; Ronquillo and Horsfall, 1969). In parallel, the internal reproductive organs develop as well, which stay connected with the genital disc via the corresponding genital ducts.

The wings have received more attention. In Lepidoptera, wing precursor cells are specified during embryogenesis, but in larvae, they form a pair of mesothoracic and metathoracic epithelial sacs or discs, located near the lateral spine just below the dorsal epidermis. The discs, which are attached to the surface by a thin pedicel, will give rise to the corresponding adult forewings and hindwings (Kango-Singh et al., 2001; Miner et al., 2000). During the last larval instar, the wing discs grow continuously, and shape and size differences between the mesothoracic and metathoracic discs appear. During metamorphosis, the wing discs behave like presumptive wing buds opening out as wing blades, not through complex rearrangement patterns and evagination, as occurs in *D. melanogaster* (see below) (Kango-Singh et al., 2001). In tenebrionid beetles, disc-like clusters of wing precursor cells located in the mesothoracic and metathoracic lateral margins give rise to the adult wing (Fig. 5.2) (Quennedey and Quennedey, 1999, 1990). In *T. molitor*, the wing precursor cells actively proliferate during the final larval instar until the prepupal commitment, when approximately 40% of these cells disappear through apoptosis, whereas the others form the adult wing (Quennedey and Quennedey, 1990).

The general integument, in particular that of the abdomen, also differs between the larva and the adult. As shown in *M. sexta*, adult integument, especially the abdominal epithelial cells, derives from a larval monolayer of

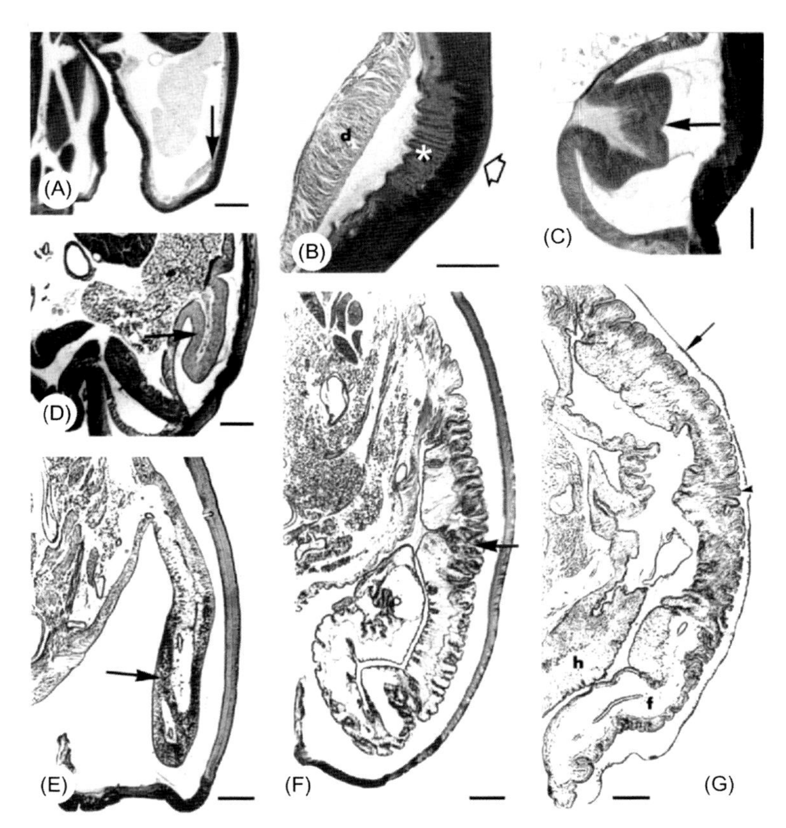

FIGURE 5.2 Disc-like clusters of wing precursor cells in the metathorax of the last larval instar of the coleopteran *Zophobas atratus*. (A) Eight days after the larval molt, the apolysis and thickening of a small bundle of epidermal cells (arrow) are clearly recognizable. (B) Detail of the wing disc (d) adjacent to the endocuticle (asterisk) of the lateral brown line of the body (open arrow). (C) By day 15 (wandering period), the disc increases in size, taking on a puffball shape (arrow). (D) One day later, the whole epidermis apolyses while the mitotic crisis becomes generalized and not restricted to the disc (arrow). (E) By day 19, the disc (arrow) has quickly grown, adopting a paddle-like shape. (F) By day 21, the disc (arrow) has finished lengthening and upward motion and adopts a scalloped appearance. (G) By day 23, the preecdysial pupal cuticle (arrowhead) is secreted, while the old larval cuticle (arrow) is almost fully digested; a hindwing (h) and a forewing (f) can be observed. Scale bars: 50 μm (A), 100 μm (B and C), 200 μm (D–G). From Quennedey and Quennedey (1999), with permission.

cells that show different ploidy levels, and there is no segregation of diploid cells destined to form adult structures. During metamorphosis, polyploid cells sort from cells of equal or smaller sizes in the tergite epithelial monolayer and form rosettes that act as foci for the aggregation of other polyploid cells at larva–pupa apolysis. These aggregated polyploid cells later disappear through apoptosis, and the result in the adult is a thinner cuticle with densely packed scale cells (Fig. 5.3) (Nardi et al., 2018).

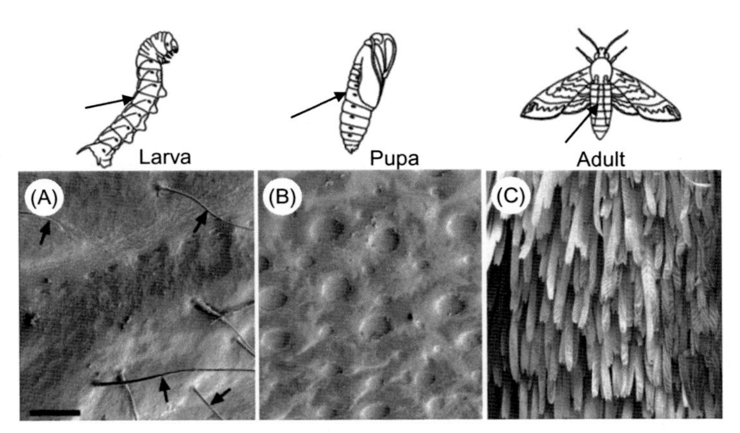

FIGURE 5.3 Transformation of the integument in the transition from larva to adult in the lepidopteran *Manduca sexta*. The fourth abdominal segment (arrows) of the last larval instar (left), pupa (middle), and adult (right) transforms over 25 days. (A) The tergite of the wandering larva (day 0) has characteristic long sensory bristles (arrows). (B) The pupal tergite on day 4 has characteristic invaginated patches of cuticle (pock marks). (C) The tergite of the newly emerged adult on day 25 has characteristic scale cells. Scale bar: 200 μm. From Nardi et al. (2018), with permission.

The precedent data refer to early branching Endopterygota, from Hymenoptera to less modified Diptera. In Diptera Cyclorrhapha, the process of formation of adult structures is much more modified, as shown in *D. melanogaster*, the model par excellence on which the biology and definition of "canonical" imaginal discs were established (Cohen, 1993; Held, 2002; Lawrence, 1995). In *D. melanogaster,* imaginal precursor cells are set aside during embryogenesis and internalized into oval epithelial sheets called imaginal discs. The imaginal discs segregate in the embryo, are patterned during the larval instars, and culminate in the formation of the corresponding adult structure during the pupal stage.

In *D. melanogaster*, there are 19 imaginal discs: the epidermis of the head, thorax, and limbs of the fly come from nine bilateral pairs of discs and the genitalia from a single disc. Clypeolabral and labial discs give rise to the mouthparts. Eye-antennal discs will form the compound eyes and the antennae. In each thoracic segment, there are also a pair of dorsal imaginal discs, namely humeral (or dorsal prothoracic), wing, and haltere discs. The genitalia derive from a medial disc extending between the 8th and 10th abdominal segments; it is sexually dimorphic and develops differently in males and females (Fig. 5.4).

Each disc is composed by a single columnar epithelial layer, but most of them also contain some adepithelial cells (mesodermal myoblasts), as well as tracheal cells and a few neurons placed between the epithelium and the basal lamina. The genital disc is the only one where cells from the mesoderm are

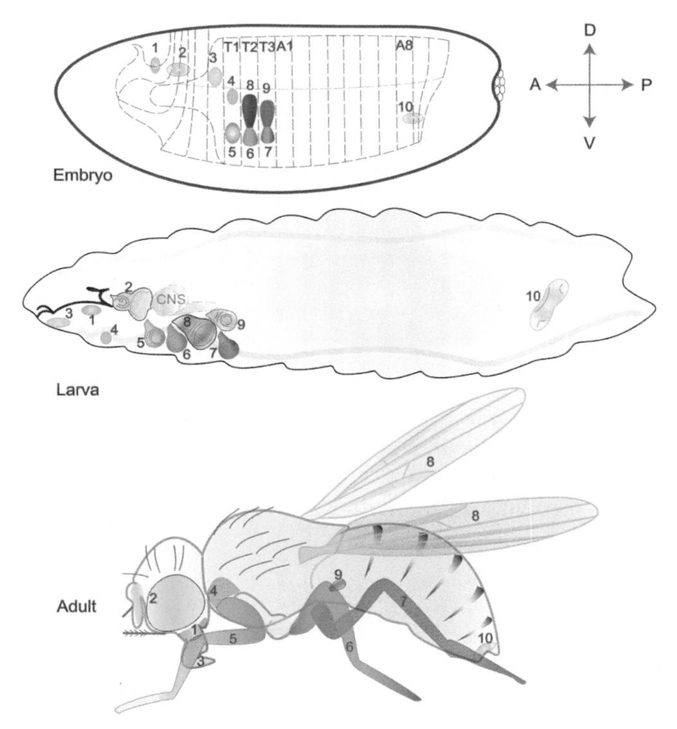

FIGURE 5.4 The imaginal discs of the dipteran *Drosophila melanogaster*. The location of imaginal tissue primordia is represented at the cellular blastoderm stage (top), with corresponding numbering in larval (middle) and adult (bottom) stages. Axes orientation is indicated by the perpendicular arrows (A, anterior; P, posterior; D, dorsal; V, ventral). T1–T3 correspond to thoracic segments, and A1–A8 represent abdominal segments. The epidermis of adult structures like the head, thorax, and appendages comes from nine pairs of bilateral discs (here, only one of each pair is shown in the larva), and the genitalia derive from a middle disc (19 discs in total): (1) clypeolabral, (2) eye-antennal, (3) labial, (4) humeral (or prothoracic), (5) first leg, (6) second leg, (7) third leg, (8) wing, (9) haltere, and (10) genital. Some portions of the head and thorax also originate from imaginal discs. For instance, the wing discs contribute both to the wings and the notum in the adult fly, which is not represented here for simplicity. From Beira and Paro (2016), with permission.

recruited into the epithelium. Discs have an outer layer, the peripodial membrane, with squamous cells that also contribute to the cuticular structures in the adult. In freshly emerged first larval instar, the biggest discs (wing, leg, and eye-antennal) have between 20 and 70 cells. Between mid and late in the instar, disc cells resume mitosis and continue dividing exponentially during the second and the third larval instars. During the third larval instar, a considerable number of cells are formed, their number doubling about every 10 hours. Thus just before pupariation, each disc contains between 10,000 and 50,000 cells, having undertaken patterning processes. During metamorphosis, mature discs undergo a dramatic morphogenetic process, by everting through their stalk in a complex deployment. Finally the peripodial membrane disintegrates through apoptosis (Aldaz and Escudero, 2010; Beira and Paro, 2016).

In *D. melanogaster*, precursors of imaginal structures are not limited to the imaginal discs. The precursor cells of the abdomen and the internal organs of the adult, such as the gut, salivary glands, and brain, arise from nests or rings of cells intimately associated with larval corresponding structures. For example, the salivary gland imaginal rings are embedded in the larval salivary glands; the midgut imaginal histoblast nests arise in the larval midgut and the abdominal histoblast nests form among the cells of the larval abdomen (Held, 2002; Lawrence and Peter, 1995). Regarding the formation of the adult abdomen, each adult abdominal segment develops from four pairs of histoblast nests as follows. The anterior and posterior dorsal pairs form the tergites, the ventral pair forms the sternites and pleurites, and the spiracular pair produces the spiracle and the pleurite tissues around it. Each anterior dorsal and ventral histoblast nest has about 16 cells; each posterior dorsal histoblast nest has about five cells, whereas each spiracle histoblast nest has about three cells. The abdominal histoblasts begin to divide within the first 3 hours after pupariation, and they continue dividing until about 15 hours of pupal development but without displacing the larval cells. Then at approximately 15 hours of pupal life, the abdominal histoblast cells begin to migrate and displace the larval cells, which are then histolyzed. Following proliferation and migration, the cells of adjacent segments fuse at the dorsal/ventral and segmental borders. Toward the end of the process, the adult cells differentiate to produce epidermal tissues, like the microchaetae and macrochaetae, and to secrete the adult cuticle (Ninov et al., 2007; Ninov and Martín-Blanco, 2009).

THE LARVA

Larval types show a spectacular diversity, as most of the peculiar morphologies result from secondary adaptations to diverse particular modes of life and interaction with the environment. Several authors have proposed different classifications of holometabolan larvae, but the most frequently used distinguish four main types: protopod, polypod, oligopod, and apodous. This classification is based on the work of Berlese (1913), who proposed that the evolution of metamorphosis is based on a more or less premature eclosion of the embryo. Thus the three original larval types categorized by Berlese, protopod, polypod, and oligopod, fit with the three classic stages of embryogenesis in which the corresponding larvae would have been enclosed. Thus this larval classification is very convenient for his metamorphosis theory (Berlese, 1913), which became very popular, especially after being adopted by Imms in his textbook of entomology (Imms, 1930). The four categories can be defined as follows:

- Protopod larva: Without abdominal segmentation and with dramatically reduced cephalic and thoracic appendages (Fig. 5.5A). The digestive, nervous, and respiratory systems are also reduced. This type of larva represents a high specialization common to endoparasitoid species (like hymenopterans belonging to the families Platygastridae, Scelionidae, Figitidae, and Dryinidae) that live immersed in the host used as food.

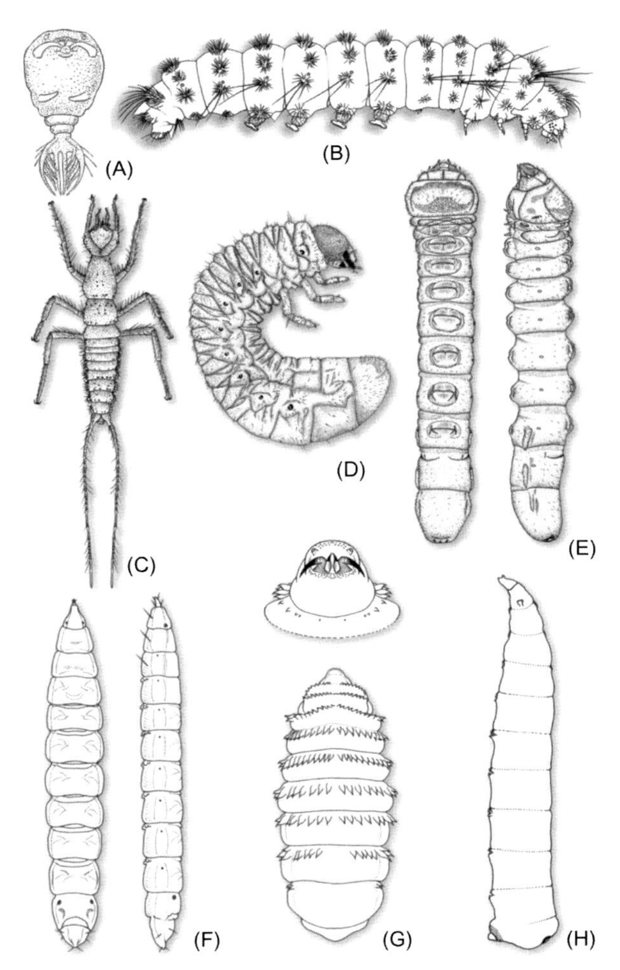

FIGURE 5.5 Larval types. (A) Protopod (*Platygaster instricator*, Hymenoptera Platygastridae). (B) Polypod (Lepidoptera Zygaenidae). (C) Olygopod campodeiform (*Galerita brasiliense*, Coleoptera Carabidae). (D) Olygopod scarabaeiform (*Macraspis cincta*, Coleoptera Scarabaeidae). (E) Apodous eucephalous (*Parandra glabra*, Coleoptera Cerambycidae, dorsal and lateral view). (F) Apodous hemicephalous (Diptera Mydidae, dorsal and lateral view). (G and H) Apodous acephalous (G: Diptera Gasterophilidae, dorsal view and detail of the head showing the mouth hooks; H: Diptera Muscidae). From Kulagin (1898) (A) and Costa et al. (2006) (B–H), with permission.

- Polypod larva: This type can be defined as a modified eruciform larva. The segmentation is well marked, but the antennae and thoracic legs are underdeveloped. Characteristically, it possesses abdominal locomotor appendages (prolegs) (Fig. 5.5B). This larval type is characteristic of most Mecoptera, Lepidoptera, and Hymenoptera Symphita.
- Oligopod larva: Probably represents the most primitive type of holometabolan larva. It has well-developed thoracic legs, but it does not have abdominal locomotor appendages. The head and cephalic appendages are well developed. Two main subtypes can be distinguished: campodeiform and scarabaeiform. Campodeiform oligopod larvae show an elongated and depressed body, well-developed thoracic ambulatory legs, and a head that tends to be prognathous (Fig. 5.5C). This subtype is typical of Megaloptera, Raphidioptera, Neuroptera, many Coleoptera, Strepsiptera, and some Trichoptera. Scarabaeiform oligopod larvae have robust, subcylindrical, C-shaped, or U-shaped body, with short thoracic legs (Fig. 5.5D). This is the typical larva of Coleoptera Scarabaeidae but also of other Coleoptera families, such as Ptinidae, Bostrychidae, and Scolytidae.
- Apodous larva: A dramatically derived type characterized by the complete absence of thoracic appendages. Three subtypes can be distinguished: eucephalous, hemicephalous, and acephalous. The acephalous subtype has a well-sclerotized cephalic capsule and a slight reduction of the chewing or biting mouthparts (Fig. 5.5E); examples are provided by some Coleoptera (namely in Buprestidae, Cerambycidae, and Curculionidae families), Diptera Nematocera, and Hymenoptera Apocrita. The hemicephalous larvae are characterized by having the head and appendages markedly reduced, the head being more or less retracted into the prothorax, with the mandibles moving in the vertical plane (Fig. 5.5F); this subtype is represented by some Diptera (Nematocera and Brachycera). In the acephalous larvae, the head has become dramatically reduced and withdrawn into the thorax, and the mouthparts form a pair of mouth hooks (Fig. 5.5G and H); the larvae of *D. melanogaster* and other Diptera Cyclorrhapha belong to this subtype.

Of course, such a simple classification in four types does not exhaust the enormous larval diversity. Thus many other larval morphologies have been described by various authors, especially in coleopterans, where many subtypes have been categorized, like the caraboid larva characterized by a narrow, elongate body with long legs; the cerambycoid, morphologically similar to the caraboid, but legless; or the curculionoid, with a reduced and grublike body (Costa et al., 2006; Stehr, 1987, 1991). The general four-type classification is useful in its simplicity, but it has no phylogenetic or evolutionary basis. In this sense, an attempt to frame the different larval types in an evolutionary context was made by Chen (1946), but the lack of a robust phylogenetic framework at that time did not make it possible to form strong conclusions. However, among other interesting reflections, Chen proposed that the oligopod larva represents the less modified type, followed by the polypod type, whereas the apodous and protopod types would be very modified because of their specialization, reflections that are worthy of consideration.

In holometabolan species, growth occurs during the larval instars, whereas the pupal stage is generally quiescent. The length of the larval period and the number of molts is quite variable depending on each species group, and sometimes the number of molts can vary within a given species as a function of the environmental conditions. However, the duration is generally much shorter and the number of molts is lower than in hemimetabolan species. A duration between days and months, and a number of molts between 3 and 7 may fit with many holometabolan species (Sehnal, 1985). As in hemimetabolan nymphs, Dyar's rule also applies for holometabolan larval growth (Kivelä et al., 2016; Sehnal, 1985). The transition from the last larval instar to the pupa can be initiated when the larva reaches a critical size, which is the size with a minimal weight in which further feeding and growth are not required for a normal time course to metamorphosis and pupation (Nijhout et al., 2014; Nijhout and Callier, 2015). The mechanism by which critical size is internally sensed, and hence the formation of the pupa is triggered, is unclear. In *M. sexta*, it has been proposed (Nijhout et al., 2006) that critical size triggers events similar to those that initiate molting in previous instars. In *D. melanogaster*, an alternative hypothesis postulates that the prothoracic gland is used to assess size (Mirth et al., 2005) (see Chapter 9: Molting: the basis for growing and for changing the form).

THE PUPA

The pupa is a generally quiescent, nonfeeding juvenile stage, typical of holometabolan species, that bridges the morphological divergence between the larval stage and the adult. For this the pupa undertakes important processes of destruction of many larval structures through cell death and the construction of proadult structures by means of the imaginal precursor cells, imaginal discs, and histoblasts (Costa, 1984; Hinton, 1948, 1963; Tettamanti and Casartelli, 2019).

Linnaeus (1758) already recognized three types of pupa: the incomplete (with appendages free), the obtect (appendages attached to the body), and the coarctate (represented by the puparium of Diptera Brachycera). Packard (1898) used the term libera to name the incomplete pupa of Linnaeus and those that were not enclosed in a cocoon. To avoid confusion, Imms (1930) coined the term exarate to characterize exclusively the pupa in which the appendages are not attached to the body. Finally, Hinton (1946) clarified that the pupa coarctate was not strictly a type of pupa, but an exarate pupa covered by the puparium, that is, the tanned cuticle of the last larval instar. With this background, Hinton (1946, 1948) established the nomenclature and classification most used today, which recognizes two main pupal types, as follows.

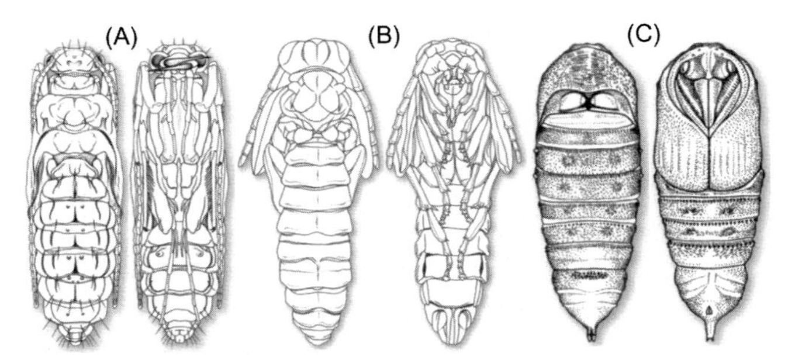

FIGURE 5.6 Pupal types. (A) Decticous (Trichoptera Philopotamidae, note the strong free mandibles). (B) Exarate adecticous (Hymenoptera Tenthredinidae). (C) Obtect adecticous (Lepidoptera Saturniidae). Dorsal and ventral view in all cases. From Costa et al. (2006), with permission.

- Decticous pupa: Type defined by having large functional mandibles operated by the pharate adult muscles (Fig. 5.6A). The decticous pupa uses the functional mandibles to cut a hole in the cocoon before emerging as adult. All decticous pupa are exarate (appendages not attached to the body). This type is presumed to be the closest to the ancestral endopterygote pupa, and it occurs in Megaloptera, Neuroptera, Raphidioptera, Trichoptera, Lepidoptera Micropterigidae (mandibulate archaic moths), and Mecoptera.
- Adecticous pupa: Characterized by having reduced nonfunctional mandibles or lacking them entirely. This type can be subclassified into two subtypes: exarate and obtect. In the exarate adecticous pupae, the appendages are not attached to the body, as in decticous pupae (Fig. 5.6B). It is a subtype characteristic of male Strepsiptera and certain Coleoptera, Diptera, and Siphonaptera. The pupae of Diptera Cyclorrhapha are exarate adecticous enclosed in the puparium; the use of the term "coarctate" for this pupa, although still used by some authors, should be discouraged. The second subtype, obtect adecticous, is characterized by having the appendages attached to the body, with the corresponding external surfaces considerably tanned (Fig. 5.6C). This subtype is observed in certain Coleoptera, Lepidoptera, and Diptera Nematocera.

HYPERMETAMORPHOSIS

The variety of forms of larvae and pupae in the different holometabolan species is overwhelming, which demonstrates the powerful capacity of natural selection and adaptation to generate morphological diversity. However, still more amazing is the occurrence of a diversity of forms within the larval period of the same species. It is the phenomenon known as hypermetamorphosis, a kind of holometaboly in which at least one of the larval instars of the life cycle differs considerably from the others (Pinto, 2003; Snodgrass, 1954). The alternative term "heteromorphosis" used by some authors (notably by Snodgrass, 1954) is much more generic, since it also refers to the more or less important morphological changes that occur throughout the life cycle of any insect. Hypermetamorphosis is more frequent among parasitoid species but may also occur in nonparasitoids. According to Snodgrass (1954), two main categories of hypermetamorphosis can be distinguished. In the most widespread form, the adult female lays the eggs in the open; thus the first larval instar must search for a food source (the host in the case of parasitoid species) by their own means. Thus they have a wandering first larval instar. In the second category, oviposition takes place in the "food" itself (the body or the egg of the host in parasitoid species). Thus they have a sedentary first larval instar. Both categories are discussed below.

Hypermetamorphosis in parasitoids with a wandering first larval instar

In these species, the first larval instar must actively find a host, thus it is specially adapted for that function. This larva can find a host directly, by its own means, or indirectly by phoresy, attaching itself to a carrier (usually the host species) that will transport it to the nest where it can find the host eggs or larvae (Clausen, 1976). This first larval instar is generally flattened, notably tanned, has functional legs or other means of locomotion, and is quite mobile; in some species, it has eyes, whereas in others not. This larva is called a planidium, a term derived from the Greek (πλανής, planis) that means "wanderer." The term "planidium" has a functional rather than a morphological sense, as planidia of different species differ variously from each other in form, as illustrated by Snodgrass (1954). It has been alternatively called triungulin by some authors, because of the story of a synonymy. In 1828, Dufour described what he supposed to be a new insect genus and species, *Triungulinus andrenetarum*, for an previously unidentified larva showing the pretarsus empodial claw flanked by two strong spines, thus forming a tridentate structure. However, Newport later discovered that the *Triungulinus* were in fact the first larval instar of a Coleoptera Meloidae. He observed that they were being carried into *Anthophora* bees nests by phoresy and described the older instars in the nest's cells (Newport, 1851). Fabre, who by the way coined the term "hypermetamorphosis" (Fabre, 1857), recorded this synonymy, but the term "triungulin larvae" became applied to the first mobile larval instar of hypermetamorphic species. However, the use of the alternative term planidium has been generalized, while triungulin has been relegated to refer to some Meloidae species that have the typical tridentate structure in the pretarsi. This type of hypermetamorphosis with a planidium stage occurs in Hymenoptera, Neuroptera, Coleoptera, Strepsiptera, Lepidoptera, and Diptera.

Hymenoptera

Eucharitidae and Perilampidae (and the Ichneumonidae genus *Euceros*) have a legless planidium larvae, which is followed by successive soft-bodied, relatively immobile larval instars. The Eucharitidae are endoparasitic in mature larva or pupae of ants (Hymenoptera Formicidae) and are phoretic. They oviposit on vegetation, and in general the planidia of most groups attach to foraging worker ants that transport them to the nest (Clausen, 1941). The morphology of the different stages in the life cycle was thoroughly studied in *Stilbula cyniformis* (=*Schizaspidia tenuicornis*) by Clausen (1923) (Fig. 5.7A–C). Eucharitidae are one of the few parasitoid groups that have been able to use ants as hosts, despite ants' defense systems against most parasitoids. This is possible because during development within the ant nest, the parasitoid acquires the colony odor of the host, as shown in the Eucharitidae *Orasema* sp. that parasitize nests of the fire ant *Solenopsis invicta* (Vander Meer et al., 1989). The family Perilampidae is closely related to the Eucharitidae, but they parasitize insect species of several orders. Classical and detailed reports of Smith (1912) on the life cycle of *Perilampus*-parasitizing larvae of *Chrysopa* (Neuroptera) or *Hypanthria* (Lepidoptera) are commented by Snodgrass (1954) (Fig. 5.7D and E). In the

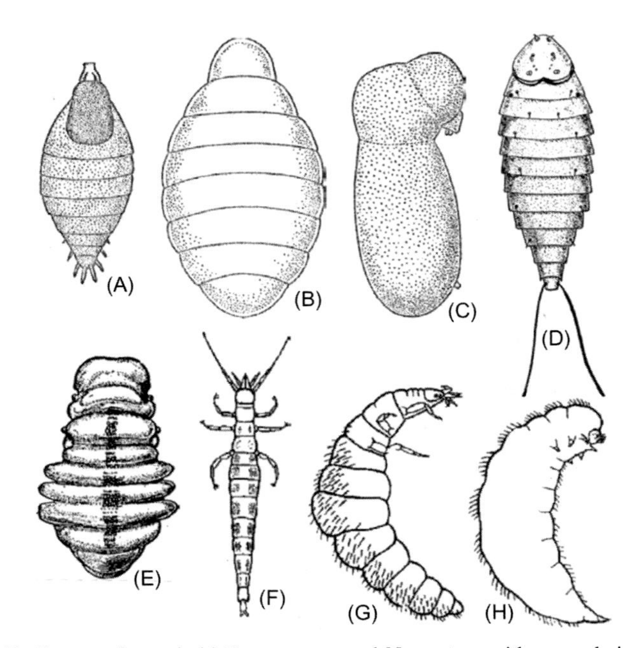

FIGURE 5.7 Larvae of parasitoid Hymenoptera and Neuroptera with a wandering first larval instar. First (A), second (B), and third (C) larval instars of *Stilbula cyniformis* (=*Schizaspidia tenuicornis*). First larval instar (D) and mature larva (E) of *Perilampus hyalinus*. Freshly emerged first larval instar (F), fully fed first larval instar (G), and mature larva (H) of *Mantispa styriaca*. (A–C) From Clausen (1923), (D and E) from Smith (1912) , and (F–H) from Brauer (1869).

Ichneumonidae *Euceros frigidus*, the planidium attaches to the integument of the larvae of the sawfly *Neodiprion swainei* and transfers to feed on the larva of other ichneumonid species, which are primary parasites of the sawfly (Tripp, 1961).

Neuroptera

The order Neuroptera is eminently predatory, but the family Mantispidae offers the best examples of hypermetamorphosis with a planidium instar. The first studies on this order were carried out by Brauer (1869), who described the life cycle of *Mantispa styriaca* feeding on spider eggs in the spider's cocoon. *M. styriaca* belongs to the subfamily Mantispinae, which is the subfamily whose life cycles are best known (Redborg, 1998). The immatures are exclusively spider egg predators during their development, and the first larval instar is a host-finding planidium, while subsequent larval instars are scarabaeiform (Fig. 5.7F–H). Larvae locate and attach to a spider and enter the spider egg sac either upon its construction or afterward. Once inside, the larvae pierce and drain the spider eggs, undergoing three larval instars within the sac (Redborg, 1998).

Compared to Mantispinae, knowledge of the morphology, biology, and ecology of immatures of the other subfamilies of Mantispidae (Symphrasinae, Calomantispinae, and Drepanicinae) is fragmentary (Redborg, 1998), although in all cases, there are more or less dramatic levels of hypermetamorphosis. The feeding habits of Symphrasinae and Calomantispinae larvae appear to be broader than the spider egg association of Mantispinae. Larvae of *Plega* spp. (Symphrasinae) have been reared in the laboratory on immature Lepidoptera, Hymenoptera, and Coleoptera (Redborg, 1998). In some species, the larvae predate upon larvae of bees or wasps, either social or solitary (Maia-Silva et al., 2013). Larvae of Calomantispinae appear to predate on other arthropods. *Nolima pinal* larvae have been reared to adulthood on hosts that included larval Diptera, Hymenoptera, Lepidoptera, and Coleoptera, in addition to spider eggs (Redborg, 1998). The Drepanicinae are the least known, and only a recent report describes the planidium of the Australian species *Ditaxis biseriata*, which follows a subterranean life; nothing is known about the following larval instars, nor about their diet (Dorey and Merritt, 2017).

Coleoptera

The family Meloidae has been thoroughly studied and offers examples of the two systems of reaching the host, directly and by phoresy. Snodgrass (1954) relates in detail both behaviors following the classical literature that deals with American species of the genus *Epicauta* and the European species *Mylabris variabilis*, which feed on grasshopper eggs. *M. variabilis* oviposits in the ground of an area where the grasshoppers live and from them emerges the planidium (Fig. 5.8A and B), which searches actively until finding the host eggs. After feeding on them, the planidium molts into a short-legged, soft-bodied second larval instar (Fig. 5.8C), which continues feeding, and grows and molts into two more instars, becoming a robust scarabaeiform larva (Fig. 5.8D). The fifth larval instar resembles the fourth, but it leaves the grasshopper egg nest, does not feed anymore, and molts into the sixth larval instar, also referred to as a coarctate, which has a heavily tanned cuticle and reduced mandibles and legs (Fig. 5.8E). This larval instar, which is quiescent and partially enveloped by remains of the exuvia of the preceding instar, passes the winter hidden in the ground. In spring, it molts into the seventh larval instar that has again taken the shape of a robust scarabaeiform larva (Fig. 5.8F). The seventh larval instar burrows upward to near the surface of the ground, where it molts into the pupa (Fig. 5.8G), from which the adult (Fig. 5.8H) finally emerges. This description corresponds to the classical observations of Paoli (1938) on *M. variabilis* but generally applies to other hypermetamorphic species of meloid beetles including several which are parasitoids in the nests of Apoidea (bees) (Pinto, 2003; Pinto and Selander, 1970).

Meloid beetles afford also examples of phoresy in those species that are parasitoids in the nests of several families of Apoidea (Hymenoptera) (Erickson et al., 1976). Detailed descriptions of the behavior of these species have been reported by classical entomologists, like Fabre (1857) for the Meloidae of the genus *Sitaris* infesting the underground nests of *Anthophora* bees, and those of Parker and Böving (1924) for the meloid *Tricrania sanguinipennis* infesting the nests of *Colletes rufithorax* bees. In both cases, the meloid female oviposits on the ground, near the nesting places of the host bees. These well-known species, however, are rather an exception. The majority of meloids with phoresy related to these genera oviposit on vegetation. Then the larvae are picked up by bees from flowers, as in the subfamily Nemognathinae. In most species of the other major subfamily, Meloinae, the eggs are laid in the ground, and larvae find the bee nests directly (as do the *Epicauta* feeding on grasshopper eggs) (Pinto, 2003). In any case, the hatched first larval instar is a planidium that once it reaches the host and molts into a second larval instar that has a soft and smooth boat-shaped body. Then, after further molts, it becomes a robust scarabaeiform larva, remaining in this form until the last larval instar. *T. sanguinipennis* goes through six larval instars in this way, and there are no overwintering or estivation instars, although the fifth and sixth larval instars remain within the unbroken fourth and fifth exuvia, which will cover also the pupa. The adult *T. sanguinipennis* emerges in the autumn but remains within the nest of the host bees until the following spring (Parker and Böving, 1924). It is worth mentioning, however, that counting instars can be complicated. Seven larval instars have been recorded in *T. sanguinipennis*, but the sixth and the seventh remain within the exuvia of the fifth, which is the last feeding instar. However, Parker and Böving (1924) counted only six larval instars instead of seven. Selander (1986) considered that counts of three rather than four feeding instars in the literature are due to missing the molt between the third and fourth instars.

Other coleopteran families have parasitoid species that show hypermetamorphic life cycles with a planidium first larval instar. These include Carabidae, with the species *Lebia scapularis*, a parasitoid of the larvae of

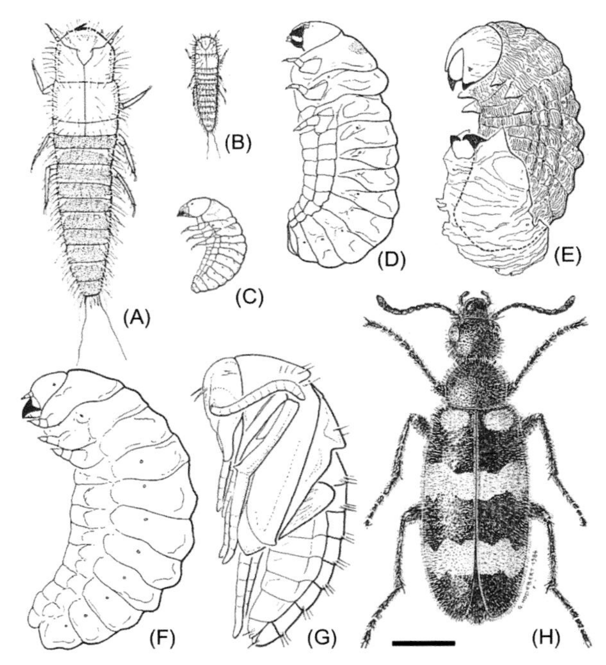

FIGURE 5.8 Life cycle stages of the coleopteran *Mylabris variabilis*. (A) First larval instar, planidium, at high magnification. (B) The same at scale. (C) Second larval instar. (D) Fourth larval instar (the third and fifth are very similar to the fourth). (E) Sixth, quiescent, larval instar, with remains of the exuvia of the preceding instar in the apical part of the abdomen. (F) Seventh (last) larval instar. (G) Pupa. (H) Adult. Scale bar: 3 mm. Modified from Paoli (1938) and Paulian (1988), with permission.

another Coleoptera (the Chrysomelidae *Galerucella luteola*), already studied by Silvestri in 1904 (Snodgrass, 1954). Also, the family Staphylinidae, which contains species of *Aleochara* that predate cabbage flies and constitute economically important natural enemies of cabbage fly pests (Broatch et al., 2008). More thoroughly studied have been the Ripiphoridae, which include parasitoids of immature Hymenoptera (Ripiphorinae) and Blattaria (Ripidiinae). Both groups follow the phoresy strategy to penetrate host colonies as a planidium attached to an adult of their host. After being transported to the host colony, the planidium burrows into the body of the host and then molts into a scarabaeiform larva that emerges to feed externally. The planidium of Ripidiinae attaches directly to the cockroach host and molts into a legless larva, which penetrates the host to feed. The last larval instar, which possesses poorly developed legs, exits the host and pupates at a certain distance. The adult of *Metoecus paradoxus* (a Ripiphorinae parasitoid of the common wasp, *Vespula vulgaris*) that emerges in the wasp colony, avoids detection by mimicking the cuticular hydrocarbon profile of the host (Van Oystaeyen et al., 2015). It is possible that this camouflage strategy is used also by other species of hypermetamorphic parasitoids whose adults emerge in the host's nest.

Strepsiptera

The order contains 640 described species and is divided into two suborders: Mengenillidia and Stylopidia. The early-branching suborder Mengenillidia is the sister group to all Stylopidia and is represented by a single family, Mengenillidae. The suborder Stylopidia includes, among others, the families Myrmecolacidae and Stylopidae (Kathirithamby, 2009, 2018; McMahon et al., 2011; Pohl and Beutel, 2008). Strepsipterans are endoparasitoids of other insects, and 34 families distributed across seven orders (from Zygentoma to Diptera) have been recorded as hosts. The planidium encounters the host where it molts into an endoparasitic legless, grub-like larva, which may undertake several molts. Species parasitizing social hymenoptera are phoretic, and the first larval instar is carried to the host nesting site, where the immature stages are attacked. Very characteristic of strepsipterans is the dramatic differences between sexes. Females live within the host from infection as first larval instar until the end of the life cycle. Thus their morphology and reproductive strategy are adapted to this mode of life, having lost eyes, antennae, legs, and wings. Instead the mature females are neotenic, retaining a larva-like appearance during the reproductive period. In contrast, adult males emerge after pupation in the host as free-flying insects ready to mate. Mating is exceptionally bizarre as it occurs with the female inside the host. The female extrudes the cephalothorax, through which the male inserts the sperm. Females are viviparous, and the progeny emerges outside through a canal that opens in the cephalothorax, and then they actively search a new host. An exception to this mode of life that is general in strepsipterans is provided by the family Mengenillidae, in which pupation takes place outside the host, and the female is free living (Kathirithamby, 2009). The family Myrmecolacidae is of special interest because males and females parasitize hosts belonging not only to different species but also to different insect orders. This complex and extreme form of behavior was called heterotrophic heteronomy by Walter in 1983, applied to Hymenoptera Aphelinidae. In Myrmecolacidae strepsipterans, the males develop as primary parasitoids in ants (Hymenoptera Formicidae), and the females develop as primary parasitoids in grasshoppers, crickets, and mantids (Orthoptera Tettigoniidae, Gryllidae; Mantodea) (Kathirithamby, 2009).

Lepidoptera

In the order Lepidoptera, the only family that shows hypermetamorphic life cycles is the Epipyropidae. Representatives of this family parasitize various homopteran species, which is unique in an eminently phytophagous insect order such as the Lepidoptera. The active first larval instar seeks out a host and the scarabaeiform larval instar that follows occurs on the body of a single homopteran. Their life cycle has often been studied in the context of their role as natural enemies of homopteran pests. This is the case, for example, of the species *Fulgoraecia melanoleuca*, a parasitoid of the fulgorid *Pyrilla perpusilla*, a serious pest of sugarcane in Asia, whose life cycle stages have been recently reported (Kumar et al., 2015).

Diptera

Hypermetamorphosis has been reported in the families Acroceridae, Nemestrinidae, and many species of Bombylidae. Some examples have been also recorded in some Tachinidae and Sarcophagidae. All these species have legless, well-tanned, and active planidia that search for the host that are followed by soft-bodied vermiform larvae. The Acroceridae are internal parasitoids of spiders. Detailed descriptions of the life cycle of *Oncodes pallipes* and *Pterodontia flavipes*, reported by Millot and King, respectively, are related in detail by Snodgrass (1954). More recent studies include the description of the species *Exetasis jujuyensis* from South America, with data on the effect of fly larval development on the behavior of the host spider, the Theraphosidae *Acanthoscurria sternalis* (Barneche et al., 2013). In *Acrocera orbicula*, a parasitoid of the wolf spider, *Pardosa prativaga*, the planidium attaches itself to the host spider with its mouthparts, cutting a tiny hole through the integument, but without penetrating the interior. A week later, the parasitoid molts into a small and flexible second larval instar, which invades the host through the hole made by the first larval instar (Overgaard Nielsen et al., 1999). This peculiar invasion strategy might reduce physical damage to the host, thus enhancing host (and parasitoid) survival. The Nemestrinidae are endoparasitoids of grasshoppers and beetles. They can have strong impacts on grasshopper populations, as shown in managed tallgrass prairies (Laws and Joern, 2012). The vast majority of Bombyliidae are ectoparasitoids, whereas only a few examples of endoparasitoid species are known in the tribes Gerontini, Systropodini, and Villini. The recorded host range of Bombyliidae spans seven insect orders and some spiders although almost half of all records are from Hymenoptera. They have not evolved structures to inject eggs directly into the host, as usual in parasitoid Hymenoptera. Hypermetamorphic Bombyliid larvae usually contact the host while it is in the larval stage and consume it when it reaches a quiescent stage, such as the mature larva, prepupa, or pupa (Yeates and Greathead, 1997).

Hypermetamorphosis in nonparasitoids with a wandering first larval instar. The case of *Micromalthus debilis*

The minute coleopteran *Micromalthus debilis* constitutes a singular case of hypermetamorphosis in a nonparasitoid insect. This species is the only representative of the family Micromalthidae (belonging to the archaic Coleoptera suborder Archostemata), known not only as extant species but also from amber fossil samples from different origins dated from Cretaceous to Miocene (Hörnchemeyer et al., 2010). Although *M. debilis* displays archaic morphological characters, its life cycle is so complex and singular that the first observations on this species, made by Barber in 1913, were received with remarkable disbelief (Pollock and Normark, 2002). Larvae of *M. debilis* live in decaying wood, feeding on fungi that develop in it. They are an occasional pest in wooden structures, such as railway sleepers or telegraph poles, which has made its distribution practically cosmopolitan by passive transport, although the species is apparently native to eastern North America.

The life cycle of *M. debilis* includes a typical planidum instar, different types of mature larvae, the pupa, and the adult (Fig. 5.9). Moreover, reproduction combines haplodiploidy, male uniovipartity (one male egg laid at a time), female polyviviparity (several female larvae delivered at a time), including internal and external matriphagy (the offspring eats the mother) (Perotti et al., 2016). During a significant part of the year, a given population of *M. debilis* is entirely formed by female planidia, cerambycoid, and paedogenetic larvae. The latter can reproduce by giving birth to about 10 small larvae of the planidium type, which are very agile and can disperse easily in the decaying wood. The planidium larva undertakes an unknown number of molts to become the cerambycoid larva, which often molts into a

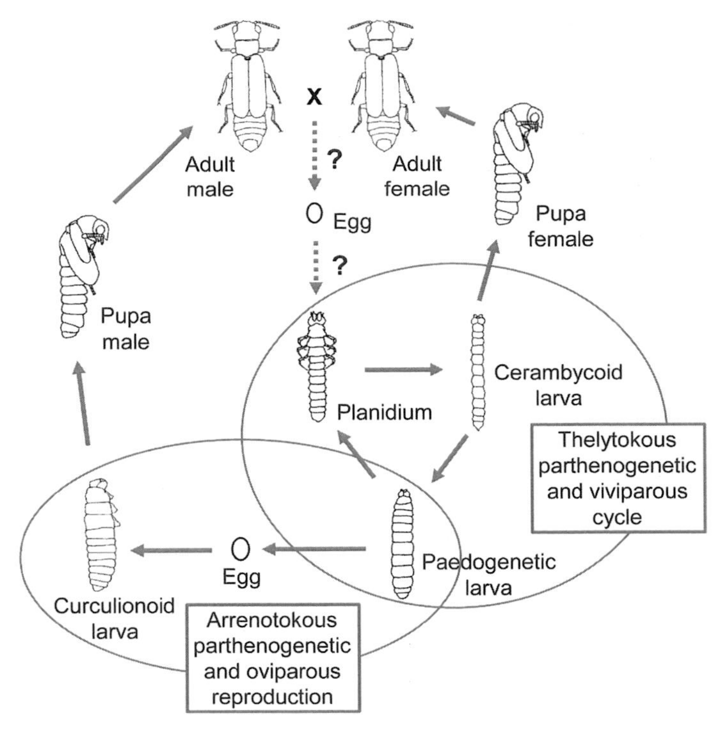

FIGURE 5.9 Schematic representation of the stages and reproduction modes in the life cycle the coleopteran *Micromalthus debilis*. The thelytokous parthenogenetic and viviparous cycle is common most of the year. The arrenotokous parthenogenetic and oviparous reproduction is rarer, as is the production of male and female pupa and adults, which can occur if the conditions become drier and hotter. It is not certain whether the adults can reproduce sexually.

paedogenetic larva; then this thelytokous parthenogenetic and viviparous cycle can start again (Fig. 5.9). However, in late summer (or earlier if the decaying wood becomes drier), some of the cerambycoid female larvae can develop into pupae, which eventually molt into winged adult females (Fig. 5.9). Moreover, the paedogenetic larvae can produce a single unfertilized egg, from which a male larva emerges, which is a typical arrenotokous parthenogenetic and oviparous type of reproduction (Fig. 5.9). Following an extraordinary behavior, the emerged male larva inserts its head into the genital atrium of his mother and begins to feed upon her. It takes a male larva about a week to devour his mother entirely and then it becomes a curculionoid larva, which can eventually pupate and then molt into a winged adult male (Fig. 5.9). It is not completely certain whether the adult beetles can reproduce sexually. However, recent experiments based on raising the temperature of the *M. debilis* colonies to 55°C to favor the production of pupae and adults (Perotti et al., 2016) have provided valuable results. For example, the observations indicate that adults exhibit a female-biased sex ratio. Heat and drought experiments produced 1000 females and 59 males, of which only 17 were useful for mating behavior observations. Intriguingly, these observations indicated that there is a sex-role reversal, the female mounting the male. However, no true copulation was observed, and none of the 1000 adult females was seen laying an egg or producing progeny, either by sexual or parthenogenetic reproduction. Interestingly, the life span of males (13 hours on average) was notably shorter than that of the adult females (148 hours on average) (Perotti et al., 2016). While these observations suggest that adults of *M. debilis* are unable to reproduce, the artificiality of the experiment does not rule out that under natural conditions adult sexual reproduction might occur.

Hypermetamorphosis in parasitoids with a sedentary first larval instar

This case is represented by parasitoids that oviposit directly in the host or in the host egg. Therefore the first larval instar emerges in the material that will serve as food supply and is adapted to these circumstances, not needing then a first active stage in the life cycle. Examples are provided by parasitoids belonging to the orders Hymenoptera and Diptera.

Hymenoptera

The first larval stage of most parasitoid Hymenoptera has diverse morphologies, some of them being so bizarre that they do not really look like insect larvae. In his classic book about entomophagous insects, Clausen (1940) describes 14 different types of extravagant first larval instars, although in the respective species, most of them transform to typical Hymenopteran larvae after the subsequent molts. A classic example is the Braconidae *Aridelus* (=*Helorimorpha*) *antestiae* that parasitizes pentatomid Hemiptera of the genus *Antestia*, which was studied by Kirkpatrick in 1918 (see Snodgrass, 1954). The first larval instar is very simplified, showing a vermiform unsegmented body with a characteristically big cephalic region and a kind of tail in the distal end of the body (Fig. 5.10A). The second larval instar still has a vermiform body, but it is markedly segmented, with the cephalic region smaller and the distal tail shortened. Finally, the third larval instar adopts the typical morphology of a nonparasitic Hymenoptera (Fig. 5.10B).

Within the family Platygastridae, the genus *Platygaster*, with some 600 described parasitoid species, includes valuable natural enemies of insect pests. In some cases, they can parasitize both eggs and first instar larvae of the host, as in the case of *Platygaster demades*, parasitoid of the Cecidomyid fly *Dasineura mali*, an important pest of apple trees (He and Wang, 2015). Some species of *Platygaster* provide conspicuous examples of a morphologically divergent first larval instar, like in the case of *Platygaster instricator* (Fig. 5.5A), *Platygaster herrickii* (Fig. 5.10C), or *Platygaster marchali* (Fig. 5.10D). Also within the Platygastridae, the genus *Synopeas*, with some 250 described species that are parasitic, also shows a very divergent, and characteristic first larval instar (Fig. 5.10E), that would conform to the "teleaform" type categorized by Clausen (1940). Also corresponding to the teleaform type of first larval instar is that of *Hadronotus ajax* (a Scelionidae, closely related to the Platygastridae) whose life cycle has been described in detail (Schell, 1943). *H. ajax* contributes to the biological control of the hemipteran *Anasa tristis*, the squash bug, a pest of cucurbitaceous crops (Doughty et al., 2016). The first larval instar (Fig. 5.10F) has a vermiform body showing a caudal horn-like structure curved anteriorly; the body is unsegmented but shows a constriction between a large cephalothoracic anterior region and an elongated posterior part. The cephalothoracic region shows a pair of large mandibles and a kind of "labial projection." Apparently, the caudal horn-like structure contributes to feeding mechanisms. The second larval instar of *H. ajax* adopts an oval sacciform shape, still unsegmented, whereas the third instar (Fig. 5.10G) is completely segmented, showing the larval morphology of a typical nonparasitoid hymenopteran.

Diptera

The most dramatic specializations of the first larval instar have been described in species of the genus *Cryptochaetum*, with some 40 described species that are parasitoids of scale insects, thus susceptible to being used as pest control agents. Indeed the introduction of the Australian species *Cryptochaetum iceryae* into California in the 1880s provided complete

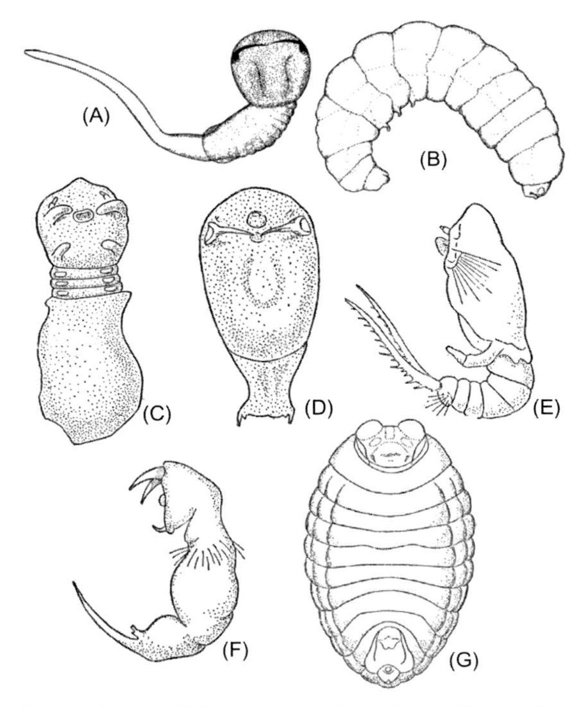

FIGURE 5.10 Larvae of parasitoid Hymenoptera with a sedentary first larval instar. First (A) and third (B) larval instars of *Aridelus* (=*Helorimorpha*) sp. (C) First larval instar of *Platygaster herrickii*. (D) First larval instar of *Platygaster marchali*. (E) First larval instar of *Synopeas* sp. First (F) and third (G) larval instars of *Hadronotus ajax*. From Snodgrass (1954), with permission.

control of the Cottony Cushion Scale (*Icerya purchasi*) (Thorpe, 1930). *Cryptochaetum* larvae have been reared from scale insects (Coccidae and Margarodidae), which has allowed the observation of the elaborate endoparasitic life cycle in a number of species (Thorpe, 1930, 1934, 1941; Yang and Yang, 1996). The *Cryptochaetum* oviposits in a medium-sized nymph of a scale insect when their teguments have not completely tanned. The freshly emerged first larval instar looks like an opaque sac, with rudimentary or absent mouthparts, which absorbs the nutrients mainly through the tegument while literally immersed in the liquid food source provided by the host. The more conspicuous feature of this larva is a pair of hemolymph-filled, finger-like appendages at the distal part of the body. In some species, like *C. iceryae*, the first larval instar is unsegmented, but in others, like *Cryptochaetum grandicorne*, segments appear when the larva matures. The second larval instar is completely segmented, with the posterior segments armed with short spines and the mouth showing tanned mandibles and other mouth pieces. The third larval instar resembles the second, but the finger-like appendages have become considerably longer. However, it progressively transforms into a barrel-shaped, yellowish maggot formed by a head followed by 10 distinct segments; the finger-like appendages are still present but have become fragile and easily broken (Thorpe, 1930, 1934). The seemingly simple structure of the *Cryptochetum* larva would seem to fit Berlese's protopod larva and his theory of the premature eclosion of the embryo (Berlese, 1913). However, Thorpe (1930) stated that "the theory obviously cannot be pushed too far, for there are many truly adaptive characters which arise de novo in insect larvae, and cannot in any way be described as embryonic."

CONTRACTED LIFE CYCLES

Another type of special postembryonic development that deserves attention is the contracted life cycle undertaken by very specialized cave-dwelling Coleoptera, which is characterized by a dramatic reduction in the number of larval instars. Initial observations were due to the remarkable work of Sylvie Deleurance (née Sylvie Glaçon) in the subterranean laboratory of Moulis, in southern France. She focused the observations on Coleoptera Leiodidae belonging to the tribe Leptodirini (subfamily Bathysciinae, in classical literature) and also on a number of species of Carabidae Trechini (Deleurance-Glaçon, 1963a,b). She presented her results in the first international congress of speleology held in Paris in 1953, but the results were received with skepticism, if not with incredulity. However, further research fully confirmed their observations, which were subsequently expanded with ecophysiological and reproductive data (Delay, 1978).

The Leptodirini is a quite large group (around 900 species described) with many representatives that live in subterranean habitats. The most specialized show all the adaptations associated with subterranean life, like depigmentation, lack of eyes and membranous wings, and elongation of antennae and legs, although there is a continuum from less modified species, living on dark and humid habitats, to the most modified, which are strictly cavernicolous (Belles, 1987). The life cycles show also this kind of gradation, from less to more modified, and have been classified into the following three categories. The first category is represented by the life cycle of Leptodirini living in mosses, forest litter, or in shallow soil generally, which comprises three larval instars, like in the muscicolous species *Bathysciola schiodtei*. The female of *B. schiodtei* produces small oligolecital eggs and the emerging larvae undergo two molts, thus passing through three larval instars, and then pupate. The larvae feed throughout their life except for the last period of the third instar. This life cycle is similar to that of a current epigean coleopteran, except by the significant detail that the larva builds a clay chamber where it hides at the time of molting. The second category is exemplified by cavernicolous species that are not dramatically adapted to cave life, like *Parvospeonomus delarouzeei*. The female of this species also produces small eggs, from which the first larval instar emerges and actively feeds. At the end of the first instar, the larvae construct the typical clay chamber in which to hide for molting to a second larval instar, which does not feed, and encloses itself in another clay chamber at the time of pupation. Inside this cell, the larva undergoes a quiescent period that lasts between 3 and 4 months. A similar example of a life cycle with only two larval instars is provided by the more modified cavernicolous *Diaprisius serullazi*, but in this species, neither of the two larval instars feeds. Finally, the third category corresponds to the more cave-specialized Leptodirini, like the species *Speonomus longicornis*, whose female oviposits a single large egg at a time. From this macrolecital egg emerges a larva that has an extremely short free life (it can be as short as a few hours), during which it does not feed and soon proceeds to build the typical clay chamber hiding in it. The larva remains within the chamber in a quiescent state during 5–6 months and then pupates in the same chamber. Thus, this "contracted" cycle comprises a single larval instar, during which it does not feed at all (Fig. 5.11) (Delay, 1978; Deleurance-Glaçon, 1963a; reviewed in Belles, 1987 and Vandel, 1964).

Although much less studied, due to the higher difficulty for laboratory rearing and manipulation, the data available on the life cycles of cavernicolous Trechini beetles fit with those of Leptodirini. In the little modified species *Trichaphaenops gounellei*, there are three larval instars, and feeding is observed during the first part of the first and second larval instars. The more modified *Hydraphaenops ehlersi* has two larval instars, and feeding occurs in the first part of both. Finally, the highly modified species *Aphaenops cerberus* also has two larval instars but in none of them was feeding activity observed (Deleurance-Glaçon, 1963b). More recent studies comparing the female internal reproductive system for six species of Trechini, including subterranean species of the genus *Aphaenops*, show again a correlation

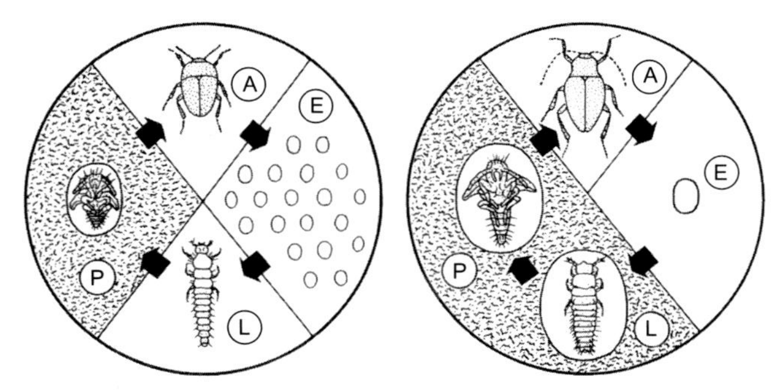

FIGURE 5.11 Extreme models of life cycle in cave-dwelling Coleoptera Leptodirini. (A) Normal cycle, characteristic of poorly modified species; the adult female oviposits many small eggs (E); the larval period (L) involves three free stages with feeding activity (although the molts occur inside a clay chamber built by the larva); at the end of the final larval stage, the larva builds again a clay chamber where it will live until molting to the pupal stage (P). (B) Contracted cycle, characteristic of the most modified species; the adult female oviposits a single large egg at a time (E); the larval period (L) consists of a single instar that lives in a clay chamber and is practically quiescent, like the pupal stage (P). From Belles (1987), with permission.

between the reduction in ovariole number and size and the degree of adaptation to the subterranean life (Faille and Pluot-Sigwalt, 2015). A similar correlation between ovariole number and degree of adaptation to the subterranean life has been reported in hemimetabolan species, like the cavernicolous cockroaches of the genus *Loboptera* of Canary Islands (Izquierdo et al., 1990). This led to the prediction that cavernicolous *Loboptera* with fewer and bigger ovarioles will have reduced the number of nymphal instars.

The data thus indicate that in beetles highly adapted to live in caves the first larval instar emerges from a macrolecital egg, where the abundant yolk provides enough nutrients to complete larval development without feeding. The contracted cycles, although they do not shorten the juvenile life, minimize the number of molts and the foraging time to search for food, thereby reducing the chances of being predated. The construction of clay chambers in the Leptodirini helps to reinforce the protection against predators. Interestingly, the contracted cycles, with a reduced number of larval instars in which the insect does not increase its body weight, raise questions of the applicability of the concept and mechanisms of critical size, which contributes to triggering the metamorphic molt (Nijhout et al., 2014; Nijhout and Callier, 2015), to these species. It is possible that in cavernicolous beetles the weight decreases during the larval period instead of increasing. In that case, a critical size could also be reached, but going from more to less weight, which might be assessed by sensors adapted to these weight dynamics. In any case, the study of the mechanisms that trigger the pupal molt in these Coleoptera is challenging.

REFERENCES

Aldaz, S., Escudero, L.M., 2010. Imaginal discs. Curr. Biol. 20, R429–R431. Available from: https://doi.org/10.1016/j.cub.2010.03.010.

Allee, J.P., Pelletier, C.L., Fergusson, E.K., Champlin, D.T., 2006. Early events in adult eye development of the moth, *Manduca sexta*. J. Insect Physiol. 52, 450–460. Available from: https://doi.org/10.1016/j.jinsphys.2005.12.006.

Anderson, D.T., 1972. The development of Holometabolous insects. In: Counce, S.J., Waddington, C.H. (Eds.), Developmental Systems: Insects. Academic Press, London, pp. 165–242.

Aspiras, A.C., Smith, F.W., Angelini, D.R., 2011. Sex-specific gene interactions in the patterning of insect genitalia. Dev. Biol. 360, 369–380. Available from: https://doi.org/10.1016/j.ydbio.2011.09.026.

Barneche, J.A., Gillung, J.P., González, A., 2013. Description and host interactions of a new species of *Exetasis* Walker (Diptera: Acroceridae), with a key to species of the genus. Zootaxa 3664, 525–536.

Beira, J.V., Paro, R., 2016. The legacy of *Drosophila* imaginal discs. Chromosoma 125, 573–592.

Belles, X., 1987. Fauna cavernićola i intersticial de la Peninśula Iberica i les Illes Balears. Consell Superior d'Investigacions Cientifiques-Editoral Moll, Madrid-Palma de Mallorca.

Berlese, A., 1913. Intorno alle metamorfosi degli insetti. Redia 9, 121–136.

Beutel, R.G., Pohl, H., 2006. Endopterygote systematics – where do we stand and what is the goal (Hexapoda, Arthropoda)? Syst. Entomol. 31, 202–219. Available from: https://doi.org/10.1111/j.1365-3113.2006.00341.x.

Brauer, F., 1869. Beschreibung der Verwandlungsgeschichte der *Mantispa styriaca* Poda und Betrachtungen über die sogenannte Hypermetamorphose Fabre's. Verh. Zool. Bot. Ges. Wien 19, 831–840.

Broatch, J.S., Dosdall, L.M., Yang, R.-C., Harker, K.N., Clayton, G.W., 2008. Emergence and seasonal activity of the entomophagous rove beetle *Aleochara bilineata* (Coleoptera: Staphylinidae) in canola in Western Canada. Environ. Entomol. 37, 1451–1460.

Campos-Ortega, J.A., Hartenstein, V., 1997. The Embryonic Development of *Drosophila melanogaster*, second ed. Springer Berlin Heidelberg, Berlin, Heidelberg. Available from: https://doi.org/10.1007/978-3-662-02454-6.

Champlin, D.T., Truman, J.W., 1998. Ecdysteroids govern two phases of eye development during metamorphosis of the moth, *Manduca sexta*. Development 125, 2009–2018.

Chen, S.H., 1946. Evolution of the insect larva. Trans. R. Entomol. Soc. London 97, 381–404. Available from: https://doi.org/10.1111/j.1365-2311.1946.tb00270.x.

Chen, Q., Li, T., Hua, B., 2012. Ultrastructure of the larval eye of the scorpionfly *Panorpa dubia* (Mecoptera: Panorpidae) with implications for the evolutionary origin of holometabolous larvae. J. Morphol. 273, 561–571. Available from: https://doi.org/10.1002/jmor.20001.

Clausen, C.P., 1923. The biology of *Schizaspidia tenuicornis* Ashm. a eucharid parasite of *Camponotus*. Ann. Entomol. Soc. Am 16, 32–69.

Clausen, C.P., 1940. Entomophagous Insects. McGraw-Hill, New York.

Clausen, C.P., 1941. The habits of the Eucharidae. Psyche (Stuttg) 48, 57–69.

Clausen, C.P., 1976. Phoresy among entomophagous insects. Annu. Rev. Entomol. 21, 343–368. Available from: https://doi.org/10.1146/annurev.en.21.010176.002015.

Cohen, S.M., 1993. Imaginal disc development. In: Bate, M., Martinez Arias, A. (Eds.), The Development of *Drosophila melanogaster*. Cold Spring Harbor Laboratory Press, pp. 747–841.

Costa, C., 1984. Origem e função da pupa dos Endopterygota. 1. Revisão histórica e considerações gerais. Cienc. e Cult. São Paulo 36, 1126–1134.

Costa, C., Ide, S., Simonka, C.E. (Eds.). 2006. Insectos inmaduros. Metamorfosis e identificación. Monografías del tercer milenio, vol. 5. SEA, Zaragoza.

Davis, G.K., Patel, N.H., 2002. Short, long, and beyod: molecular and embryological approaches to insect segmentation. Annu. Rev. Entomol. 47, 669–699. Available from: https://doi.org/10.1146/annurev.ento.47.091201.145251.

Delay, B., 1978. Milieu souterrain et écophysiologie de la reproduction et du développement des Coléoptères Bathysciinae hypogés. Mém. Biospéol. 5, 1–349.

Deleurance-Glaçon, S., 1963a. Recherches sur les Coléoptères troglobies de la sous-famille des Bathysciinae. Ann. Sci. Nat. Zool. 5, 1–172.

Deleurance-Glaçon, S., 1963b. Contribution à l'étude des coléoptères cavernicoles de la sous-famille des Trechinae. Ann. Speleol. 18, 227–265.

Dorey, J.B., Merritt, D.J., 2017. First observations on the life cycle and mass eclosion events in a mantis fly (Family Mantispidae) in the subfamily Drepanicinae. Biodivers. Data J. 5, e21206. Available from: https://doi.org/10.3897/BDJ.5.e21206.

Doughty, H.B., Wilson, J.M., Schultz, P.B., Kuhar, T.P., 2016. Squash Bug (Hemiptera: Coreidae): biology and management in cucurbitaceous crops. J. Integr. Pest Manag. 7, 1. Available from: https://doi.org/10.1093/jipm/pmv024.

Erickson, E.H., Enns, W.R., Werner, F.G., 1976. Bionomics of the bee-associated Meloidae (Coleoptera); bee and plant hosts of some Nearctic Meloid beetles—a synopsis. Ann. Entomol. Soc. Am. 69, 959–970. Available from: https://doi.org/10.1093/aesa/69.5.959.

Fabre, J.H., 1857. Mémoire sur l'hypermétamorphose et les moeurs des Méloïdes. Ann. Sci. Nat. Zool. 7 (4), 299–365.

Faille, A., Pluot-Sigwalt, D., 2015. Convergent reduction of ovariole number associated with subterranean life in beetles. PLoS One 10, e0131986. Available from: https://doi.org/10.1371/journal.pone.0131986.

Friedrich, M., 2003. Evolution of insect eye development: first insights from fruit fly, grasshopper and flour beetle. Integr. Comp. Biol. 43, 508–521. Available from: https://doi.org/10.1093/icb/43.4.508.

Friedrich, M., 2006. Continuity versus split and reconstitution: exploring the molecular developmental corollaries of insect eye primordium evolution. Dev. Biol. 299, 310–329. Available from: https://doi.org/10.1016/j.ydbio.2006.08.027.

Grbić, M., Strand, M.R., 1998. Shifts in the life history of parasitic wasps correlate with pronounced alterations in early development. Proc. Natl. Acad. Sci. U.S.A. 95, 1097–1101.

He, X.Z., Wang, Q., 2015. Ability of *Platygaster demades* (Hymenoptera: Platygastridae) to parasitize both eggs and larvae makes it an effective natural enemy of *Dasineura mali* (Diptera: Cecidomyiidae). J. Econ. Entomol. 108, 1884–1889. Available from: https://doi.org/10.1093/jee/tov116.

Held, L.I., 2002. Imaginal Discs: The Genetic and Cellular Logic of Pattern Formation. Cambridge University Press, Cambridge.

Heming, B.S., 2003. Insect Development and Evolution. Comstock Pub. Associates.

Hinton, H.E., 1946. A new classification of insect pupae. Proc. Zool. Soc. London 161, 282–328.

Hinton, H.E., 1948. On the origin and function of the pupal stage. Trans. R. Entomol. Soc. London 99, 395–409.

Hinton, H.E., 1963. The origin and function of the pupal stage. Proc. R. Entomol. Soc. London 38, 77–85.

Hörnchemeyer, T., Wedmann, S., Poinar, G., 2010. How long can insect species exist? Evidence from extant and fossil *Micromalthus* beetles (Insecta: Coleoptera). Zool. J. Linn. Soc. 158, 300–311. Available from: https://doi.org/10.1111/j.1096-3642.2009.00549.x.

Horsfall, W.R., Ronquillo, M.C., 1970. Genesis of the reproductive system of mosquitoes. II. Male of *Aedes stimulans* (Walker). J. Morphol. 131, 329–357.

Huet, C., Lenoir-Rousseaux, J.J., 1976. Étude de la mise en place de la patte imaginale de *Tenebrio molitor*. 1. Analyse experimentale des processus de restauration au cours de la morphogenèse. J. Embryol. Exp. Morph. 35, 303–321.

Imms, A.D., 1930. A General Textbook of Entomology. Methuen and Co. Ltd., London.

Izquierdo, I., Oromi, P., Belles, X., 1990. Number of ovarioles and degree of dependence with respect to the underground environment in the Canarian (North Atlantic Ocean) species of the genus *Loboptera* Brunner (Blattaria, Blattellidae). Mém. Biospéol. 17, 107–112.

Kango-Singh, M., Singh, A., Gopinathan, K.P., 2001. The wings of *Bombyx mori* develop from larval discs exhibiting an early differentiated state: a preliminary report. J. Biosci. 26, 166–177.

Kathirithamby, J., 2009. Host-parasitoid associations in Strepsiptera. Annu. Rev. Entomol. 54, 227–249. Available from: https://doi.org/10.1146/annurev.ento.54.110807.090525.

Kathirithamby, J., 2018. Biodiversity of Strepsiptera. In: Foottit, R.G., Adler, P.H. (Eds.), Insect Biodiversity: Science and Society, vol. 2. John Wiley & Sons Ltd, New York, pp. 673–703.

Kim, C.-W., 1959. The differentiation centre inducing the development from larval to adult leg in *Pieris brassicae* (Lepidoptera). Development 7, 572–582.

Kivelä, S.M., Friberg, M., Wiklund, C., Leimar, O., Gotthard, K., 2016. Towards a mechanistic understanding of insect life history evolution: oxygen-dependent induction of moulting explains moulting sizes. Biol. J. Linn. Soc. 117, 586–600. Available from: https://doi.org/10.1111/bij.12689.

Konopová, B., Zrzavý, J., 2005. Ultrastructure, development, and homology of insect embryonic cuticles. J. Morphol. 264, 339–362.

Kulagin, N., 1898. Beiträge zur Kenntnis der Entwicklungsgeschichte von *Platygaster*. Ztschr. Wiss. Zool. 63, 195–235.

Kumar, R., Mittal, V., Chutia, P., Ramamurthy, V.V., 2015. Taxonomy of *Fulgoraecia melanoleuca* (Fletcher, 1939), (Lepidoptera: Epipyropidae) in India, a biological control agent of *Pyrilla perpusilla* (Walker) (Hemiptera: Lophopidae). Zootaxa 3974, 431–439.

Lawrence, P.A., 1995. The Making of a Fly: The Genetics of Animal Design. Blackwell Science, Oxford.

Laws, A.N., Joern, A., 2012. Variable effects of dipteran parasitoids and management treatment on grasshopper fecundity in a tallgrass prairie. Bull. Entomol. Res. 102, 123–130. Available from: https://doi.org/10.1017/S0007485311000472.

Linnaeus, C., 1758. Systema Naturae per Regna Tria Naturae secundum Classes, Ordines, Genera, Species, cum Characteribus, Differentiis, Synonymis, Locis. Vol. 1, Editio dec. ed. L. Salvii, Holmiae.

Maia-Silva, C., Hrncir, M., Koedam, D., Machado, R.J.P., Imperatriz-Fonseca, V.L., 2013. Out with the garbage: the parasitic strategy of the mantisfly *Plega hagenella* mass-infesting colonies of the eusocial bee *Melipona subnitida* in northeastern Brazil. Naturwissenschaften 100, 101–105. Available from: https://doi.org/10.1007/s00114-012-0994-1.

McMahon, D.P., Hayward, A., Kathirithamby, J., 2011. Strepsiptera. Curr. Biol. 21, R271–R272. Available from: https://doi.org/10.1016/j.cub.2011.02.038.

Mehta, D.R., 1933. On the development of the male genitalia and the efferent genital ducts in Lepidoptera. Quart. J. Microscop. Sci. 76, 35–61.

Miner, A.L., Rosenberg, A.J., Frederik Nijhout, H., 2000. Control of growth and differentiation of the wing imaginal disk of *Precis coenia* (Lepidoptera: Nymphalidae). J. Insect. Physiol. 46, 251–258.

Mirth, C., Truman, J.W., Riddiford, L.M., 2005. The role of the prothoracic gland in determining critical weight for metamorphosis in *Drosophila melanogaster*. Curr. Biol. 15, 1796–1807. Available from: https://doi.org/10.1016/j.cub.2005.09.017.

Monsma, S.A., Booker, R., 1996. Genesis of the adult retina and outer optic lobes of the moth, *Manduca sexta*. I. patterns of proliferation and cell death. J. Comp. Neurol. 367, 10–20.

Nardi, J.B., Bee, C.M., Wallace, C.L., 2018. Remodeling of the abdominal epithelial monolayer during the larva-pupa-adult transformation of *Manduca*. Dev. Biol. 438, 10–22. Available from: https://doi.org/10.1016/j.ydbio.2018.03.017.

Newport, G., 1851. On the natural history, anatomy and development of the oil beetle, *Meloe*, more especially of *Meloe cicatricosus* Leach. First memoir. The natural history of *Meloe*. Trans. Linn. Soc. London 20, 297–320.

Nijhout, H.F., Callier, V., 2015. Developmental mechanisms of body size and wing-body scaling in insects. Annu. Rev. Entomol. 60, 141–156. Available from: https://doi.org/10.1146/annurev-ento-010814-020841.

Nijhout, H.F., Davidowitz, G., Roff, D.A., 2006. A quantitative analysis of the mechanism that controls body size in *Manduca sexta*. J. Biol. 5, 16. Available from: https://doi.org/10.1186/jbiol43.

Nijhout, H.F., Riddiford, L.M., Mirth, C., Shingleton, A.W., Suzuki, Y., Callier, V., 2014. The developmental control of size in insects. Wiley Interdiscip. Rev. Dev. Biol. 3, 113–134. Available from: https://doi.org/10.1002/wdev.124.

Ninov, N., Martín-Blanco, E., 2009. Changing gears in the cell cycle: histoblasts and beyond. Fly (Austin) 3, 286–289.

Ninov, N., Chiarelli, D.A., Martin-Blanco, E., 2007. Extrinsic and intrinsic mechanisms directing epithelial cell sheet replacement during *Drosophila* metamorphosis. Development 134, 367–379. Available from: https://doi.org/10.1242/dev.02728.

Overgaard Nielsen, B., Funch, P., Toft, S., 1999. Self-injection of a dipteran parasitoid into a spider. Naturwissenschaften 86, 530–532.

Packard, A.S., 1898. A Textbook of Entomology. Macmillan, New York.

Panfilio, K.A., 2008. Extraembryonic development in insects and the acrobatics of blastokinesis. Dev. Biol. 313, 471–491. Available from: https://doi.org/10.1016/j.ydbio.2007.11.004.

Paoli, G., 1938. Note sulla biologia e sulla filogenesi dei Meloidi (Coleoptera). Mem. Soc. Ent. Ital. 16, 71–97.

Parker, J.B., Böving, A.G., 1924. The blister beetle *Tricrania sanguinipennis*—biology, descriptions of different stages, and systematic relationship. Proc. U.S. Nat. Mus. 64, 1–40.

Paulian, R., 1988. Biologie des coléoptères. Lechevalier, Paris.

Paulus, H.F., 1989. Das Homologisieren in der Feinstrukturforschung: Das Bolwig-Organ der hoeheren Dipteren und seine Homologisierung mit Stemmata und Ommatidien eines urspruenglichen Facettenauges der Mandibulata. Zool. Beitr. N.F. 32, 437–478.

Peel, A.D., 2008. The evolution of developmental gene networks: lessons from comparative studies on holometabolous insects. Philos. Trans. R. Soc. Lond. B. Biol. Sci. 363, 1539–1547. Available from: https://doi.org/10.1098/rstb.2007.2244.

Perotti, M.A., Young, D.K., Braig, H.R., 2016. The ghost sex-life of the paedogenetic beetle *Micromalthus debilis*. Sci. Rep. 6, 27364. Available from: https://doi.org/10.1038/srep27364.

Pinto, J.D., 2003. Hypermetamorphosis. In: Resh, V.H., Cardé, R. (Eds.), Encyclopedia of Insects. Academic Press, Amsterdam, pp. 484–486.

Pinto, J.D., Selander, R.B., 1970. The bionomics of blister beetles of the genus *Meloe* and a classification of the New World species. Illinois Biol. Monogr. 42, 1–250. Available from: https://doi.org/10.5962/bhl.title.50239.

Pohl, H., Beutel, R.G., 2008. The evolution of Strepsiptera (Hexapoda). Zoology 111, 318–338. Available from: https://doi.org/10.1016/j.zool.2007.06.008.

Pollock, D.A., Normark, B.B., 2002. The life cycle of *Micromalthus debilis* LeConte (1878) (Coleoptera: Archostemata: Micromalthidae): historical review and evolutionary perspective. J. Zool. Syst. Evol. Res. 40, 105–112. Available from: https://doi.org/10.1046/j.1439-0469.2002.00183.x.

Quennedey, A., Quennedey, B., 1990. Morphogenesis of the wing Anlagen in the mealworm beetle *Tenebrio molitor* during the last larval instar. Tissue Cell 22, 721–740.

Quennedey, A., Quennedey, B., 1999. Development of the wing discs of *Zophobas atratus* under natural and experimental conditions: occurrence of a gradual larval-pupal commitment in the epidermis of tenebrionid beetles. Cell Tissue Res. 296, 619–634.

Redborg, K.E., 1998. Biology of the Mantispidae. Annu. Rev. Entomol. 43, 175–194. Available from: https://doi.org/10.1146/annurev.ento.43.1.175.

Reinecke, L.H., Reinecke, J.P., Adams, T.S., 1983. Morphology of the male reproductive tract of mature larval, pupal, and adult tobacco hornworms (Lepidoptera: Sphingidae), *Manduca sexta*. Ann. Entomol. Soc. Am. 76, 365–375.

Ronquillo, M.C., Horsfall, W.R., 1969. Genesis of the reproductive system of mosquitoes. I. Female of *Aedes stimulans* (Walker). J. Morphol. 129, 249–280.

Rottmar, B., 1966. Über Züchtung, Diapause und postembryonale Entwicklung von *Panorpa communis* L. Zool. Jahrb. Anat. Ontog. Tiere 83, 497–570.

Schell, S.C., 1943. The biology of *Hadronotus ajax* Girault (Hymenoptera-Scelionidae), a parasite in the eggs of squash-bug (*Anasa tristis* DeGeer). Ann. Entomol. Soc. Am. 36, 625–635. Available from: https://doi.org/10.1093/aesa/36.4.625.

Sehnal, F., 1985. Growth and life cycles. In: Kerkut, G.A., Gilbert, L. (Eds.), Comprehensive Insect Physiology, Biochemistry and Pharmacology. Pergamon Press, Oxford, pp. 1–86.

Selander, R.B., 1986. Rearing blister beetles (Coleoptera, Meloidae). Insecta Mundi 1, 209–218.

Smith, H.S., 1912. The chalcidoid genus *Perilampus* and its relation to the problem of parasitic introduction. Tech. Ser. Bur. Ent. US 19, 32–69.

Snodgrass, R.E., 1954. Insect metamorphosis. Smithson. Misc. Collect. 122, 1–124.

Stehr, F.W. (Ed.), 1987. Immature Insects, vol. 1. Kendall/Hunt Publishing, Dubuke.

Stehr, F.W. (Ed.), 1991. Immature Insects, vol. 2. Kendall/Hunt Publishing, Dubuke.

Sucena, É., Vanderberghe, K., Zhurov, V., Grbić, M., 2014. Reversion of developmental mode in insects: evolution from long germband to short germband in the polyembrionic wasp *Macrocentrus cingulum* Brischke. Evol. Dev. 16, 233–246. Available from: https://doi.org/10.1111/ede.12086.

Švácha, P., 1992. What are and what are not imaginal discs: reevaluation of some basic concepts (Insecta, Holometabola). Dev. Biol. 154, 101–117.

Tanaka, K., Truman, J.W., 2005. Development of the adult leg epidermis in *Manduca sexta*: contribution of different larval cell populations. Dev. Genes Evol. 215, 78–89. Available from: https://doi.org/10.1007/s00427-004-0458-5.

Tanaka, K., Truman, J.W., 2007. Molecular patterning mechanism underlying metamorphosis of the thoracic leg in *Manduca sexta*. Dev. Biol. 305, 539–550. Available from: https://doi.org/10.1016/j.ydbio.2007.02.042.

Tettamanti, G., Casartelli, M., 2019. Cell death during complete metamorphosis. Philos. Trans. R. Soc. B 374, 20190065.

Thorpe, W.H., 1930. The biology, post-embryonic development and economic importance of *Cryptochaetum iceryae* (Diptera, Agromyzidae) parasitic on *Icerya purchasi* (Coccidae, Monophlebinae). Proc. Zool. Soc. Lond. 60, 929–971. Available from: https://doi.org/10.1111/j.1096-3642.1930.tb01007.x.

Thorpe, W.H., 1934. Memoirs: the biology and development of *Cryptochaetum grandicorne* (Diptera), an internal parasite of *Guerinia serratulae* (Coccidae). J. Cell Sci. s2–77, 273–304.

Thorpe, W.H., 1941. A description of six new species of the genus *Cryptochaetum* (Diptera-Agromyzidae) from east Africa and east indies; together with a key to the adults and larvae of all known species. Parasitology 33, 131–148.

Tikhomirova, A.L., 1983. On homology of the joints of larval and imaginal legs in insects with complete metamorphosis (*Tenebrio molitor* taken as an example). Zool. Zhurn. 62, 530–539 (in Russian, with English summary).

Tripp, H.A., 1961. The biology of a hyperparasite, *Euceros frigidus* Cress. (Ichneumonidae) and description of the planidial stage. Can. Entomol. 93, 40–58. Available from: https://doi.org/10.4039/Ent9340-1.

Vandel, A., 1964. Biospeologie: la biologie des animaux cavernicoles. Gauthier-Villars, Paris.

Vander Meer, R.K., Jouvenaz, D.P., Wojcik, D.P., 1989. Chemical mimicry in a parasitoid (Hymenoptera: Eucharitidae) of fire ants (Hymenoptera: Formicidae). J. Chem. Ecol. 15, 2247–2261. Available from: https://doi.org/10.1007/BF01014113.

Van Oystaeyen, A., van Zweden, J.S., Huyghe, H., Drijfhout, F., Bonckaert, W., Wenseleers, T., 2015. Chemical strategies of the beetle *Metoecus paradoxus*, social parasite of the wasp *Vespula vulgaris*. J. Chem. Ecol. 41, 1137–1147. Available from: https://doi.org/10.1007/s10886-015-0652-0.

Yang, C.K., Yang, C.Q., 1996. Cryptochetidae, Flies of China, vol. 1. Liaoning Science and Technology Press, Shenyang, pp. 224–233.

Yeates, D.K., Greathead, D., 1997. The evolutionary pattern of host use in the Bombyliidae (Diptera): a diverse family of parasitoid flies. Biol. J. Linn. Soc. 60, 149–185. Available from: https://doi.org/10.1006/BIJL.1996.0097.

Chapter 6

Hormones involved in the regulation of metamorphosis

The earliest evidence for the existence of hormones in insects is due to Stephan Kopeć, who between 1917 and 1922 showed that caterpillars of the lepidopteran *Lymantria dispar* with the brain removed were unable to pupate. On the basis of these results, Kopeć proposed the existence of a humoral factor of cerebral origin that governed the metamorphosis. In the early 1930s, the pioneering observations of Kopeć were followed by researchers such as Vincent B. Wigglesworth, working with the bloodsucking bug *Rhodnius prolixus*, and Dietrich Bodenstein, with the cockroach *Periplaneta americana*. The results obtained using classical techniques in endocrinology, such as extirpation and reimplantation of organs and tissues, and parabiosis (Fig. 6.1), suggested the existence of a humoral factor originated in the brain that triggered molting. In addition, and also using parabiosis experiments in *R. prolixus*, Wigglesworth demonstrated in 1936 the existence of a humoral factor that inhibited the metamorphosis. This factor, later called juvenile hormone (JH), was present in young nymphs, and when it diffused into the hemolymph of last instar nymphs, these nymphs molted into adults but showing a number of nymphal characters.

The observations of Wigglesworth and Bodenstein on the existence of a molting hormone were confirmed in dipterans of the genus *Calliphora* by Gottfried Fraenkel in 1935 and in the lepidopteran *Galleria mellonella* by Alfred Kühn and Hans Piepho in 1936. Shortly after, between 1940 and 1944, Soichi Fukuda showed in the silk moth *Bombyx mori* that the molting hormone was not released from the brain but from the prothoracic gland (PG). With regard to the JH, Wigglesworth's works were followed by those of Otto Pflugfelder's between 1937 and 1938 in the phasmid *Carausius morosus*, those of Jean Jacques Bounhiol in 1938 in *B. mori*, and those of Piepho at the beginning of the 1940s in *G. mellonella*. In all these species, the existence of JH and their inhibitory properties upon metamorphosis was demonstrated (see Karlson, 1996; Wigglesworth, 1985).

These data were soon integrated into a general scheme on endocrine control of development and metamorphosis. According to this scheme (Fig. 6.1), the molting hormone would induce the molting process from one instar to

Insect Metamorphosis. DOI: https://doi.org/10.1016/B978-0-12-813020-9.00006-5

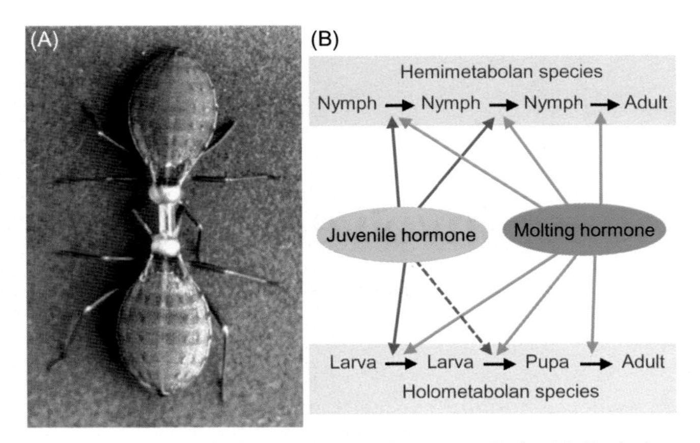

FIGURE 6.1 Endocrine control of development and metamorphosis. (A) Typical experiment of parabiosis using *Rhodnius prolixus* as a model. (B) General scheme on endocrine control of development and metamorphosis by juvenile hormone and molting hormone. (A) Photo courtesy of the late Vincent B. Wigglesworth.

the following one, whereas the JH would inhibit metamorphosis. Thus, the joint action of the molting hormone and the JH would determine a conservative molt, from a juvenile to another juvenile instar, whereas the molting hormone in the absence of JH would determine the metamorphic process.

This simple scheme has remained essentially unchanged, and even today it is used to explain in a few words how insect metamorphosis is regulated by hormones. Subsequently, the chemical structure of these hormones was elucidated, and the PG and corpora allata (CA, the glands that produce JH) were thoroughly studied. More modernly, the study of mechanisms of action at the cellular and molecular level has been deepened. The monographs of Nijhout (1994) and Reynolds (2013) cover the historical, chemical, and physiological aspects of insect hormones in a comprehensive way.

THE NATURE OF THE MOLTING HORMONE

Karl Butenandt and Peter Karlson were the first to purify the molting hormone. In 1952, these authors were able to isolate 25 mg in crystalline form from about 5 tons of *B. mori* pupae (see Karlson, 1996). To follow the biological activity of the extracts, the test of pupa formation in *Calliphora* flies, which had been designed by Fraenkel in 1935, was used as a bioassay. It is based on subjecting a larva about to pupate to a ligature that isolates the PG in the anterior region of the body. When the puparium is formed in this region, the extract is injected into the posterior part to check whether the puparium is formed there, out of the influence of the PG.

More than 10 years later, in 1965, the chemical analyses of Karlson and his group, and the structural studies of X-ray diffraction made by Huber and

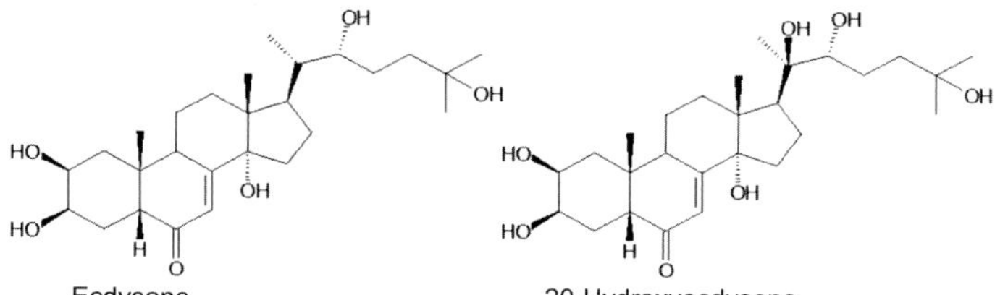

FIGURE 6.2 Chemical structure of ecdysone and 20-hydroxyecdysone.

Hoppe, led to the elucidation of the chemical structure of the molting hormone. It turned out to be a polyhydroxylated steroid, or ecdysteroid, that was called ecdysone (Fig. 6.2). Shortly thereafter, from the same *B. mori* pupae extracts, the Karlson group identified the 20-hydroxyecdysone derivative, which was designated ß-ecdysone. In the following years, about 30 ecdysteroids isolated from various species of insects have been identified although we know today that most of them are precursors or metabolites of the two mentioned, which are true hormonal products. 20-Hydroxyecdysone (20E) is the biologically active hormone in most of the insects studied (Fig. 6.2).

Ecdysone is synthesized in the PG from dietary sterols, like cholesterol and other steroids (Lavrynenko et al., 2015). Then, ecdysone is secreted to the hemolymph and oxidized to 20E in peripheral tissues, such as the epidermis, midgut, Malpighian tubules, ovaries, and fat body. The first step of the ecdysone biosynthesis is the conversion of cholesterol to 7-dehydrocholesterol, which is mediated by a 7,8-dehydrogenase encoded by the gene *neverland*. Conversion of 7-dehydrocholesterol to the Δ^4-diketol constitutes the so-called Black Box, because no intermediates have been reported yet. However, the so-called Halloween genes *spook*, *spookier*, and *spookiest* have been identified as being involved in the Black Box reactions in the fly *Drosophila melanogaster*. Moreover, *nonmolting glossy* in *B. mori* or its ortholog *shroud* in *D. melanogaster*, which encode a short-chain dehydrogenase/reductase, appear to participate in Black Box reactions although the precise function of these enzymes is still unclear. Then 5β-reduction and 3β-reduction steps convert the Δ^4-diketol into the 5β-ketodiol. The products of Halloween genes *phantom*, *disembodied*, and *shadow* sequentially convert 5β-ketodiol into 2,22-dideoxyecdysone, 2-deoxyecdysone, and ecdysone. Once ecdysone is secreted to the hemolymph, it is oxidized to 20E by the product of the last Halloween gene *shade* (see Gilbert et al., 2002; Niwa and Niwa, 2016; Ou et al., 2016).

THE PROTHORACIC GLAND AND THE SYNTHESIS OF ECDYSTEROIDS

The PG is the endocrine organ best known as a producer of ecdysone. The gland is located in the anterior part of the prothorax although in some species it may also occupy the basal region of the cephalic capsule and the cervical region. Morphological studies carried out in species of different groups revealed a significant diversity of models, which have been classified into the following five basic types (Joly, 1968):

- Massive type, in which the glands are constituted by lobular cellular groups, elongated, and laterally compressed. It is observed in early-branching insects, such as Apterygota and Palaeoptera.
- Long lamellar type, in which the cells form a thin and long lamina that can localize in different parts of the head and prothorax, depending on the species. This type is observed in Polyneoptera and, among the Endopterygota, in Coleoptera.
- Type in X, which is similar to the previous one, but with the lamellar structures forming an X, showing very fine muscular fibers along the axes (Fig. 6.3A). This type is typical of Blattaria.
- Diffuse type, in which the cells are arranged in small clusters scattered over the prothoracic tracheae. It has been observed in Hemiptera and in numerous species of Endopterygota, such as Lepidoptera and Hymenoptera (Fig. 6.3B).
- Annular type, in which the cells corresponding to the PG fuse with those of the corpora cardiaca (CC) and the CA, forming an annular structure (the Weismann ring or ring gland) that surrounds the aorta. It is typical of Diptera Brachycera (Fig. 6.3C).

At the ultrastructural level, the PG cells are characterized by very apparent intercellular spaces, an inconspicuous Golgi complex, and a poorly developed smooth endoplasmic reticulum, features that are typical

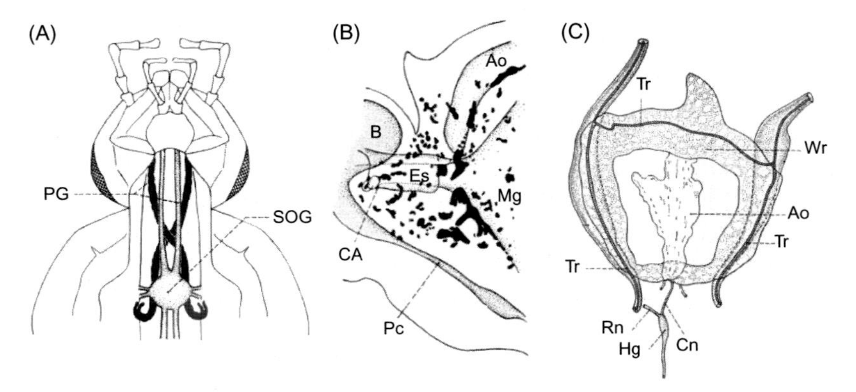

FIGURE 6.3 Different types of prothoracic gland (PG). (A) Type in X, characteristic of Blattaria. (B) Diffuse type, observed in Hemiptera and in many species of Endopterygota, such as Lepidoptera and Hymenoptera (the structures in black correspond to PG tissue). (C) Annular type characteristic of Diptera Brachycera, where the PG tissues are integrated in the Weissman ring (Wr). Ao, aorta; B, brain; CA, corpora allata; Cn, cardiohypocerebral nerve; Hg, hypocerebral ganglion; Mg, midgut; Es, esophagus; Pc, perioesophagic connective; Rn, recurrent nerve; SOG, suboesophageal ganglion; Tr, tracheae. From Joly (1968), with permission.

of steroid-producing cells. The visualization of other organelles (nucleoli, mitochondria, and lysosomes, in particular) depends on the secretory activity of the gland. Importantly, in hemimetabolan and holometabolan insects, the PG disintegrates in the days following the imaginal ecdysis. The process often begins with the apparition of autophagic vacuoles in the cytoplasm, followed by cell fragmentation and rupture of the nuclear membrane, with the release of picnotic nuclear material.

THE PROTHORACICOTROPIC HORMONE

The prothoracicotropic hormone (PTTH) is a neuropeptide produced by lateral neurosecretory cells in the brain, which promotes the synthesis of ecdysteroids in the PG (see Smith and Rybczynski, 2012). Indeed, PTTH was the hormone discovered by Kopeć in 1922 when he removed the brain in last instar larvae of *L. dispar* and observed that pupation was prevented. In the 1930s, Wigglesworth, using the famous experiments of parabiosis and brain implantations, showed that the anterior region of the brain induced molting activity. Toward the end of the 1940s, Carroll Williams, using experiments of ligature and brain and PG implantations in the silk moth *Hyalophora cecropia*, elucidated the complete endocrine circuit. Williams concluded that a factor from the brain (the brain hormone now known as PTTH) triggered the production of the actual molting hormone (the ecdysone) in the PG (see Williams, 1952). An important milestone was the identification of two pairs of neurosecretory cells in the brain of the lepidopteran *Manduca sexta* as the source of PTTH (Agui et al., 1979).

Preliminary purifications and characterization of active extracts suggested that PTTH could form dimers. However, the first report describing the complete amino acid sequence and dimeric structure of PTTH was reported by Kataoka et al. (1991) in *B. mori*. All PTTH reported sequences have seven cysteines, six of which are involved in forming intramolecular bonds, and one that forms an intermolecular bond needed for dimerization. The only exception is the PTTH of Homoptera that has only six (Barberà and Martínez-Torres, 2017). The six cysteines of aphid PTTH appear to be needed to form the three usual intramolecular disulfide bonds, whereas the absence of a seventh cysteine could prevent these PTTHs from establishing intermolecular bonds. This implies that aphid PTTH would be monomeric (Barberà and Martínez-Torres, 2017). A monomeric structure has been also proposed for the *D. melanogaster* PTTH (Kim et al., 1997).

INSULIN-LIKE PEPTIDES

In general, insulin-like peptides (ILPs) of insects are mainly involved in metabolism and growth (see Mizoguchi and Okamoto, 2013; Wu and Brown, 2006), contributing to regulating the achievement of critical size for molting and metamorphosis (see Nijhout and Callier, 2015). However, ILPs have been also involved in the regulation of PG activity. The most consistent data on ILPs regarding this function have been obtained in *B. mori* and *D. melanogaster*.

The first ILP sequence was obtained from *B. mori*. Brain extracts gave a peptide that activated the PG, which was partially sequenced, showing that the N-terminal region was very similar to vertebrate insulin (Nagasawa et al., 1984). It was finally named bombyxin, and, at present, more than 30 *bombyxin* (or *ILP*) genes have been identified in *B. mori* (see Mizoguchi and Okamoto, 2013). Most of these *ILP* genes encode polypeptides similar to preproinsulin, consisting of the signal peptide, B-chain, C-peptide, and A-chain (Fig. 6.4). All six cysteines and some hydrophobic residues in the

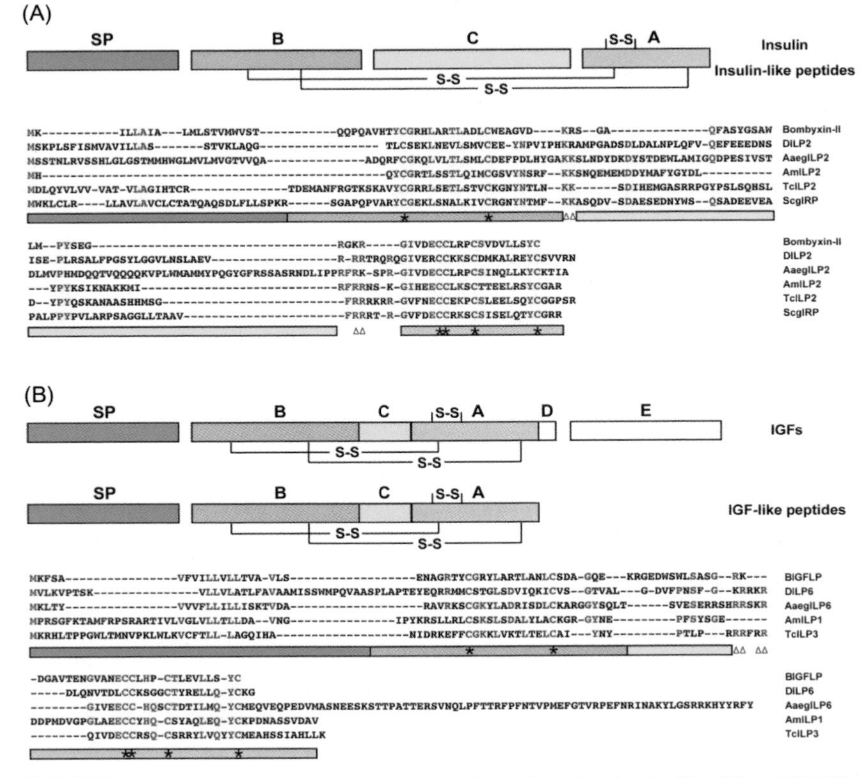

FIGURE 6.4 Amino acid sequences of representatives of putative insulin-like (A) and IGF-like (B) peptides. The sequences aligned are from *Bombyx mori* (bombyxin-II and BIGFLP), *Drosophila melanogaster* (DILP2 and DILP6), *Aedes aegypti* (AaegILP2 and AaegILP6), *Apis mellifera* (AmILP2 and AmILP1), *Tribolium castaneum* (TcILP2 and TcILP3), and *Schistocerca gregaria* (ScgIRP). Highly conserved amino acid residues are shown in red. Color bars below the alignment indicate the predicted domains in the precursor peptides as follows: Green: signal peptide; red: B-domain; yellow: C-domain; blue: A-domain. Asterisks represent cysteine residues, and paired triangles potential cleavage sites (dibasic amino acids). From Mizoguchi and Okamoto (2013), with permission.

A- and B-domains responsible for the formation of a hydrophobic core are conserved in them, as in insulin. The C-domains of all *B. mori* pro-ILPs are flanked by dibasic or monobasic sites, suggesting that the C-domain is removed to generate mature ILPs, as occurs in insulin maturation. This is the case of the ILPs bombyxin-II and bombyxin-IV, obtained from brain extracts, which consist of A- and B-chains, like insulin. However, an ILP containing the C-domain was purified from the hemolymph of *B. mori* pupae. It is mainly produced in the fat body and was named *B. mori* IGF-like peptide (BIGFLP) by analogy to vertebrate insulin-like growth factors (IGFs) (Okamoto et al., 2009). The work on *B. mori* has led to the classification of the insulin-related peptides into two types: ILPs (bombyxins), as a counterpart of vertebrate insulin; and BIGFLPs, or IGF-like peptides (Fig. 6.4) (see Mizoguchi and Okamoto, 2013). Regarding the action on the PG, it has been also reported that bovine insulin stimulates ecdysteroidogenesis in *B. mori* PG in long-term incubations (Gu et al., 2009). In vivo, an injection of insulin in day 6 of last larval instar increases both the hemolymph ecdysteroid titers, and the ecdysteroidogenic activity of the PG explanted 24 hours after the injection and incubated in vitro. The results suggest that some endogenous ILP of *B. mori* may exert these actions on the PG. Measurements of circulating BIGFLP indicate that titers increase to remarkably high levels early in the pupal stage (Okamoto et al., 2011), which suggests a role of BIGFLP in metamorphosis. Experiments incubating genital imaginal discs with BIGFLP support this hypothesis, as the size, protein content, and cell number of the genital discs increase after 5 days of incubation (Okamoto et al., 2009). This suggests that BIGFLP acts as a growth hormone regulating adult tissue development during metamorphosis of *B. mori*.

D. melanogaster has eight ILPs (DILP-1 to DILP-8), which have been extensively studied from a molecular and functional points of view (Brogiolo et al., 2001; Colombani et al., 2012; Garelli et al., 2012). DILP-2, DILP-3, and DILP-5 are mainly expressed in the median neurosecretory cells (MNCs) of the brain; thus they have been considered as functional homologues of *B. mori* bombyxins (Mizoguchi and Okamoto, 2013). Overexpression of individual DILPs (DILP-1 to DILP-7) triggers a proportionate body size increase in adult flies (Brogiolo et al., 2001; Ikeya et al., 2002). Conversely, genetic ablation of the DILP-producing MNCs in the brain triggers a severe delay of larval growth and puparium formation and a reduction of the adult size. All these phenotypes can be rescued by ubiquitously expressing DILP-2 using a heat-shock promoter (Rulifson et al., 2002). These results show that the MNCs-secreted DILPs efficiently regulate systemic growth in *D. melanogaster*. Another important aspect of growth is how organs scale with other body parts. Using the imaginal discs of *D. melanogaster* as model, Boulan et al. (2019) have shown that when the growth of one disc domain is perturbed, other parts of the disc and other discs slow down their growth, thus keeping proper interdisc and intradisc proportions. Interestingly, DILP8 is crucial for this inter-organ coordination. Growth and metamorphosis are closely linked, as adult body size is determined by the size of the premetamorphic stage, which must be precisely regulated. This has been investigated in *D. melanogaster*, where 20E triggers molting in larval stages, while larval growth rate is increased by DILPs. Data available suggest that there is a complex crosstalk between the ecdysteroid and insulin pathways, which coordinates growth and developmental timing, thus determining the final size of the adult fly (Colombani et al., 2005, 2012) (see Chapter 9: Molting: the basis for growing and for changing the form). In *D. melanogaster*, DILPs, acting through the insulin receptor, trigger a kinase cascade that contains phosphatidylinositol 3-kinase (PI3K) (see Chapter 7: Molecular mechanisms regulating hormone production and action). A clear link between DILPs and ecdysone production was shown when the expression of two ecdysteroidogenic genes, *disembodied* and *phantom*, was upregulated upon PI3K activation in the ring gland (Colombani et al., 2005).

OTHER PEPTIDES AFFECTING ECDYSONE SYNTHESIS IN THE PROTHORACIC GLAND

A remarkable number of factors, most of them peptidic in nature, have been shown to stimulate or inhibit ecdysone production by the PG (Marchal et al., 2010; Tanaka, 2011). The best characterized are three prothoracicostatic factors: prothoracicostatic peptide (PTSP) (Hua et al., 1999), bommo-myosuppressin (BMS) (Yamanaka et al., 2005), and bommo-FMRF-amide-related peptides (BRFas) (Yamanaka et al., 2006), as well as the prothoracicotropic activity of the pigment-dispersing factor (PDF) (Iga et al., 2014). All these data have been reported in *B. mori*.

PTSP peptides (Fig. 6.5) belong to the $W(X)_6$Wamide peptide family, which is widespread in insects. They have been also referred to as

Prothoracicostatic peptides (PTSPs)

Peptide	Sequence
PTSP-I:	AWQDLNSAW-NH_2
PTSP-II:	GWQDLNSAW-NH_2
PTSP-III:	APEKWAAFHGSW-NH_2
PTSP-IV:	GWNDISSVW-NH_2
PTSP-V:	AWQDMSSAW-NH_2
PTSP-VI:	AWSALHGTW-NH_2
PTSP-VII:	AWQDLNSVW-NH_2
PTSP-VIII:	AWSSLHSGW-NH_2

Bommo-myosuppressin (BMS)

pEDWHSFLRF-NH_2

Bommo-FMRF-amide-related peptides (BRFa)

Peptide	Sequence
BRFa-I:	SAIDRSMIRF-NH_2
BRFa-II:	SASFVRF-NH_2
BRFa-III:	DPSFIRF-NH_2
BRFa-IV:	ARNHFIRL-NH_2

Pigment dispersing factor (PDF)

NADLINSLLALPKDMNDA-NH_2

FIGURE 6.5 Peptides affecting the activity of the prothoracic gland. Prothoracicostatic peptides: prothoracicostatic peptides (PTSPs) reported by Hua et al. (1999) and Yamanaka et al. (2010), bommo-myosuppressin (BMS) reported by Yamanaka et al. (2005), and bommo-FMRF-amide-related peptides (BRFas) reported by Yamanaka et al. (2006). Prothoracicotropic peptide: pigment-dispersing factor (PDF) reported by Iga et al. (2014). All data refer to *B. mori*.

myoinhibitory peptides (MIPs) or as allatostatin-B (AST-B), owing to their originally identified biological activities (see Marchal et al., 2010; Tanaka, 2011). The first PTSP (PTSP-I, Fig. 6.5) was isolated from larval brains of *B. mori* (Hua et al., 1999). It showed to have the same sequence as a MIP previously isolated from the lepidopteran *M. sexta* and named Mas-MIP-I. PTSP-I inhibits in a dose-dependent manner both basal and PTTH-stimulated ecdysteroidogenesis in PG incubated in vitro from fifth larval instar of *B. mori* (in spinning and feeding stages) (Hua et al., 1999).

The BMS isolated from *B. mori* (Yamanaka et al., 2005) (Fig. 6.5) turned out to be identical to the FLRFamide I peptide previously isolated from *M. sexta*. BMS dose-dependently inhibited both basal and PTTH-stimulated ecdysone production in PGs from fifth larval instar of *B. mori* incubated in vitro. Comparison of BMS and PTSP activities revealed that BMS inhibits PG activity at much lower concentrations. Yamanaka et al. (2005) also identified the BMS receptor (see Chapter 7: Molecular mechanisms regulating hormone production and action), which was predominantly expressed in the PG but also in some other tissues including the midgut, hindgut, and Malpighian tubules, suggesting that this factor may play other roles in *B. mori*. The expression pattern of BMS in the brain during the fifth larval instar suggests that it operates mainly in the first half of the instar, during the feeding period, when the PG show low activity (Yamanaka et al., 2005).

Four BRFa peptides (Yamanaka et al., 2006) (Fig. 6.5) isolated from *B. mori* inhibit ecdysone production in PGs incubated in vitro. The four peptides are encoded by the same *B. mori FMRFamide* gene, which is predominantly expressed in neurosecretory cells of the thoracic ganglia. Firing activity of BRFa neurons increases in periods with low PG activity of the last larval instar. Although these BRFa use the same receptor as BMS peptides, they are transported to the PG surface through direct innervation, which contrasts to BMS, which is transported through the hemolymph (Yamanaka et al., 2006).

PDF (Fig. 6.5) was discovered in crustaceans, where it regulates pigment distribution in the eye and the epithelial chromatophores (Rao and Riehm, 1993). The peptide is localized in a subset of clock neurons in the brain and in neurons of the abdominal ganglia (Meelkop et al., 2011). In *B. mori*, PDF is involved in ecdysone production by the PG (Iga et al., 2014). The authors identified an orphan G protein-coupled receptor, which was named BNGR-B2 (from *Bombyx* neuropeptide G protein–coupled receptor), whose expression pattern in the PG correlated with the ecdysteroid titer in the hemolymph. Subsequently, PDF was identified as a ligand for BNGR-B2. PDF stimulates the production of ecdysone in PG from last larval instar of *B. mori* incubated in vitro. Intriguingly, the PG is sensitive to the stimulatory action of PDF only immediatley before the production of the edcysone pulse (Iga et al., 2014).

OTHER FACTORS AFFECTING THE PROTHORACIC GLAND ACTIVITY

Given that JH and ecdysone-20E are the two most important hormones in the regulation of metamorphosis, and have to a certain extent opposing actions, it has always been suspected that there must be crosstalk between the two. Classical studies using hormonal treatments have shown different effects of JH on PG activity in various species and stages (see Marchal et al., 2010). Much data have been obtained in *B. mori*, which suggest that JH inhibits ecdysone production in the first days of the fifth (last) larval instar. Removal of the CA on the first day of the last larval instar shortens the time between larval ecdysis and gut purge, whereas administration of JH early to the last larval instar delays the onset of pupal transformation (Sakurai, 2005). Furthermore, the expression of the steroidogenic gene *spook* is inhibited in the PG of early fifth larval instar of *B. mori* coincubated in vitro with the CA. More recent reports have confirmed that JH can inhibit ecdysone production in *B. mori* and *D. melanogaster*, and that the effect is mediated by the transcription factor Krüppel homolog 1, a typical transducer of the JH signal (see Chapter 7: Molecular mechanisms regulating hormone production and action), which represses the expression of steroidogenic genes (Liu et al., 2018; Zhang et al., 2018).

Biogenic amines have been also involved in the regulation of PG activity. Knockdown of tyramine biosynthesis genes expressed in the PG of *D. melanogaster* results in ecdysone depletion and metamorphosis arrest. Similar defects are observed if the β3-octopamine receptor (a monoamine G protein-coupled receptor) is depleted (Ohhara et al., 2015). The results suggest that monoaminergic autocrine signaling in the PG of *D. melanogaster* regulates ecdysone biosynthesis in conjunction with PTTH, ILPs, and the other prothoracicotropic and prothoracicostatic factors.

THE JUVENILE HORMONES

The first extracts enriched in JH were obtained by Williams in 1956 from adult males of *H. cecropia*. At the same time, the tests needed to monitor the biological activity were developed. One of the most used was the *Tenebrio* test designed by Wigglesworth, which consists of applying the bioactive extract to *Tenebrio molitor* pupae, and then quantifying the pupal characters retained in the resulting adult after the imaginal molt. Following these procedures, the group of Herbert Röller was able to isolate the first practically pure JH in 1965, and this same group elucidated its structure 2 years later, which turned out to be that of an acyclic sesquiterpenoid. To date, seven JHs have been identified, and all of them have a similar structure (Fig. 6.6).

The identification by Röller and collaborators of the JH I was made from extracts of *H. cecropia* (Röller et al., 1967). The proposed structure was

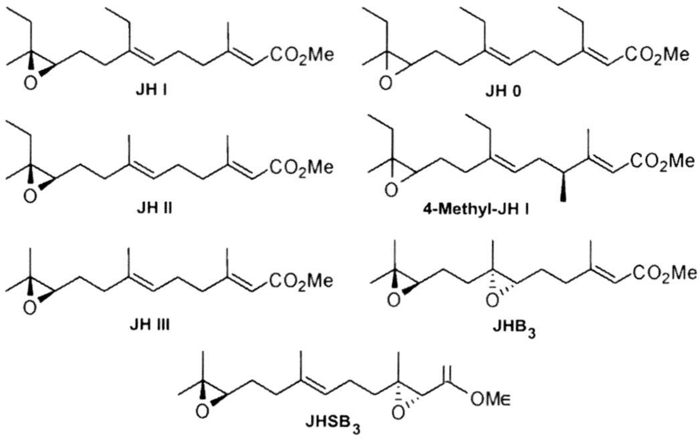

FIGURE 6.6 Chemical structure of the seven known natural juvenile hormones.

methyl methyl (2*E*,6*E*)-10,11-epoxy-3,11-dimethyl-7-ethyl-2,6-tridecadienoate. Subsequently, the stereochemistry of the oxirane ring and the C-6 double bond were elucidated as *cis* and *trans*, respectively, and the absolute configuration of the C-10 and C-11 chiral centers was established as 10*R*, 11*S*. Shortly thereafter, Meyer et al. (1968) found that in *H. cecropia* extracts, there was also a small proportion of a second JH, JH II, and a third compound, JH III, was subsequently identified from the medium where *M. sexta* CA had been incubated in vitro (Judy et al., 1973). Seven years later, JH 0 and 4-methyl-JH I were identified in embryos of the same species by Bergot et al. (1980), and almost 10 years after the discovery of these two structures, a JH bisepoxide compound was isolated from the medium where ring glands from *D. melanogaster* had been incubated in vitro. The compound was identified as methyl 6,7;10,11-bisepoxy-3,7,11-trimethyl-(2*E*)-dodecenoate and was named JHB_3 (Richard et al., 1989). The same authors showed that JHB_3 is produced solely by the CA portion of the ring gland, and that ring glands from larvae of other flies (Diptera Brachycera) also produce JHB_3 almost exclusively. Moreover, it was found that CA from larvae of mosquitoes (Diptera Nematocera) only produce JH III. Finally, a seventh JH structure was discovered in the Hemiptera *Plautia stali* that had the same molecular formula as that of JHB_3 but with a different bisepoxide arrangement, which was named JH III skipped bisepoxide ($JHSB_3$) (Kotaki et al., 2009) (Fig. 6.6). JH III is the most common JH, having been identified in several species of Orthoptera, Blattaria, Hemiptera, Hymenoptera, Coleoptera, Lepidoptera, and Diptera. JH0, 4-methyl-JH I, JH I, and JH II have only been found in Lepidoptera. JHB_3 appears to be exclusive to Diptera Brachycera, whereas $JHSB_3$ seems specific to Hemiptera. It is not known for certain whether this molecular diversity of JHs has any physiological or functional sense.

The JH is produced in the CA (see below) through a biosynthetic pathway that comprises 13 enzymatic steps. The early steps correspond to the conserved mevalonate pathway, which in mammals leads to cholesterol, but the five steps after the formation of farnesyl pyrophosphate are arthropod-specific. These five steps successively lead to the formation of farnesol (catalyzed by farnesyl diphosphate pyrophosphatase), farnesal (farnesol oxidase), farnesoic acid (farnesal dehydrogenase), methyl farnesoate (JH acid methyl transferase), and JH (JH epoxidase) (Belles et al., 2005). The genes encoding the enzymes involved in the JH pathway are expressed coordinately (Nouzova et al., 2011), which suggests that mechanisms involved in regulating their transcription affect the entire pathway.

THE CORPORA ALLATA

The CA are glands responsible for the biosynthesis and secretion of JH. Together with the CC, they are part of the retrocerebral complex that is connected to the posterior region of the brain and stands on the aorta and the digestive tube. CA morphology is relatively diverse depending on the species (Sedlak, 1985). Therefore different classifications have been proposed on a morphological basis. One of the most exhaustive works in this regard is that of Cazal (1948), who established five morphological types, defined as follows:

- Lateralized type, characterized by presenting two CA glands arranged symmetrically, one on each side of the digestive tube, and a pair of CC well differentiated. It is common in Polyneoptera orders, such as Isoptera and Phasmatodea.
- Distal-lateralized type, in which the CC are located more distant from the brain and are juxtaposed with CA. This type is characteristic of some Diptera and Coleoptera.
- Semicentralized type, with CC fused and attached to the aorta, and two CA well differentiated. It is typical of Palaeoptera and early-branching Neoptera (like Blattaria and Orthoptera).
- Centralized type, characteristic for presenting a single corpus allatum in direct contact with the CC, which is typical of certain terrestrial Heteroptera.
- Annular type, in which the CA are incorporated into the Weismann ring, mentioned in the section on the PG. This type is characteristic of Diptera Brachycera.

The nervous connections between the CA and the CC and the brain and the subesophageal ganglion also show different variants depending on the species. In general, the CA show two major nerves, the nervi corporis allati I, which connects them to the CC, and the nervi corporis allati II, which connects them to the subesophageal ganglion. Regarding CC, in most studied species, there are two or three nervi corporis cardiaci that connect to the brain (Fig. 6.7A).

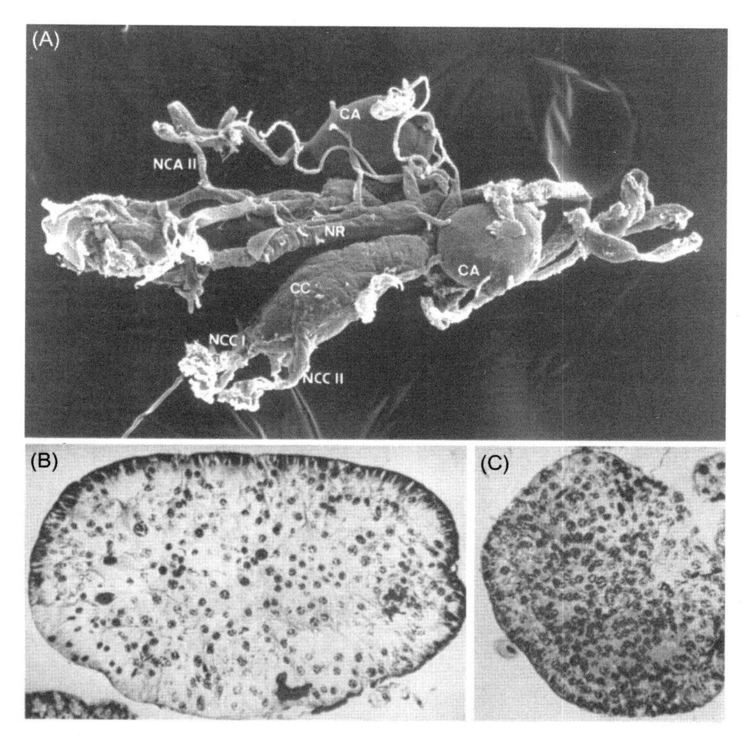

FIGURE 6.7 The nervous connections of the corpora allata and changes related to glandular activity. (A) Micrography showing the main nervous connections in the cockroach *Blattella germanica*. CA, corpora allata; CC, corpora cardiaca; NCAI and NCAII, nervi corporis allati I and II; NCCII, nervi corporis cardiaci II. NR: recurrent nerve and its connections. (B–C) Aspect of an active (B) and an inactive (C) CA of the locust *Locusta migratoria* taken at the same magnification. (A) Photo courtesy of Maria-Dolors Piulachs; (B and C) from Joly (1968), with permission.

Structural studies of the CA have shown that the more general model consists of a cluster of more or less well-packed parenchymal cells, which are surrounded by a thin acellular sheath called the basal lamina. Research at the ultrastructural level has revealed the detailed features of the basal lamina, glandular cells, and neurosecretory fibers that connect the CA and the brain (Cassier, 1979). Numerous studies have focused on clarifying the role of different organelles in JH production, by comparing active and inactive CA. A good model in this sense is provided by locusts and cockroaches, which have a CA resting phase between two reproductive cycles. Thus the CA of females during the vitellogenesis period are clearly bigger than those in the resting phase, especially due to the higher volume showed by the secretory cells, which have a well-developed ergastoplasm, a high number of ribosomes, and glycogen accumulation (Fig. 6.7B). In contrast, in the females during the intervitellogenic phase, when CA are inactive, the smooth endoplasmic reticulum is more developed, the number of mitochondria increases, while dense bodies appear (Fig. 6.7C).

ALLATOTROPINS AND ALLATOSTATINS

JH production is regulated by two types of neuropeptides, stimulating (allatotropins or ATs) or inhibiting (allatostatins or ASTs) the CA activity. Although most of the members of both types were discovered due to their activity in relation to the CA, as both, ATs and ASTs, are pleiotropic, having also functions unrelated to the CA and JH regulation.

Allatotropins

The first neuropeptide reported with stimulatory activity on JH was the AT of the lepidopteran *M. sexta* (Fig. 6.8). This AT was isolated from nervous system tissues and stimulated JH production in CA incubated in vitro (Kataoka et al., 1989). Subsequently, a number of ATs have been described in other insect species, and structural comparisons revealed that most of them have a conserved C-terminal pentapeptide, TARGF-NH_2. Exceptionally, an hymenopteran AT has a TAYGF-NH_2 C-terminus. Intriguingly, ATs have not been found in *D. melanogaster* or in any *Drosophila* species (see Verlinden et al., 2015). As stated above, a number of functions other than stimulation of JH production have been ascribed to ATs in different insect species. For example, in Lepidoptera, they include myostimulation of the dorsal vessel, inhibition of ion transport in the midgut, stimulation of contractions in the foregut, stimulation of lateral oscillation of the ventral nerve cord, inhibition of food

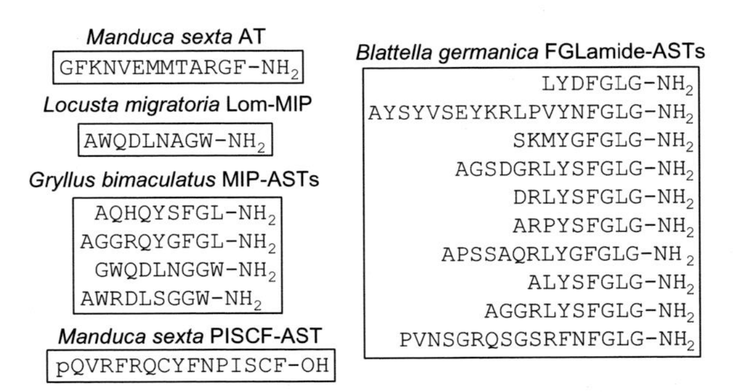

FIGURE 6.8 Allatotropins and allatostatins. Allatotropin (AT) of *Manduca sexta*, identified by Kataoka et al. (1989). Ten canonical FGLamide allatostatins (ASTs) of *Blattella germanica*, reported by Belles et al. (1999). Lom-MIP of *Locusta migratoria*, identified by Schoofs et al. (1991). Four MIP-ASTs identified by Lorenz et al. (1995) in *Gryllus bimaculatus*. PISCF allatostatin of *Manduca sexta*, discovered by Kramer et al. (1991).

intake, and regulation of digestive enzymes release in the midgut (see Verlinden et al., 2015). In a number of hemimetabolan insects, AT stimulates myotropic activity in different muscular systems, is involved in the photic entrainment of the circadian clock, and stimulates the secretion of saliva and the contractions of the muscles surrounding the salivary glands (see Verlinden et al., 2015). As AT myotropic activity is common to most of the species studied, it has been suggested that this would be the ancestral function of ATs, whereas allatotropic activity would have evolved secondarily (Elekonich and Horodyski, 2003).

Allatostatins

Three types of ASTs, known as FGLamide, MIP, and PISCF ASTs (see Coast and Schooley, 2011), have been reported in insects. As in the case of ATs, ASTs are pleiotropic peptides with functions beyond JH regulation, which have been observed in many insect orders. Nevertheless, the inhibitory role on JH production has been demonstrated in a limited group of hemimetabolan species. In these cases, ASTs originate from brain neurosecretory cells and are axonally transported to the CA.

FGLamide allatostatins (FGLa-ASTs) constitute a family of peptides with a conserved pentapeptide C-terminal sequence Y/FXFGL/I-amide (Fig. 6.8). The first FGLa-ASTs were isolated from the cockroach, *Diploptera punctata*, and elicited a rapid and reversible inhibition of JH biosynthesis by the CA incubated in vitro (Pratt et al., 1989; Woodhead et al., 1989). In cockroaches, the precursor contains 13–14 distinct FGLa-AST sequences (Belles et al., 1999). Similar peptides have been identified in other insect orders, but the inhibitory activity on JH biosynthesis has been determined only in cockroaches and termites (Verlinden et al., 2015). In Blattaria, Orthoptera, Dermaptera, Hemiptera, and Lepidoptera, FGLa-ASTs have myoinhibitory effects on visceral muscles in different organs, including gut and oviduct (see Verlinden et al., 2015). In Blattaria, immunocytochemical studies have revealed that FGLa-ASTs-specific neurons of the brain project immunoreactive axons toward the CC–CA complex. In the German cockroach, *Blattella germanica*, additional FGLa-AST-like immunoreactive neurons have been observed in the frontal ganglion, the suboesophageal ganglion, thoracic ganglia, abdominal ganglia, with axons projecting toward the antennal pulsatile organ, hindgut, heart, and ovaries, as well as in midgut endocrine cells (Maestro et al., 1998).

MIP-ASTs were first isolated from the brain–CC–CA–suboesophageal ganglion complex of the locust *Locusta migratoria* (Schoofs et al., 1991), and named Lom-MIP (Fig. 6.8). MIP-ASTs are characterized by a C-terminal W(X6/7)Wamide sequence, which is important for biological activity (see Coast and Schooley, 2011). MIP-ASTs inhibit spontaneous hindgut and oviduct contractions in vitro in *L. migratoria*, as well as hindgut contractions in the cockroach *Leucophaea maderae* (Schoofs et al., 1991). Four new MIP-ASTs were subsequently isolated from brain extracts of the cricket *Gryllus bimaculatus* (Fig. 6.8), which elicit allatostatic activity on CA incubated in vitro (Lorenz et al., 1995). In coleopterans, MIP-ASTs show allatostatic activity in vitro in *T. molitor* and in vivo in *Tribolium castaneum*. MIP-ASTs have also been found in many other insect species, and in *R. prolixus*, the precursor encodes peptides with an extra nonconserved amino acid in between the flanking conserved tryptophans W(X)7Wamide (see Verlinden et al., 2015). MIP-ASTs occur also in arthropods other than insects, such as ticks and crustaceans, as well as in mollusks and annelids. In insects, the precursor mRNA of MIP-AST is widely distributed throughout the body and developmental stages; however, it is most abundant in the central nervous system (Vandersmissen et al., 2013).

The first PISCF allatostatin (PISCF-AST) was characterized from head extracts of *M. sexta* pupae (Kramer et al., 1991). It resulted in being a nonamidated 15-residue peptide with a pyroglutamate-blocked N-terminus (Fig. 6.8). It forms a disulfide bridge between the Cys residues at positions 7 and 14 and is characterized by a C-terminal pentapeptide PISCF (Veenstra, 2009). Structure–activity studies in *M. sexta* with alanine substitutions of all amino acid residues showed the importance of the disulfide bond. The aromatic side chains proved to be also important as well, since alanine substitution of these residues reduced the peptide biological activity (see Verlinden et al., 2015). Similar or identical PISCF-ASTs have been found in Orthoptera (*L. migratoria*), Coleoptera (*T. castaneum*), Lepidoptera (*Pseudaletia unipuncta*, *Spodoptera frugiperda*, and *Samia cynthia ricini*), and Diptera (*D. melanogaster* and *Aedes aegypti*). Unlike FLGa-ASTs or MIP-ASTs, where the precursor contains multiple related peptides, the PISCF-AST precursor contains a single peptide. An additional precursor gene similar to the *PISCF-AST* has been found in the genome of a number of insect species, which encodes a PISCF-AST-like peptide that has been named ASTCC (from allatostatin double C). ASTCCs contain a ring structure very similar to that of PISCF-ASTs and display sequence similarity with PISCF-ASTs. In a few insect species, such as *D. melanogaster*, the predicted precursors encoded by the ASTCCs have some unusual features, like the absence of a signal peptide, having instead a peptide anchor (Veenstra, 2009). In Lepidoptera and Diptera, PISCF-ASTs elicit allatostatic activity (Li et al., 2006), and in a number of species, these peptides can have both allatostatic and allatotropic properties depending on the age of the insect (see Verlinden et al., 2015). Like other allatoregulatory peptides, PISCF-ASTs elicit myotropic effects, promoting gut contractions. In this context, it has been suggested that the ancestral function of PISCF-ASTs was related to myo- and/or neuroregulation, which later evolved to allatomodulatory functions in a number of species (Stay and Tobe, 2007).

PEPTIDES REGULATING ECDYSIS AND TANNING

Successful ecdysis, or shedding off the exuviae in the last part of the molting process (see Chapter 9: Molting: the basis for growing and for changing the form), requires a complex multistep behavior, which is regulated by peptide hormones produced in Inka cells, and neuropeptides from the central nervous system. The first effector is the ecdysis triggering hormone (ETH), produced in Inka cells, which acts on crustacean cardioactive peptide (CCAP) neurons in the central nervous system. These neurons, in addition to expressing the ETH receptor, produce peptides that contribute to the regulation of the sequential ecdysis steps (see Mena et al., 2016, and references therein). Among them, and in addition to CCAP, they produce the eclosion hormone (EH) and corazonin. Moreover, the regulatory peptide bursicon is an important player in the tanning process that follows the ecdysis.

Ecdysis triggering hormone

The first ETH was identified in *M. sexta* as a 26-amino acid amidated peptide (Fig. 6.9) that can induce precocious ecdysis (Zitnan and Adams, 2012). The ETH precursor was subsequently cloned and sequenced from the same species, which showed that it is processed into three peptides, ETH, preecdysis triggering hormone (PETH), and an ETH-associated peptide that has no known biological activity. ETH and PETH have a C-terminal PRXamide motif (Fig. 6.9). The precursor of *B. mori* also contains an ETH and a PETH sequence, whereas in *D. melanogaster*, the ETH precursor contains two active peptides (ETH1 and ETH2), both appearing more similar in sequence to ETH than to PETH. A similar structure has been observed in the precursor of the mosquito *Anopheles gambiae* and the hemimetabolan *L. migratoria* (see Zitnan and Adams, 2012).

ETH peptides are secreted by single-celled glands, the Inka cells, which are associated with exocrine epitracheal glands placed on particular sites on the trachea. In Lepidoptera, epitracheal glands are composed of four cells: large endocrine Inka cells, smaller narrow cells with possible endocrine function, as well as exocrine and duct cells. Similar glands are only found in the more derived holometabolan insects (Diptera, Lepidoptera, and some Coleoptera and Hymenoptera), whereas other holometabolan and hemimetabolan species possess a rather large number of Inka cells scattered on the tracheal system (see Zitnan et al., 2003).

Eclosion hormone

EH activity was demonstrated with the experiments of Truman and Riddiford (1970), using brain transplantation in the giant silk moths

ETH and PETH

M. sexta ETH:	SNEAISPFDQGMMGYVIKTNKNIPRM-NH$_2$
B. mori ETH:	SNEA---FDEDVMGYVIKSNKNIPRM-NH$_2$
M. sexta/B. mori PETH:	SFIKPNNVPRV-NH$_2$
D. melanogaster ETH1:	DDSSPGFFLKITKNVPRL-NH$_2$
D. melanogaster ETH2:	GENFAIKNLKTIPRI-NH$_2$
A. gambiae ETH1:	SESPGFFIKLSKSVPRI-NH$_2$
A. gambiae ETH2:	GDLENFFLKQSKSVPRI-NH$_2$
L. migratoria ETH1:	SDFFLKTAKSVPRI-NH$_2$
L. migratoria ETH2:	SDLFLKSAKSVPRI-NH$_2$

Corazonin

P. americana:	pQTFQYSRGWTN-NH$_2$
S. gregaria:	pQTFQYSHGWTN-NH$_2$
A. mellifera:	pQTFTYSHGWTN-NH$_2$

CCAP

PFCNAFTGC-NH$_2$

EH

M. sexta:	NP--AIATGYDPMEICIENCAQCKKMIGAWFEGPLCAESCIKFKGKLIPECEDFASIAPFLNKL
D. melanogaster:	FDSMGGIDFVQVCLNNCVQCKTMLGDYFQGQTALSCLKFKGKAIPDCEDIASIAPFLNAE
A. mellifera:	NA--EVRSGYIDDGVCIRNCAQCKKMFGPYFLGQKCADSCFKNKGKLIPDCEDEDSIQPFLQAL
T. castaneum:	SS--LLVVDANPIGVCIRNCAQCKKMFGPYFEGQLCADACVKFKGKIIPDCEDITSIAPFLNKF
A. aegypti:	NPQLDILGGYDMLSVCINNCAQCKRMFGEFFEGRLCAEACIQFKGKMVPDCEDINSIAPFLTKLN

***Drosophila melanogaster* bursicon**

QPDSSVAATDNDITHIGDDCQVTPVIHVLQYPGCVPKPIPSFACVGRCASYIQVSGSKIWQMERCMCCQESGEREAAVSIF
CPKVKPGERKFKKVITKAPLECMCRPCTSIEEGIIPQEIAGYSDEGPLNNHFRRIALQ

***Drosophila melanogaster* partner of bursicon**

RYSQGTGDENCETLKSEIHLIKEEFDELGRMQRTCNADVIVNKCEGLCNSQVQPSVITPTGFLKECYCCRESFLKEKVITL
THCYDPDGTRLTSPEMGSMDIRLREPTECKCFKCGDFTR

FIGURE 6.9 Ecdysis and tanning peptides. Ecdysis-triggering hormone (ETH) and preecdysis-triggering hormone (PETH) of diverse hemimetabolan (*Locusta migratoria*) and holometabolan (*Manduca sexta*, *Bombyx mori*, *Drosophila melanogaster*, *Anopheles gambiae*) insects. Eclosion hormone (EH) of diverse holometabolan insects (*D. melanogaster*, *Apis mellifera*, *Tribolium castaneum*, *Aedes aegypti*). Corazonin of *Periplaneta americana* (sequence identical in *Gryllus bimaculatus*, *B. mori*, *A. gambiae*, and *D. melanogaster*), *Schistocerca gregaria* (sequence identical in *L. migratoria* and *Carausius morosus*) and *A. mellifera*. Crustacean cardioactive peptide (CCAP) discovered in the Crustacean *Carcinus maenas* but found with identical sequence in hemimetabolan (like *L. migratoria* and *P. americana*) and holometabolan (like *M. sexta* and *D. melanogaster*) insects. Bursicon and partner of bursicon of *D. melanogaster*. Sequence data from Truman (2005) and Zitnan and Adams (2012).

H. cecropia and *Antheraea pernyi*. In these moths, adult emergence (also called eclosion) is under a strong circadian control, with the adult of each species emerging at a different time of the day: *H. cecropia* early in the day, and *A. pernyi* in the late evening. Brain extirpation and implantation experiments within and between species showed that the brain controlled this timing and the associated behavior in each species (Truman, 2005; Truman and Riddiford, 1970). The first EH was isolated and sequenced from brain extracts in two other moth species, *M. sexta* (Kataoka et al., 1987; Marti et al., 1987) and *B. mori* (Kono et al., 1987). In *M. sexta,* EH turned out to be a 62-amino acid peptide with three disulfide bridges (Fig. 6.9). Two years later, the *EH* gene of this species was cloned and sequenced (Horodyski et al., 1989), showing that it encodes a precursor with a signal peptide of 26 amino acids followed by a copy of the EH sequence. *EH* has been later identified by molecular cloning in *D. melanogaster* (Horodyski et al., 1993) or through screening in silico in several other insects and even in crustaceans and tardigrades (see Zitnan and Adams, 2012) (Fig. 6.9).

Corazonin

Corazonin was discovered due to its myotropic activity on heart preparations of the cockroach *P. americana* (Veenstra, 1989). Only later was corazonin related to ecdysis regulation (Kim et al., 2004), contributing to keep ETH secretion at low levels and promoting the preecdysis behavioral sequence. Corazonin immunoreactivity has been observed in lateral neurosecretory cells of the brain in a number of species belonging to most major insect orders, except Coleoptera. Characterization of corazonin in a significant number of very diverse insect species (Zitnan et al., 2007; see also Zitnan and Adams, 2012) has shown that it is a very conserved blocked undecapeptide (Fig. 6.9).

Crustacean cardioactive peptide

Crustacean cardioactive peptide (CCAP) is a nonapeptide amidated on its C-terminus, which shows a singular cyclic structure with an intramolecular disulfide bridge between the cysteines (Fig. 6.9). It was first sequenced from crustaceans due to its stimulatory activity on heart rate (Stangier et al., 1998). Later it was observed that CCAP also regulates cardiac function in insects, such as *M. sexta* (Loi et al., 2001). The *CCAP* gene in *D. melanogaster* encodes a 155-amino acid precursor, which is cleaved to yield one copy of CCAP and three other peptides of unknown function (Ewer et al., 2001). In *M. sexta,* CCAP promotes ecdysis behavior and is produced in a number of central nervous system neurons; two pairs of cells located in most abdominal ganglia are especially relevant in relation to the stimulation of ecdysis (Ewer et al., 1994; Loi et al., 2001).

Bursicon

The existence of an insect tanning neurohormone was independently demonstrated by two groups early in the 1960s (Cottrell, 1962; Fraenkel and Hsiao, 1962). Both groups used the "ligated fly bioassay," in which blowflies ligated around their neck immediately after ecdysis remained white and soft, whereas if hemolymph from a fly that had just darkened was then injected into similarly ligated recipients, then their cuticle tanned. This suggested the existence of a hormonal factor originating in the head region that acted on the rest of the body promoting sclerotization and hardening, which was named bursicon. The first step to elucidating its structure was a partial purification of bursicon from extracts of the cockroach *P. americana*. Then, the sequences of five peptidic fragments were used to search for similar sequences in the *D. melanogaster* genome. Three of the five partial peptides independently identified the gene *CG13419* from the *D. melanogaster* genome as the putative *bursicon* gene (Dewey et al., 2004). After removing the 32-amino acid signal peptide, the predicted mature protein resulted to have 141 amino acids (Fig. 6.9), and a molecular weight of c.15 kDa, which is one-half the molecular weight of bioactive bursicon. Earlier experiments that purified bursicon under reducing conditions indicated that the compound might be a dimer, and additional evidence indicated that *CG13419* encodes a subunit of bursicon (Honegger et al., 2008). Moreover, in situ hybridization and immunohistochemical studies showed that *CG13419* is expressed in a subset of neurons whose ablation eliminated bursicon bioactivity (Dewey et al., 2004). One year after the identification of the *CG13419* gene, Luo et al. (2005) used the available bursicon sequence as a query and identified *CG15284* by sequence search in *D. melanogaster*. Interestingly, *CG15284* encodes the only other cystine-knot protein present in the *D. melanogaster* genome, and the sequence contains the partial bursicon sequence of *P. americana*, *ESFLR* that was not contained in *CG13419*. Removal of the 21-amino acid signal peptide results in an approximately 15-kDa protein known as partner of bursicon (Fig. 6.9). Subsequent experiments showing that only the conditioned media of cells cotransfected with *CG13419* and *CG15284* initiated tanning of flies, and the fact that the bursicon receptor was only activated by the media of cotransfected cells, demonstrated that bursicon must function as a heterodimer formed by bursicon and partner of bursicon (see Honegger et al., 2008). This heterodimeric cystine-knot protein was the first to be found in invertebrates. Genetic analyses have shown that the subunits of partner of bursicon also contribute to the regulation of ecdysis (Lahr et al., 2012).

REFERENCES

Agui, N., Granger, N.A., Gilbert, L.I., Bollenbacher, W.E., 1979. Cellular localization of the insect prothoracicotropic hormone: in vitro assay of a single neurosecretory cell. Proc. Natl. Acad. Sci. U.S.A. 76, 5694–5698.

Barberà, M., Martínez-Torres, D., 2017. Identification of the prothoracicotropic hormone (PTTH) coding gene and localization of its site of expression in the pea aphid *Acyrthosiphon pisum*. Insect Mol. Biol. 26, 654–664. Available from: https://doi.org/10.1111/imb.12326.

Belles, X., Graham, L.A., Bendena, W.G., Ding, Q., Edwards, J.P., Weaver, R.J., et al., 1999. The molecular evolution of the allatostatin precursor in cockroaches. Peptides 20, 11–22. Available from: https://doi.org/10.1016/S0196-9781(98)00155-7.

Belles, X., Martín, D., Piulachs, M.-D., 2005. The mevalonate pathway and the synthesis of juvenile hormone in insects. Annu. Rev. Entomol. 50, 181–199.

Bergot, B.J., Jamieson, G.C., Ratcliff, M.A., Schooley, D.A., 1980. JH zero: new naturally occurring insect juvenile hormone from developing embryos of the tobacco hornworm. Science 210, 336–338.

Boulan, L., Andersen, D., Colombani, J., Boone, E., Léopold, P., 2019. Inter-organ growth coordination is mediated by the Xrp1-Dilp8 axis in *Drosophila*. Dev. Cell 49, 811–818. e4. Avaliable from: https://doi.org/10.1016/j.devcel.2019.03.016.

Brogiolo, W., Stocker, H., Ikeya, T., Rintelen, F., Fernandez, R., Hafen, E., 2001. An evolutionarily conserved function of the *Drosophila* insulin receptor and insulin-like peptides in growth control. Curr. Biol. 11, 213–221.

Cassier, P., 1979. The corpora allata of insects. Int. Rev. Cytol. 57, 1–74.

Cazal, P., 1948. Les glandes endocrines rétro-cérébrales des insectes (Étude morphologique). Bull. Biol. Fr. Belgique Suppl. 32, 1–227.

Coast, G.M., Schooley, D.A., 2011. Toward a consensus nomenclature for insect neuropeptides and peptide hormones. Peptides 32, 620–631. Available from: https://doi.org/10.1016/j.peptides.2010.11.006.

Colombani, J., Bianchini, L., Layalle, S., Pondeville, E., Dauphin-Villemant, C., Antoniewski, C., et al., 2005. Antagonistic actions of ecdysone and insulins determine final size in *Drosophila*. Science 310, 667–670. Available from: https://doi.org/10.1126/science.1119432.

Colombani, J., Andersen, D.S., Léopold, P., 2012. Secreted peptide Dilp8 coordinates *Drosophila* tissue growth with developmental timing. Science 336, 582–585. Available from: https://doi.org/10.1126/science.1216689.

Cottrell, C.B., 1962. The imaginal ecdysis of blowflies. Detection of the blood-borne darkening factor and determination of some of its properties. J. Exp. Biol. 39, 413–430.

Dewey, E.M., McNabb, S.L., Ewer, J., Kuo, G.R., Takanishi, C.L., Truman, J.W., et al., 2004. Identification of the gene encoding bursicon, an insect neuropeptide responsible for cuticle sclerotization and wing spreading. Curr. Biol. 14, 1208–1213. Available from: https://doi.org/10.1016/j.cub.2004.06.051.

Elekonich, M.M., Horodyski, F.M., 2003. Insect allatotropins belong to a family of structurally-related myoactive peptides present in several invertebrate phyla. Peptides 24, 1623–1632.

Ewer, J., De Vente, J., Truman, J.W., 1994. Neuropeptide induction of cyclic GMP increases in the insect CNS: resolution at the level of single identifiable neurons. J. Neurosci. 14, 7704–7712.

Ewer, J., Del Campo, M.L., Clark, A.C., 2001. Neuroendocrine control of ecdysis. J. Neurogenet. 15, 19.

Fraenkel, G., Hsiao, C., 1962. Hormonal and nervous control of tanning in the fly. Science 138, 27–29.

Garelli, A., Gontijo, A.M., Miguela, V., Caparros, E., Dominguez, M., 2012. Imaginal discs secrete insulin-like peptide 8 to mediate plasticity of growth and maturation. Science 336, 579–582. Available from: https://doi.org/10.1126/science.1216735.

Gilbert, L.I., Rybczynski, R., Warren, J.T., 2002. Control and biochemical nature of the ecdysteroidogenic pathway. Annu. Rev. Entomol. 47, 883–916. Available from: https://doi.org/10.1146/annurev.ento.47.091201.145302.

Gu, S.-H., Lin, J.-L., Lin, P.-L., Chen, C.-H., 2009. Insulin stimulates ecdysteroidogenesis by prothoracic glands in the silkworm, *Bombyx mori*. Insect Biochem. Mol. Biol. 39, 171–179. Available from: https://doi.org/10.1016/j.ibmb.2008.10.012.

Honegger, H.-W., Dewey, E.M., Ewer, J., 2008. Bursicon, the tanning hormone of insects: recent advances following the discovery of its molecular identity. J. Comp. Physiol. A. Neuroethol. Sens. Neural. Behav. Physiol. 194, 989–1005. Available from: https://doi.org/10.1007/s00359-008-0386-3.

Horodyski, F.M., Riddiford, L.M., Truman, J.W., 1989. Isolation and expression of the eclosion hormone gene from the tobacco hornworm, *Manduca sexta*. Proc. Natl. Acad. Sci. U.S.A. 86, 8123–8127.

Horodyski, F.M., Ewer, J., Riddiford, L.M., Truman, J.W., 1993. Isolation, characterization and expression of the eclosion hormone gene of *Drosophila melanogaster*. Eur. J. Biochem. 215, 221–228.

Hua, Y.J., Tanaka, Y., Nakamura, K., Sakakibara, M., Nagata, S., Kataoka, H., 1999. Identification of a prothoracicostatic peptide in the larval brain of the silkworm, *Bombyx mori*. J. Biol. Chem. 274, 31169–31173.

Iga, M., Nakaoka, T., Suzuki, Y., Kataoka, H., 2014. Pigment dispersing factor regulates ecdysone biosynthesis via *Bombyx* neuropeptide G protein coupled receptor-B2 in the prothoracic glands of *Bombyx mori*. PLoS One 9, e103239. Available from: https://doi.org/10.1371/journal.pone.0103239.

Ikeya, T., Galic, M., Belawat, P., Nairz, K., Hafen, E., 2002. Nutrient-dependent expression of insulin-like peptides from neuroendocrine cells in the CNS contributes to growth regulation in *Drosophila*. Curr. Biol. 12, 1293–1300.

Joly, P., 1968. Endocrinologie des Insectes. Masson & Cie, Paris.

Judy, K.J., Schooley, D.A., Dunham, L.L., Hall, M.S., Bergot, B.J., Siddall, J.B., 1973. Isolation, structure, and absolute configuration of a new natural insect juvenile hormone from *Manduca sexta*. Proc. Natl. Acad. Sci. U.S.A. 70, 1509–1513.

Karlson, P., 1996. On the hormonal control of insect metamorphosis. A historical review. Int. J. Dev. Biol. 40, 93–96. Available from: https://doi.org/10.1387/IJDB.8735917.

Kataoka, H., Troetschler, R.G., Kramer, S.J., Cesarin, B.J., Schooley, D.A., 1987. Isolation and primary structure of the eclosion hormone of the tobacco hornworm, *Manduca sexta*. Biochem. Biophys. Res. Commun. 146, 746–750.

Kataoka, H., Toschi, A., Li, J.P., Carney, R.L., Schooley, D.A., Kramer, S.J., 1989. Identification of an allatotropin from adult *Manduca sexta*. Science 243, 1481–1483. Available from: https://doi.org/10.1126/science.243.4897.1481.

Kataoka, H., Nagasawa, H., Isogai, A., Ishizaki, H., Suzuki, A., 1991. Prothoracicotropic hormone of the silkworm, *Bombyx mori*: amino acid sequence and dimeric structure. Agric. Biol. Chem. 55, 73–86.

Kim, A.J., Cha, G.H., Kim, K., Gilbert, L.I., Lee, C.C., 1997. Purification and characterization of the prothoracicotropic hormone of *Drosophila melanogaster*. Proc. Natl. Acad. Sci. U.S. A. 94, 1130–1135.

Kim, Y.-J., Spalovska-Valachova, I., Cho, K.-H., Zitnanova, I., Park, Y., Adams, M.E., et al., 2004. Corazonin receptor signaling in ecdysis initiation. Proc. Natl. Acad. Sci. U.S.A. 101, 6704–6709. Available from: https://doi.org/10.1073/pnas.0305291101.

Kono, T., Nagasawa, H., Isogai, A., Fugo, H., Suzuki, A., 1987. Amino acid sequence of eclosion hormone of the silkworm *Bombyx mori*. Agric. Biol. Chem. 51, 2307–2308.

Kotaki, T., Shinada, T., Kaihara, K., Ohfune, Y., Numata, H., 2009. Structure determination of a new juvenile hormone from a heteropteran insect. Org. Lett. 11, 5234–5237. Available from: https://doi.org/10.1021/ol902161x.

Kramer, S.J., Toschi, A., Miller, C.A., Kataoka, H., Quistad, G.B., Li, J.P., et al., 1991. Identification of an allatostatin from the tobacco hornworm *Manduca sexta*. Proc. Natl. Acad. Sci. U.S.A. 88, 9458–9462. o1902161x.

Lahr, E.C., Dean, D., Ewer, J., 2012. Genetic analysis of ecdysis behavior in *Drosophila* reveals partially overlapping functions of two unrelated neuropeptides. J. Neurosci. 32, 6819–6829. Available from: https://doi.org/10.1523/JNEUROSCI.5301-11.2012.

Lavrynenko, O., Rodenfels, J., Carvalho, M., Dye, N.A., Lafont, R., Eaton, S., et al., 2015. The ecdysteroidome of *Drosophila*: influence of diet and development. Development 142, 3758–3768. Available from: https://doi.org/10.1242/dev.124982.

Li, Y., Hernandez-Martinez, S., Fernandez, F., Mayoral, J.G., Topalis, P., Priestap, H., et al., 2006. Biochemical, molecular, and functional characterization of PISCF-allatostatin, a regulator of juvenile hormone biosynthesis in the mosquito *Aedes aegypti*. J. Biol. Chem. 281, 34048–34055. Available from: https://doi.org/10.1074/jbc.M606341200.

Liu, S., Li, K., Gao, Y., Liu, X., Chen, W., Ge, W., et al., 2018. Antagonistic actions of juvenile hormone and 20-hydroxyecdysone within the ring gland determine developmental transitions in *Drosophila*. Proc. Natl. Acad. Sci. U.S.A. 115, 139–144. Available from: https://doi.org/10.1073/pnas.1716897115.

Loi, P.K., Emmal, S.A., Park, Y., Tublitz, N.J., 2001. Identification, sequence and expression of a crustacean cardioactive peptide (CCAP) gene in the moth *Manduca sexta*. J. Exp. Biol. 204, 2803–2816.

Lorenz, M.W., Kellner, R., Hoffmann, K.H., 1995. A family of neuropeptides that inhibit juvenile hormone biosynthesis in the cricket, *Gryllus bimaculatus*. J. Biol. Chem. 270, 21103–21108.

Luo, C.-W., Dewey, E.M., Sudo, S., Ewer, J., Hsu, S.Y., Honegger, H.-W., et al., 2005. Bursicon, the insect cuticle-hardening hormone, is a heterodimeric cystine knot protein that activates G protein-coupled receptor LGR2. Proc. Natl. Acad. Sci. U.S.A. 102, 2820–2825. Available from: https://doi.org/10.1073/pnas.0409916102.

Maestro, J.L., Belles, X., Piulachs, M.-D., Thorpe, A., Duve, H., 1998. Localization of allatostatin-immunoreactive material in the central nervous system, stomatogastric nervous system, and gut of the cockroach *Blattella germanica*. Arch. Insect Biochem. Physiol. 37, 269–282.

Marchal, E., Vandersmissen, H.P., Badisco, L., Van de Velde, S., Verlinden, H., Iga, M., et al., 2010. Control of ecdysteroidogenesis in prothoracic glands of insects: a review. Peptides 31, 506–519. Available from: https://doi.org/10.1016/j.peptides.2009.08.020.

Marti, T., Takio, K., Walsh, K.A., Terzi, G., Truman, J.W., 1987. Microanalysis of the amino acid sequence of the eclosion hormone from the tobacco hornworm *Manduca sexta*. FEBS Lett. 219, 415–418.

Meelkop, E., Temmerman, L., Schoofs, L., Janssen, T., 2011. Signalling through pigment dispersing hormone-like peptides in invertebrates. Prog. Neurobiol. 93, 125–147. Available from: https://doi.org/10.1016/j.pneurobio.2010.10.004.

Mena, W., Diegelmann, S., Wegener, C., Ewer, J., 2016. Stereotyped responses of *Drosophila* peptidergic neuronal ensemble depend on downstream neuromodulators. Elife 5, e19686. Available from: https://doi.org/10.7554/eLife.19686.

Meyer, A.S., Schneiderman, H.A., Hanzmann, E., Ko, J.H., 1968. The two juvenile hormones from the cecropia silk moth. Proc. Natl. Acad. Sci. U.S.A. 60, 853–860.

Mizoguchi, A., Okamoto, N., 2013. Insulin-like and IGF-like peptides in the silkmoth *Bombyx mori*: discovery, structure, secretion, and function. Front. Physiol. 4, 217. Available from: https://doi.org/10.3389/fphys.2013.00217.

Nagasawa, H., Kataoka, H., Isogai, A., Tamura, S., Suzuki, A., Ishizaki, H., et al., 1984. Amino-terminal amino acid sequence of the silkworm prothoracicotropic hormone: homology with insulin. Science 226, 1344–1345. Available from: https://doi.org/10.1126/science.226.4680.1344.

Nijhout, H.F., 1994. Insect Hormones. Princeton University Press, Princeton, NJ.

Nijhout, H.F., Callier, V., 2015. Developmental mechanisms of body size and wing-body scaling in insects. Annu. Rev. Entomol. 60, 141–156. Available from: https://doi.org/10.1146/annurev-ento-010814-020841.

Niwa, Y.S., Niwa, R., 2016. Transcriptional regulation of insect steroid hormone biosynthesis and its role in controlling timing of molting and metamorphosis. Dev. Growth Differ. 58, 94–105. Available from: https://doi.org/10.1111/dgd.12248.

Nouzova, M., Edwards, M.J., Mayoral, J.G., Noriega, F.G., 2011. A coordinated expression of biosynthetic enzymes controls the flux of juvenile hormone precursors in the corpora allata of mosquitoes. Insect Biochem. Mol. Biol. 41, 660–669. Available from: https://doi.org/10.1016/j.ibmb.2011.04.008.

Ohhara, Y., Shimada-Niwa, Y., Niwa, R., Kayashima, Y., Hayashi, Y., Akagi, K., et al., 2015. Autocrine regulation of ecdysone synthesis by β3-octopamine receptor in the prothoracic gland is essential for *Drosophila* metamorphosis. Proc. Natl. Acad. Sci. U.S.A. 112, 1452–1457. Available from: https://doi.org/10.1073/pnas.1414966112.

Okamoto, N., Yamanaka, N., Satake, H., Saegusa, H., Kataoka, H., Mizoguchi, A., 2009. An ecdysteroid-inducible insulin-like growth factor-like peptide regulates adult development of the silkmoth *Bombyx mori*. FEBS J. 276, 1221–1232. Available from: https://doi.org/10.1111/j.1742-4658.2008.06859.x.

Okamoto, N., Yamanaka, N., Endo, Y., Kataoka, H., Mizoguchi, A., 2011. Spatiotemporal patterns of IGF-like peptide expression in the silkmoth *Bombyx mori* predict its pleiotropic actions. Gen. Comp. Endocrinol. 173, 171–182. Available from: https://doi.org/10.1016/j.ygcen.2011.05.009.

Ou, Q., Zeng, J., Yamanaka, N., Brakken-Thal, C., O'Connor, M.B., King-Jones, K., 2016. The insect prothoracic gland as a model for steroid hormone biosynthesis and regulation. Cell Rep. 16, 247–262. Available from: https://doi.org/10.1016/j.celrep.2016.05.053.

Pratt, G.E., Farnsworth, D.E., Siegel, N.R., Fok, K.F., Feyereisen, R., 1989. Identification of an allatostatin from adult *Diploptera punctata*. Biochem. Biophys. Res. Commun. 163, 1243–1247.

Rao, K.R., Riehm, J.P., 1993. Pigment-dispersing hormones. Ann. N. Y. Acad. Sci. 680, 78–88.

Reynolds, S., 2013. Endocrine system. in: Simpson, S.J., Douglas, A.E. (Eds.), The Insects. Structure and Function. Cambridge University Press, New York, pp. 674–707.

Richard, D.S., Applebaum, S.W., Sliter, T.J., Baker, F.C., Schooley, D.A., Reuter, C.C., et al., 1989. Juvenile hormone bisepoxide biosynthesis in vitro by the ring gland of *Drosophila melanogaster*: a putative juvenile hormone in the higher Diptera. Proc. Natl. Acad. Sci. U.S.A. 86, 1421–1425.

Röller, H., Dahm, K.H., Sweely, C.C., Trost, B.M., 1967. The structure of the juvenile hormone. Angew. Chem. Int. Ed. Engl. 6, 179–180. Available from: https://doi.org/10.1002/anie.196701792.

Rulifson, E.J., Kim, S.K., Nusse, R., 2002. Ablation of insulin-producing neurons in flies: growth and diabetic phenotypes. Science 296, 1118–1120. Available from: https://doi.org/10.1126/science.1070058.

Sakurai, S., 2005. Feedback regulation of prothoracic gland activity. In: Gilbert, L.I., Iatrou, K., Gill, S.S. (Eds.), Comprehensive Molecular Insect Science. Elsevier Pergamon, San Diego, CA, pp. 409–431.

Schoofs, L., Holman, G.M., Hayes, T.K., Nachman, R.J., De Loof, A., 1991. Isolation, identification and synthesis of locustamyoinhibiting peptide (LOM-MIP), a novel biologically active neuropeptide from *Locusta migratoria*. Regul. Pept. 36, 111–119.

Sedlak, B.J., 1985. Structure of endocrine glands. in: Kerkut, G., Gilbert, L.I. (Eds.), Comprehensive Insect Physiology, Biochemistry, and Pharmacology. Pergamon Press, Oxford, pp. 26–60.

Smith, W., Rybczynski, R., 2012. Prothoracicotropic hormone. In: Gilbert, L.I. (Ed.), Insect Endocrinology. Elsevier Science, Amsterdam, pp. 1–62.

Stangier, J., Hilbich, C., Dircksen, H., Keller, R., 1998. Distribution of a novel cardioactive neuropeptide (CCAP) in the nervous system of the shore crab *Carcinus maenas*. Peptides 9, 795–800.

Stay, B., Tobe, S.S., 2007. The role of allatostatins in juvenile hormone synthesis in insects and crustaceans. Annu. Rev. Entomol. 52, 277–299. Available from: https://doi.org/10.1146/annurev.ento.51.110104.151050.

Tanaka, Y., 2011. Recent topics on the regulatory mechanism of ecdysteroidogenesis by the prothoracic glands in insects. Front. Endocrinol. 2, 107. Available from: https://doi.org/10.3389/fendo.2011.00107.

Truman, J.W., 2005. Hormonal control of insect ecdysis: endocrine cascades for coordinating behavior with physiology. Vitam. Horm. 73, 1–30. Available from: https://doi.org/10.1016/S0083-6729(05)73001-6.

Truman, J.W., Riddiford, L.M., 1970. Neuroendocrine control of ecdysis in silkmoths. Science 167, 1624–1626. Available from: https://doi.org/10.1126/science.167.3925.1624.

Vandersmissen, H.P., Nachman, R.J., Vanden Broeck, J., 2013. Sex peptides and MIPs can activate the same G protein-coupled receptor. Gen. Comp. Endocrinol. 188, 137–143. Available from: https://doi.org/10.1016/j.ygcen.2013.02.014.

Veenstra, J.A., 1989. Isolation and structure of corazonin, a cardioactive peptide from the American cockroach. FEBS Lett. 250, 231–234.

Veenstra, J.A., 2009. Allatostatin C and its paralog allatostatin double C: the arthropod somatostatins. Insect Biochem. Mol. Biol. 39, 161–170. Available from: https://doi.org/10.1016/j.ibmb.2008.10.014.

Verlinden, H., Gijbels, M., Lismont, E., Lenaerts, C., Vanden Broeck, J., Marchal, E., 2015. The pleiotropic allatoregulatory neuropeptides and their receptors: a mini-review. J. Insect Physiol. 80, 2–14. Available from: https://doi.org/10.1016/j.jinsphys.2015.04.004.

Wigglesworth, V.B., 1985. Historical perspectives. In: Kerkut, G.A., Gilbert, L.I. (Eds.), Comprehensive Insect Physiology, Biochemistry and Pharmacology. Pergamon Press, pp. 1–24.

Williams, C.M., 1952. Morphogenesis and metamorphosis of insects. Harvey Lect. 47, 126–155.

Woodhead, A.P., Stay, B., Seidel, S.L., Khan, M.A., Tobe, S.S., 1989. Primary structure of four allatostatins: neuropeptide inhibitors of juvenile hormone synthesis. Proc. Natl. Acad. Sci. U.S.A. 86, 5997–6001.

Wu, Q., Brown, M.R., 2006. Signaling and function of insulin-like peptides in insects. Annu. Rev. Entomol. 51, 1–24. Available from: https://doi.org/10.1146/annurev.ento.51.110104.151011.

Yamanaka, N., Hua, Y.-J., Mizoguchi, A., Watanabe, K., Niwa, R., Tanaka, Y., et al., 2005. Identification of a novel prothoracicostatic hormone and its receptor in the silkworm *Bombyx mori*. J. Biol. Chem. 280, 14684–14690. Available from: https://doi.org/10.1074/jbc.M500308200.

Yamanaka, N., Hua, Y.-J., Roller, L., Spalovská-Valachová, I., Mizoguchi, A., Kataoka, H., Tanaka, Y., 2010. *Bombyx* prothoracicostatic peptides activate the sex peptide receptor to regulate ecdysteroid biosynthesis. Proc. Natl. Acad. Sci. U.S.A 107, 2060–2065. Available from: https://doi.org/10.1073/pnas.0907471107.

Yamanaka, N., Zitnan, D., Kim, Y.-J., Adams, M.E., Hua, Y.-J., Suzuki, Y., et al., 2006. Regulation of insect steroid hormone biosynthesis by innervating peptidergic neurons. Proc. Natl. Acad. Sci. U.S.A. 103, 8622–8627. Available from: https://doi.org/10.1073/pnas.0511196103.

Zhang, T., Song, W., Li, Z., Qian, W., Wei, L., Yang, Y., et al., 2018. Krüppel homolog 1 represses insect ecdysone biosynthesis by directly inhibiting the transcription of steroidogenic enzymes. Proc. Natl. Acad. Sci. U.S.A. 115, 3960–3965. Available from: https://doi.org/10.1073/pnas.1800435115.

Zitnan, D., Adams, M.E., 2012. Neuroendocrine regulation of ecdysis. Insect Endocrinology. Elsevier Science, Amsterdam, pp. 253–309.
Zitnan, D., Zitnanová, I., Spalovská, I., Takác, P., Park, Y., Adams, M.E., 2003. Conservation of ecdysis-triggering hormone signalling in insects. J. Exp. Biol. 206, 1275–1289.
Zitnan, D., Kim, Y.-J., Zitnanová, I., Roller, L., Adams, M.E., 2007. Complex steroid-peptide-receptor cascade controls insect ecdysis. Gen. Comp. Endocrinol. 153, 88–96. Available from: https://doi.org/10.1016/j.ygcen.2007.04.002.

Chapter 7

Molecular mechanisms regulating hormone production and action

Underlying the action of hormones, there are molecular mechanisms that transduce the hormonal signal. Commonly, such mechanisms begin with the interaction of the hormone with its receptor. Upon binding to the receptor, the hormone initiates a cascade of events in the cell that transduces the hormonal signal. Receptors are classified into two broad categories: transmembrane receptors, which include G protein-coupled receptors (GPCRs) and enzyme-linked hormone receptors, and intracellular receptors, which include cytoplasmic and nuclear receptors. Despite the evolutionary distance, insects possess all the types of hormones and cognate receptors that vertebrates have.

The transduction mechanisms of the GPCRs are mainly based on G proteins (Ben-Shlomo et al., 2003). Receptors of allatotropins, allatostatins, and peptides regulating ecdysis and tanning are GPCRs. In the enzyme-linked hormone receptors, the binding of an extracellular ligand causes enzymatic activity on the intracellular side. The receptors of the tyrosine kinase family (Ben-Shlomo et al., 2003), to which the prothoracicotropic hormone (PTTH) receptor and the insulin receptor (InR) belong, are enzyme-linked hormone receptors. Finally, the receptors of 20-hydroxyecdysone (20E) and juvenile hormone (JH) identified until now are intracellular, and belong, respectively, to the nuclear receptor family (Evans and Mangelsdorf, 2014) and to the basic helix-loop-helix/Per-Arnt-Sim (bHLH/PAS) family (Eben Massari and Murre, 2000; Jindra et al., 2015b).

Generally, when a hormone binds to a GPCR, it activates a G protein, which in turn activates a number of intracellular second messengers. Two main signal transduction pathways act downstream of GPCR, one based on cyclic adenosine monophosphate (cAMP) and the other on phosphatidylinositol signaling (Gilman, 1987). Activation of tyrosine kinase receptors by the ligand triggers signal transduction mechanisms that involve adaptor proteins, serine/threonine kinases, and transcription factors (Sopko and Perrimon, 2013). The interaction of a hormone with an intracellular receptor

Insect Metamorphosis. DOI: https://doi.org/10.1016/B978-0-12-813020-9.00007-7

triggers a cascade of transcription factors that transduce the hormonal signal (see Ou and King-Jones, 2013).

PRODUCTION OF ECDYSONE IN THE PROTHORACIC GLAND

A number of factors that act on the prothoracic gland (PG) stimulate or inhibit the biosynthesis of ecdysone. JH can inhibit the production of ecdysone in the PG (Liu et al., 2018; Zhang et al., 2018b), as explained later in this chapter, but most of the prothoracicotropic and prothoracicostatic factors are peptides, including the PTTH, insulin-like peptides (ILPs), and others. Their mechanism of action has been covered in a number of reviews (Niwa and Niwa, 2016; Ou and King-Jones, 2013; Tanaka, 2011).

Action of prothoracicotropic peptides

The PTTH receptor is Torso, which belongs to the receptor tyrosine kinase family and stimulates the expression of steroidogenic genes through the Ras85D/Raf/mitogen-activated protein kinase (MEK)/extracellular signal-regulated kinase (ERK) pathway (Fig. 7.1). Torso was originally described as a determinant of the anteroposterior termini in embryos of the fruit fly *Drosophila melanogaster*. However, Torso also contributes to the activation of PG cells in response to stimulation by PTTH (Rewitz et al., 2009). In larvae of *D. melanogaster*, Torso is expressed in the PG, and its depletion delays larval development due to deficient ecdysone synthesis. Moreover, stimulation of *Drosophila* S2 cells transfected with *Bombyx mori* torso and *D. melanogaster* ERK with 10^{-9} M PTTH triggers a robust phosphorylation of ERK (Rewitz et al., 2009). The mechanism of action of Torso has been also studied in the beetle *Leptinotarsa decemlineata*, where its depletion by RNA interference (RNAi) delays larval development, increases pupal weight, and impairs pupation and adult emergence (Zhu et al., 2015).

ILPs contribute to stimulate ecdysone production in the PG via the InR and activation of phosphoinositide 3-kinase (PI3K) (Caldwell et al., 2005; Colombani et al., 2005; Mirth et al., 2005). Transgene-induced activation of PI3K in the PG causes a premature synthesis of ecdysone, thus advancing the onset of metamorphosis. Moreover, blocking the protein kinase A (PKA) pathway in insulin-producing cells, increases insulin signaling, which results in increased larval growth rate and premature ecdysone production (Walkiewicz and Stern, 2009). These observations suggest that PI3K-mediated insulin action in the PG is regulated by PKA pathway activity in the insulin-producing cells. The InR is a tetramer composed of two subunits containing the putative ligand-binding domains and two transmembrane subunits with the cytoplasmic tyrosine kinase domains. Initial contributions were based on *D. melanogaster*, which only has one *InR*. However, research in other insects has shown that most species have two *InRs* originated by a gene duplication that occurred at the origin of Pterygota. Intriguingly, transcriptomic data indicate that the Blattodea have a third *InR* that originated just before the eusocial termites diverged from cockroaches. One of the *InR* paralogs is caste-biased expressed in three termite species, which suggests that it might be involved in caste differentiation (Kremer et al., 2018). In the migratory planthopper, *Nilaparvata lugens*, two insulin receptors (InR1 and InR2) act as switches to determine alternative wing morphs, which represents a control layer regulating wing polymorphism additional to JH (Xu and Zhang, 2017).

Other peptides found in *B. mori* showing prothoracicotropic activities are orcokinins and the pigment-dispersing factor (PDF) (see Tanaka, 2011). The orcokinin receptor has not yet been identified. Regarding PDF, Iga et al. (2014) showed that this neuropeptide stimulates ecdysone biosynthesis and identified a GPCR named *Bombyx* neuropeptide GPCR-B2 (BNGR-B2), as its receptor. BNGR-B2 is predominantly expressed in steroidogenic tissues, and the expression pattern in the PGs parallels the ecdysteroid titer profile in the hemolymph (Fig. 7.1).

Tyramine signaling through the β3-octopamine receptor (Octβ3R) also contributes to ecdysone production in the PG, as shown in *D. melanogaster*. Knockdown of Octβ3R specifically in the PG results in a decrease of ecdysone production, and thus metamorphosis becomes prevented. Knockdown of tyramine biosynthesis genes expressed in the PG causes a similar phenotype. Moreover, Octβ3R knockdown impairs PTTH and ILP signaling pathways, whereas activation of these pathways rescues metamorphosis (Ohhara et al., 2015). Taken together, the observations suggest that monoaminergic autocrine signaling in the PG regulates ecdysone synthesis coordinately with the activation by PTTH, ILPs, and PDF.

Action of prothoracicostatic peptides

The first factor identified in this category was called prothoracicostatic peptide (PTSP) (Hua et al., 1999) although it belongs to the myoinhibitory peptide (MIP) allatostatin family (see below). Subsequently, Yamanaka et al. (2010) cloned the corresponding gene, *PTSP*, and analyzed its expression in *B. mori*. The authors also characterized a *B. mori* GPCR as the functional PTSP receptor, which is an ortholog of the *Drosophila* sex peptide receptor, discovered previously. This receptor responds specifically to PTSP in a heterologous expression system and is highly expressed in the PG before each larval and pupal ecdysis, when PTSPs are produced at high levels (Yamanaka et al., 2010). Functional studies, using both mammalian and insect cells, have shown that the interaction of sex peptide and PTSP operate under different conditions (He et al., 2015) (Fig. 7.1).

The receptor of *B. mori* bommo-myosuppressin (BMS) is a GPCR protein that was found using *B. mori* transcriptomic data and conducting systematic

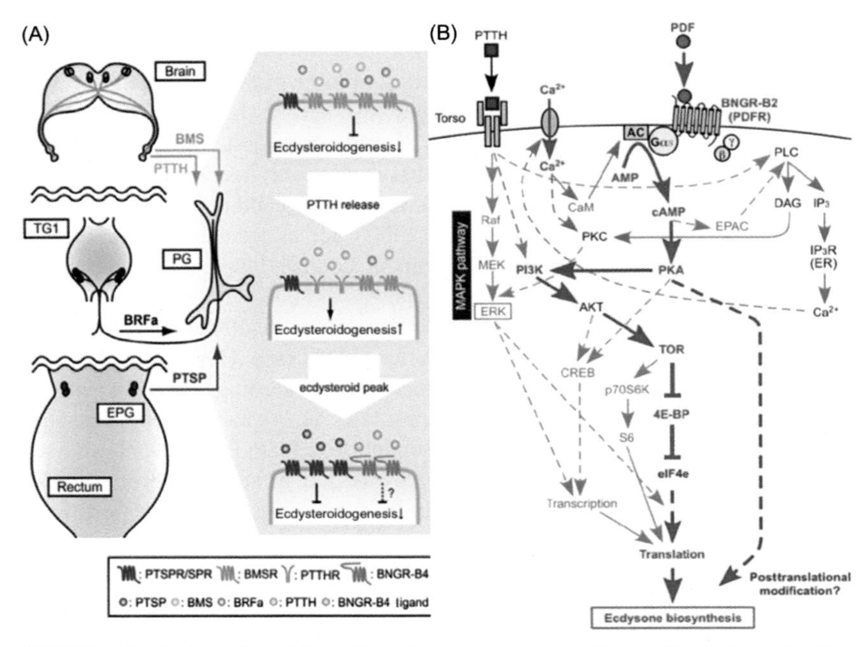

FIGURE 7.1 Action of peptides influencing ecdysone production in the prothoracic gland (PG), based on studies in *Bombyx mori*. (A) Model describing the temporal regulation of the PG activity by different neuropeptides. Bommo-myosuppressin (BMS) and Bommo-FMRFamide (BRFa) act on the receptor BMSR during the feeding stage and suppress ecdysone production; when enough prothoracicotropic hormone (PTTH) is released from the brain (which is probably coupled with the downregulation of BMSR signaling), ecdysone production increases; after the decline of the ecdysteroid titer, prothoracicostatic peptide (PTSP) and probably the BNGR-B4 neuropeptide ligand further inhibit ecdysone production. PTTHR, PTTH receptor; TG1, prothoracic ganglion. (B) Interaction of the PTTH and the pigment-dispersing factor (PDF) signaling pathways. Solid lines indicate demonstrated or highly likely pathways, and dashed lines indicate hypothetical pathways. 4E-BP, eIF4E-binding protein; AC, adenylate cyclase; AKT, protein kinase B; AMP, adenosine monophosphate; CaM, calmodulin; cAMP, cyclic AMP; CREB, cAMP response element-binding protein; DAG, diacylglycerol; eIF4e, eukaryotic translation initiation factor 4E; EPAC, exchange protein directly activated by cAMP; ERK, extracellular signal-regulated kinase; Gαs, G protein αs subunit; IP3, inositol 1,4,5-trisphosphate; IP3R, IP3 receptor; MAPK, mitogen-activated protein kinase; MEK, MAP kinase kinase; p70S6K , 70 kDa S6 kinase; PKA, protein kinase A; PKC, protein kinase C; PI3K, phosphatidylinositol 3-kinase; PLC, phospholipase C; Raf, MAP kinase kinase kinase; S6, ribosomal protein S6; TOR, target of rapamycin. (A) From Yamanaka et al. (2010), with permission and (B) from Iga et al. (2014), slightly modified, with permission.

sequence comparisons using as queries the *D. melanogaster* myosuppressin receptors. Heterologous expression experiments showed that BMS bound this receptor (BMS receptor or BMSR) at physiological concentrations. Expression studies in various tissues of wandering stage larvae of *B. mori* indicate that the highest expression of *BMSR* occurs in the PG (Yamanaka et al., 2005).

The same group reported that the *Bombyx* FMRFamide PTSPs (Bommo-FMRFamide, BRFa), which suppress steroidogenesis in the PG by reducing cAMP production, act through the BMSR (Yamanaka et al., 2006) (Fig. 7.1).

Transcription factors involved in regulating the expression of ecdysteroidogenic genes

The stimulation or inhibition of ecdysone production can be produced via transcription factors regulating the expression of ecdysteroidogenic genes, which have been discussed in Chapter 6 (Hormones involved in the regulation of metamorphosis). A number of transcription factors have been associated to the synthesis of ecdysone, such as ß-Fushi tarazu-factor 1 (ßFTZ-F1), HR4, Broad-complex (BR-C), Ventral veins lacking, the CncC-dKeap1 complex, Knirps, Molting defective (Mld), Ouija board (Ouib), and séance (Séan) (see Uryu et al., 2018). The last three, Mld, Ouib and Séan, are specially interesting due to their specific features. They belong to the zinc-finger-associated domain (ZAD)-C2H2-type zinc finger protein family, appear exclusive of drosophilid flies, and, in the case of *ouib* and *séan*, the expression is restricted to the PG region of the ring gland (Komura-Kawa et al., 2015; Uryu et al., 2018).

The studies of Uryu et al. (2018) have revealed that loss-of-function mutations in *séan*, *ouib*, or *mld* dramatically reduce the expression of the steroidogenic genes *neverland* (*nvd*), *spookier* (*spok*), or both *nvd* and *spok*, respectively. These mutations result in developmentally arrested larvae, a phenotype that can be rescued by administering intermediate metabolites of the ecdysone synthesis to mutant insects. Thus, supply of 7-dehydrocholesterol in the diet corrects the larval arrest provoked by the null mutation of *séan*, whereas 5ß-ketodiol corrects the arrest of *ouib* and *mld* mutants. The arrest of *séan*, *ouib*, or *mld* mutants can also be rescued by overexpressing in the PG *nvd* alone, *spook* (an ortholog of *spok*) alone, or *nvd* and *spo* combined, respectively. Significantly, the promoter region of *nvd* and *spok* contains specific Séan and Ouib response elements, respectively, whereas the presence of Mld has a synergistic effect with Séan and Ouib, stimulating the transcription of the steroidogenic genes. Taken together, the data suggest that the transcription factors Séan, Ouib, and Mld contribute to regulate the expression of *nvd* and *spok* in the PG region of the ring gland of *D. melanogaster* (Uryu et al., 2018).

PRODUCTION OF JUVENILE HORMONE IN THE CORPORA ALLATA

The factors regulating JH production in corpora allata (CA) are short peptides that can be stimulatory (allatotropins) or inhibitory (allatostatins). The corresponding receptors have been reviewed by Verlinden et al. (2015) and belong to the GPCR type.

Action of allatotropins

The first allatotropin receptor (ATR) was identified in *B. mori* by examining the transcript distribution of orphan receptors (Yamanaka et al., 2008). The ATR has been further reported in the lepidopterans *Manduca sexta* and *Helicoverpa armigera*, the coleopteran *Tribolium castaneum*, the dipteran *Aedes aegypti*, the hymenopteran *Bombus terrestris*, and the orthopteran *Schistocerca gregaria* (see Verlinden et al., 2015; Zhang et al., 2018). Insect ATRs have been functionally characterized with heterologous expression systems using Chinese hamster ovary (CHO) and human embryonic kidney (HEK) cells. In these cells, ATRs show dual coupling characteristics, leading to an increase in intracellular calcium ion concentrations and intracellular cAMP levels, upon activation by the corresponding ATR (see Verlinden et al., 2015).

Action of allatostatins

Allatostatins (ASTs) have been classified into three types: FGLamide, MIP, and PISCF ASTs (Coast and Schooley, 2011). All ASTs signal through GPCRs.

The first FGLamide AST receptor was characterized in *D. melanogaster*. *Xenopus* oocytes expressing the fly receptor and a mouse G-protein-gated inwardly rectifying potassium channel were used to screen against possible ligands in fly head extracts by recording inward potassium currents (Birgül et al., 1999). The receptor found, *Drosophila* allatostatin receptor 1 (DAR-1), shows similarities to the mammalian galanin receptor family. A second *D. melanogaster* FGLa/AST receptor structurally related to DAR-1 (named DAR-2) was further characterized. DAR-2 is also activated by the FGLamide ASTs of *D. melanogaster*, as well as those of the cockroach *Diploptera punctata*, when expressed in CHO cells using a calcium mobilization assay (Larsen et al., 2001). Subsequently, FGLamide AST receptors have been identified in the stick insect *Carausius morosus*, the cockroach *Periplaneta americana*, the silkworm *B. mori*, the kissing bug *Rhodnius prolixus*, and the mosquitoes *A. aegypti* and *Culex quinquefasciatus* (see Christ et al., 2018).

In cockroaches, the FGLamide AST receptor plays the role of transducer of the AST inhibitory signal. In *D. punctata*, this receptor is predominantly expressed in the CA, and its depletion by RNAi results in a significant increase in JH biosynthesis, and a significant decrease in the response of CA to AST (Huang et al., 2014; Lungchukiet et al., 2008). The FGLamide AST receptor has been functionally characterized using CHO cells stably expressing apoaequorin, transfected with the receptor. With this system, the binding affinity of different FGLamide ASTs has been tested by measuring the corresponding bioluminescent response (Huang et al., 2014).

Independent work of two laboratories working on *D. melanogaster* (Kim et al., 2010; Poels et al., 2010) showed that MIP ASTs are the ancestral ligands of the promiscuous *Drosophila* sex peptide receptor, the same receptor that transduces the signal of the PTSP mentioned above. Sex peptide receptors form a group of rhodopsine receptors specifically characterized by several substitutions of the generally most conserved residues (see Verlinden et al., 2015). Although relatively dissimilar, sex peptide and MIP ASTs share a tryptophan-rich region that is involved in receptor activation (Kim et al., 2010; Poels et al., 2010). In the cricket *Gryllus bimaculatus*, in which MIP ASTs have allatostatic activity, their action is also mediated by the corresponding sex peptide receptor ortholog (Tsukamoto and Nagata, 2016).

Heterologous expression assays with *Xenopus* oocytes and *D. melanogaster* brain extracts led to the characterization of two GPCRs activated by PISCF ASTs, named Drostar-1 and Drostar-2, which are related to the mammalian somatostatin/opioid receptor family (Kreienkamp et al., 2002). Inhibition of *Drostar-1* expression in third instar larvae of *D. melanogaster* significantly reduces the production of methyl farnesoate (MF), a distal precursor of JH, and JH bisepoxide, but does not affect JH III. Inhibition of Drostar-2 also reduces MF biosynthesis. In adult females, reduction in expression of *Drostar-1* or *Drostar-2* increases MF and JH III production (Wang et al., 2012). In the mosquito *A. aegypti,* two PISCF AST receptors have been identified with a heterologous assay using HEK cells, measuring a calcium-dependent increase in fluorescence upon binding with the corresponding PISCF AST. Quantification of mRNA levels in adult mosquitoes showed that the two receptors are highly expressed in the abdominal ganglia. However, one of them (named AedaeAS-CrA) was comparatively more expressed in the thoracic ganglia and CA, while the other (AedaeAS-CrB) was more expressed in the Malpighian tubules and in the heart (Mayoral et al., 2010).

In *B. mori*, a PISCF AST receptor that shows high expression in the corpora cardiaca (CC)–CA complex and fat body of larval stages has been identified. When expressed in HEK cells, it is activated in a dose-dependent manner by PISCF AST (Yamanaka et al., 2008). Finally, a PISCF AST receptor has been characterized in the beetle *T. castaneum*, which, when expressed in mammalian cells, is selectively activated by PISCF ASTs (but not by FGLamide and MIP ASTs) (Audsley et al., 2013).

MECHANISMS OF REGULATION OF ECDYSIS AND TANNING

The complex behavior of shedding off the exuvia is regulated by a sequential production of regulatory peptides, which in a number of species include corazonin, ecdysis-triggering hormone (ETH), eclosion hormone (EH), and the crustacean cardioactive peptide (CCAP) (Mena et al., 2016; Roller et al., 2010; Zitnan et al., 1996) (see Chapter 6: Hormones involved in the

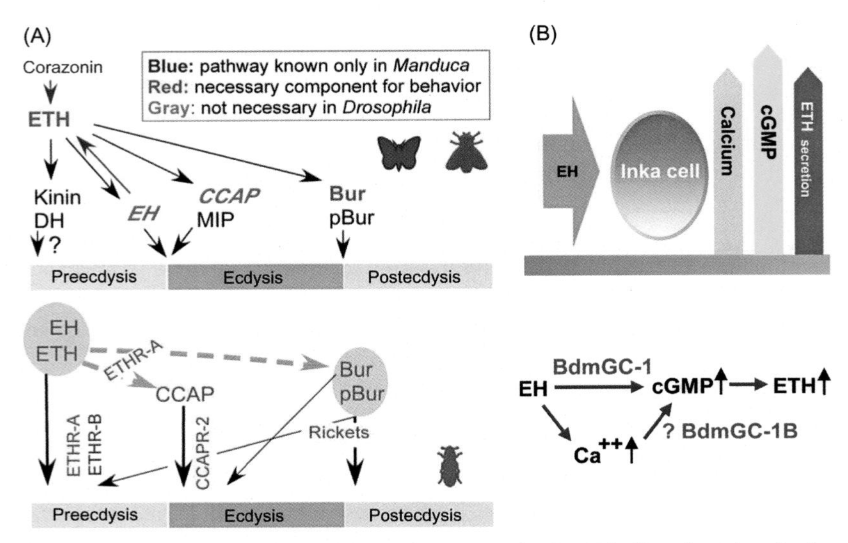

FIGURE 7.2 Action of peptides regulating ecdysis and tanning. (A) Cascade actions in three insect species. The upper panel depicts the models for *Manduca sexta* and *Drosophila melanogaster*. In blue, components observed only in *M. sexta.* Ecdysis-triggering hormone (ETH) and bursicon (Bur) were found to be necessary in experiments with *D. melanogaster* mutants; eclosion hormone (EH) and the crustacean cardioactive peptide (CCAP) were found to be nonessential in *D. melanogaster* cell-knockout and null mutants, respectively. The lower panel shows the case of *Tribolium castaneum.* In red, necessary components. Gray arrow represents the hypothetical pathway inspired by the *D. melanogaster–M. sexta* model; red background circles represent possible feedback between EH and ETH, and putative heterodimerization between Bur and partner of bursicon (pBur) based on the *D. melanogaster–M. sexta* model; the receptor ETHR-B has a mild effect on preecdysis, Bur alone also has a mild effect on ecdysis behavior, while Bur/pBur and rickets have an effect on postecdysis and only a mild effect on preecdysis. (B) Model for EH-mediated ETH secretion by Inka cells. EH triggers cGMP synthesis by binding to its receptor BdmGC-1. EH may act via two distinct pathways: a direct action on BdmGC-1 increasing cGMP content, and a second transduction pathway functioning via calcium mobilization (and regulating BdmGC-1B activity?). (A) From Arakane et al. (2008) and (B) from Chang et al. (2009), with permission.

regulation of metamorphosis). When the exuvia is detached, the tanning of the new cuticle takes place, a process that is mainly regulated by bursicon (Honegger et al., 2008) (Fig. 7.2).

Action of corazonin

The GPCR of corazonin was first reported in *D. melanogaster* when it was expressed in CHO cells and corazonin was identified as the endogenous specific ligand (Cazzamali et al., 2002). The corazonin receptor was subsequently studied in the lepidopteran *M. sexta* in the context of ecdysis initiation. The corazonin receptor is located in Inka cells and exhibits high sensitivity and selectivity for corazonin when expressed in *Xenopus* oocytes or CHO cells. Assays using the receptor as a biosensor revealed that corazonin concentrations in the hemolymph 20 minutes before natural preecdysis onset range from 20 to 80 pM and then decline over the next 30–40 minutes, which supports the role of corazonin signaling in ecdysis initiation (Kim et al., 2004) (Fig. 7.2A). The corazonin receptor has been also characterized in the mosquito *Anopheles gambiae*, the silkworm *B. mori*, the kissing bug *R. prolixus*, and cloned in the house fly *Musca domestica* (see Hamoudi et al., 2016).

Action of ecdysis-triggering hormone and eclosion hormone

Two kinds of endocrine cells are involved in coordinating the ecdysis events: the peripherally located Inka cells, which release ETH (Zitnan et al., 1996), and the centrally located neurosecretory neurons, the ventral median (VM) neurons, which release EH. EH acts on the Inka cells causing the release of ETH, and ETH, in turn, acts on the VM neurons triggering the release of EH (see Ewer et al., 1997). Functional experiments showed that the *ETH receptor* (*ETHR*) gene of *D. melanogaster* encodes two distinct subtypes of GPCRs (later named ETHR-A and ETHR-B). The two subtypes show differences in ligand sensitivity and specificity, which suggests that they play different roles in ETH signaling (Park et al., 2003). Using "Trojan exon" approaches to simultaneously mutate the *ETHR* genes and gain genetic access to the neurons that express its two isoforms, Diao et al. (2016) showed that *ETHR-A* and *ETHR-B* are expressed in different subsets of neurons. *ETHR-A*-expressing neurons are required for ecdysis at all developmental stages, whereas *ETHR-B* has an essential role in pupal and adult (but not larval) ecdysis. ETH receptors have also been characterized in *M. sexta*, *T. castaneum*, *A. aegypti*, *S. gregaria*, and the Oriental fruit fly *Bactrocera dorsalis*. As in *D. melanogaster*, the *ETHR* gene in these species encodes two ETHR subtypes (ETHR-A and ETHR-B) (see Shi et al., 2017).

EH triggers an increase of cyclic guanosine monophosphate (cGMP) in Inka cells, which leads to the massive release of ETH and to ecdysis initiation. In *B. dorsalis*, the process is mediated by a receptor guanylyl cyclase, named BdmGC-1, which is activated by EH. *BdmGC-1* is expressed in Inka cells, and heterologous expression of this gene in HEK cells leads to an increase in cGMP following exposure to picomolar concentrations of EH. An isoform of BdmGC-1 (BdmGC-1B), which is also triggered by EH, possesses the same domains and putative *N*-glycosylation sites present in BdmGC-1 but has an additional 46-amino acid insertion in the extracellular region and lacks the C-terminal tail of BdmGC-1. Intriguingly, BdmGC-1B responds only to high EH concentrations, suggesting that it has different physiological roles with respect to BdmGC-1. Given that another typical action of EH is to mobilize cytoplasmic calcium, it has been proposed that a second signal transduction pathway could function through calcium mobilization, which would regulate BdmGC-1B activity (Fig. 7.2B) (Chang et al., 2009).

Action of crustacean cardioactive peptide

The first functional characterization of a CCAP receptor was reported in *D. melanogaster*, following an approach of candidate GPCRs expressed in *Xenopus* oocytes and candidate peptides (Park et al., 2002). In the beetle *T. castaneum*, two paralog genes encoding putative CCAP receptors, *CCAPR-1* and *CCAPR-2*, have been reported. CCAP activates both receptors in a heterologous expression system, but specific RNAi experiments revealed that CCAPR-2 is essential for eclosion behavior, while depletion of CCAPR-1 does not elicit any apparent anomaly (Lee et al., 2011). CCAP signaling is indispensable for successful ecdysis in the kissing bug *R. prolixus*, as RNAi of the CCAP receptor results in high mortality, typically at the expected time of the ecdysis sequence, or considerably delays the ecdysis. CCAPRs have been characterized also in the mosquito *A. gambiae* (see Lee et al., 2013).

Action of bursicon

The receptor for bursicon is encoded by the *rickets* gene, although it had been originally named *Drosophila leucine-rich repeats containing GPCR 2* (*dLGR2*), when its role as bursicon receptor was independently reported by two groups (Luo et al., 2005; Mendive et al., 2005). Both described that the bursicon-partner of bursicon heterodimer from *D. melanogaster* bound to dLGR2, triggering the stimulation of cAMP signaling. This activates a protein kinase A that, in turn, phosphorylates tyrosine hydroxylase, the rate-limiting enzyme in the melanization/sclerotization pathway (Davis et al., 2007). Moreover, labeled bursicon heterodimer binds with high affinity and specificity to dLGR2 (Luo et al., 2005). Depletion of the bursicon receptor with RNAi in *T. castaneum* shows that it is required for wing expansion, cuticle tanning, development of integumentary structures, and adult emergence (Arakane et al., 2008; Bai and Palli, 2010).

TRANSDUCTION OF THE 20-HYDROXYECDYSONE SIGNAL

The first data suggesting that ecdysteroids regulate gene expression were based on the phenomenon of puffing in the giant polytene chromosomes of dipteran salivary glands. Some puffs (enlargements of specific loci on chromosomes) were induced rapidly after the supply of ecdysone to salivary glands of the midge *Chironomus tentans* incubated in vitro (Clever and Karlson, 1960). A number of puffs (which were interpreted as representing local transcriptional activity) appeared rapidly after ecdysone administration, so they were called early puffs, while others were delayed, being called late puffs. Elegant studies by Ulrich Clever in *C. tentans* and by Michael Ashburner in *D. melanogaster* showed that the early puffs were still triggered in the presence of protein synthesis inhibitors, leading to the prediction that the early puffs would be direct targets of ecdysone (bound to their receptor). Ashburner continued with a series of detailed studies showing that when protein synthesis was inhibited the early puffs did not regress and the late puffs never appeared. The complete model proposed that the ecdysone-receptor complex would directly induce the expression of early genes, whose respective products (early proteins) would induce the expression of late genes. At the same time, the ecdysone-receptor complex would repress the premature expression of the late genes, and the early proteins would repress the expression of their own genes (Fig. 7.3). This conceptual framework is known as the Ashburner model (Ashburner, 1974; Ashburner et al., 1974) and remains a key reference not only for the molecular action of insect ecdysteroids but also for steroids in general (see Hill et al., 2013; King-Jones and Thummel, 2005; Ou and King-Jones, 2013).

The ecdysone receptor

The identification of the *ecdysone receptor (EcR)* gene in *D. melanogaster* (Koelle et al., 1991) and that of several early ecdysone response genes (Burtis et al., 1990; DiBello et al., 1991; Segraves and Hogness, 1990), at the beginning of the 1990s, established a new era in the study of ecdysteroid action. EcR was found to be a member of the nuclear receptor superfamily, which are ligand-dependent transcription factors that contain a highly conserved DNA-binding domain (DBD) as well as a less conserved ligand-binding domain (LBD) (King-Jones and Thummel, 2005). As found

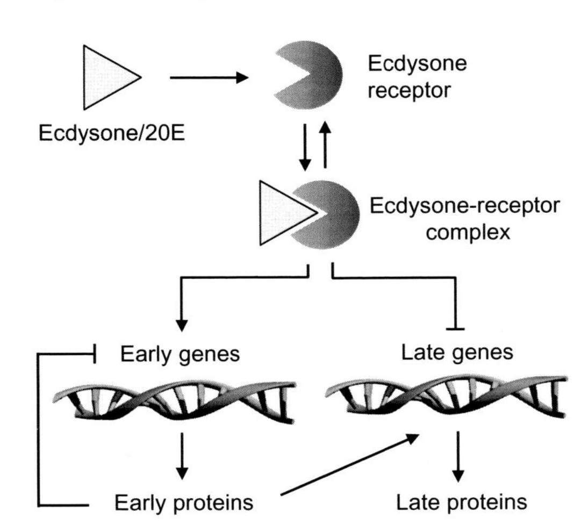

FIGURE 7.3 The Ashburner model. The ecdysone-receptor complex induces the expression of early genes, whose respective products (early proteins) induce the expression of late genes. At the same time, the ecdysone-receptor complex represses the expression of the late genes, and the early proteins repress the expression of their own genes.

out subsequently, the surprising discovery was that many early puffs encoded nuclear receptors.

In *D. melanogaster, EcR* encodes three protein isoforms, EcR-A, EcR-B1, and EcR-B2 (Talbot et al., 1993). All three are able to interact with its receptor partner ultraspiracle (USP, another nuclear receptor, see below), and all can bind 20E with similar affinity. Null mutations in the region common to all EcR isoforms are embryonic lethal, which is consistent with the observation that 20E signaling is required during germ-band retraction during *D. melanogaster* embryogenesis (Bender et al., 1997; Kozlova and Thummel, 2003). EcR-B1 is mostly produced in larval tissues that do not contribute to adult structures, and loss of its function prevents the ecdysone action in these tissues, and thus metamorphosis cannot be completed (Bender et al., 1997; Schubiger et al., 1998; Talbot et al., 1993). EcR-A is produced in imaginal discs and the ring gland, and insects mutant for EcR-A arrest development during late pupa (Davis et al., 2005; Talbot et al., 1993).

EcR requires heterodimerization with another nuclear receptor, USP, to form a functional ecdysteroid-receptor complex (Thomas et al., 1993; Yao et al., 1992, 1993). USP is homologous to vertebrate retinoic X receptor (RXR) and is required, as an essential partner of EcR, during embryogenesis and metamorphosis (Hall and Thummel, 1998; Oro et al., 1992). Genetic studies of *usp* have shown that the EcR is required to suppress premature metamorphic responses, consistent with the Ashburner model (Schubiger and Truman, 2000). Like mammalian RXR, USP is a promiscuous factor that can dimerize with other nuclear receptors, like HR38 and Seven-up (Sutherland et al., 1995; Zelhof et al., 1995). In 2001, the X-ray crystal structure of the isolated USP LBD of *D. melanogaster* was determined (see Hill et al., 2013).

Although EcR usually heterodimerizes with USP to perform its functions, a number of cases in which EcR works independently of USP have been reported. For example, disruption of endogenous EcR function in *D. melanogaster* causes premature death of larval crustacean cardioactive peptide (CCAP) neurons during metamorphosis, a phenotype that is rescued by co-expressing *EcR*. The mechanism is based on the repressive action of EcR signaling on the expression of *grim*, an essential cell death gene. However, USP is dispensable in the protection of CCAP neurons (Lee et al., 2019). This contrasts with the proapoptotic action of EcR signaling on corazonin-producing peptidergic neurons shortly after the prepupal stage of *D. melanogaster*, an action for which both EcR and USP are required (Wang et al., 2019a).

Ecdysteroids, which have a lipophilic character, were believed to enter cells by diffusion. However, genetic screens in *D. melanogaster* have led to the identification of a membrane transporter, called Ecdysone Importer (EcI), which is involved in the cellular uptake of ecdysteroids. *EcI* encodes an organic anion-transporting polypeptide that belongs to the evolutionarily conserved solute carrier organic anion superfamily. EcI depletion prevents cellular uptake of ecdysteroids, and *EcI* loss of function results in phenotypes identical to those obtained in insects deficient in ecdysteroids or EcR (Okamoto et al., 2018).

Early response genes

E74, *E75*, *BR-C*, and *E93* are examples of early response genes in the 20E signaling of *D. melanogaster*, the four encoding transcription factors, albeit belonging to different DNA-binding protein families (Fig. 7.4A) (see Sullivan and Thummel, 2003). *E74* maps to the 74EF early puff, encodes a member of the *ets* proto-oncogene family, and is directly induced by 20E. *E74* produces two protein isoforms, E74A and E74B, which share a C-terminal ETS DBD (Burtis et al., 1990). Both isoforms are precisely controlled by changes in 20E titers: E74A is produced when 20E concentration is high, while E74B becomes abundant when hormone concentrations decrease. This difference is critical for the proper timing of secondary gene responses. Mutations in *E74* confer pupal lethality and display defects in late gene induction, indicating that it plays essential roles in metamorphosis (Fletcher et al., 1995).

E75 maps to the 75B early puff and encodes a member of the nuclear receptor superfamily. *E75* generates three protein isoforms (E75A, E75B, and E75C) (Segraves and Hogness, 1990), which differ in their N-terminal sequences but share a common LBD. Mutant insects that do not produce E75A show larval lethality, molting defects, and developmental delays, while E75C is required for late pupal development and adult viability (Bialecki et al., 2002). E75B has a truncated DBD domain, and specific mutations for this isoform are viable. It binds to another nuclear receptor, HR3, in an inhibitory fashion, delaying the induction of the prepupal competence factor ß-FTZ-F1 (White et al., 1997) (see next section). An interesting property of E75 is its capacity to bind heme groups with high affinity. Caceres et al. (2011) have shown that E75 is able to sense the signaling molecule nitric oxide (NO), and that NO signaling regulates the interaction of HR3 and E75, thus controlling the transcriptional activation of *ßFTZ-F1*.

The *BR-C* gene (also known as *broad*) corresponds to the 2B5 early puff and encodes protein isoforms belonging to the Broad-complex, Tramtrack, Bric a brac/Poxvirus and zinc finger (BTB/POZ) family of C2H2 zinc finger transcription factors (DiBello et al., 1991). In *D. melanogaster*, *BR-C* expresses four different transcripts that produce four protein isoforms, depending on which zinc finger module (designated Z1–Z4) is incorporated. The zinc fingers are believed to confer target specificity (DiBello et al., 1991). Mutations that disrupt all BR-C isoforms result in larval lethality with a failure to enter metamorphosis, suggesting their essential role in pupal morphogenesis (Kiss et al., 1988). Consistent with

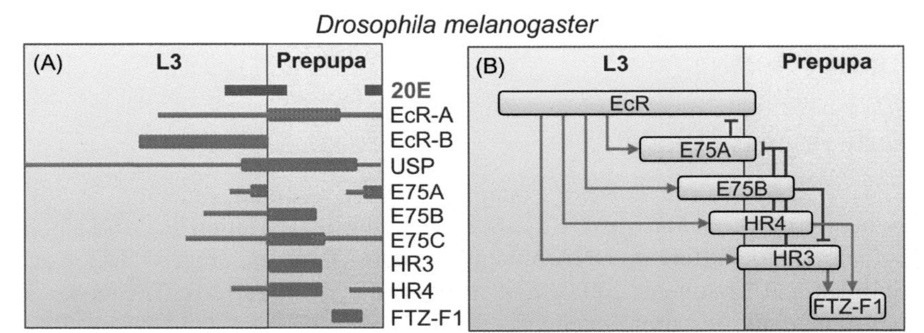

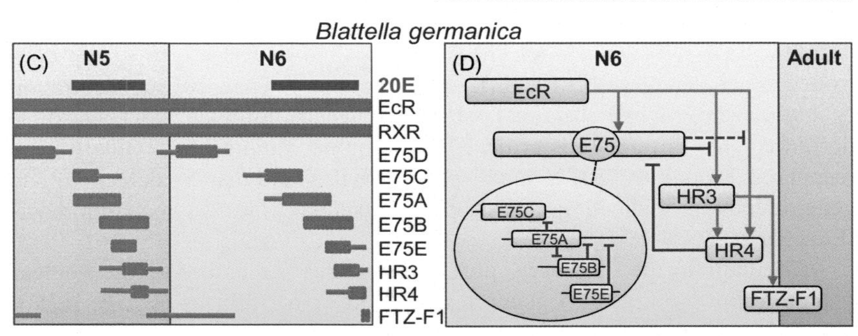

FIGURE 7.4 Ecdysone/20E signaling. (A and B): In the holometabolan *Drosophila melanogaster* in third (last) larval instar (L3) and in prepupa; expression hierarchy of early and late genes (A) and gene interactions (B). (C and D): In the hemimetabolan *Blattella germanica* in penultimate (fifth, N5) and last nymphal instar (sixth, N6), and in the transition to adult; expression hierarchy of early and late genes (C), and gene interactions (D). Original figures with data from different sources compiled by Sullivan and Thummel (2003) (A and B) and Mané-Padrós et al. (2012) (C and D).

this, *BR-C* is essential for proper ecdysone-regulated gene expression as larvae enter metamorphosis (von Kalm et al., 1994). The main function of BR-C in hemimetabolan species is to regulate wing development during nymphal stages, whereas in holometabolans, BR-C transcription factors determine the larval–pupal transition. This is described in more detail in a following subsection.

E93 maps to the 93F late prepupal stage-specific early puff and encodes a protein with RHF domains that have significant similarity to Pipsqueak motifs. 93F is directly induced by 20E in late prepupae but display no response to the hormone in late larvae. *E93* was discovered while studying the degeneration of the salivary glands during *D. melanogaster* metamorphosis (Baehrecke and Thummel, 1995). Consistent with the stage specificity of the 93F puff, *E93* is transcribed only in 10–12 hours prepupal salivary glands (Baehrecke and Thummel, 1995). Genetic studies further demonstrated that *E93* is an important regulator of adult patterning during metamorphosis (Mou et al., 2012), as discussed in detail in a subsection below.

Late response genes

Two delayed early response genes with very similar expression patterns are *HR3* and *HR4*, both coding nuclear receptors. Their expression patterns show a peak at the beginning of the prepupal stage, when the expression of early genes like *BR-C*, *E74A*, and *E75A* is declining, and that of *ßFTZ-F1* is about to be induced (Fig. 7.4A). Both proteins HR3 and HR4 are sufficient to repress the early genes and are required for maximal *ßFTZ-F1* expression in mid-prepupae (King-Jones et al., 2005; Lam et al., 1997). *HR4* mutants show precocious wandering behavior followed by precocious metamorphosis. The resulting adults are smaller than normal due to the shortened feeding period, a phenotype not observed in any other mutants associated with the ecdysone hierarchy, and which was eventually tracked to a role for HR4 in the PG (see Ou and King-Jones, 2013).

An important late response gene is *FTZ-F1*, which also encodes a nuclear receptor. Two protein isoforms have been described in *D. melanogaster*, α and ß (Lavorgna et al., 1991; Ueda et al., 1990). While αFTZ-F1 is maternally loaded and plays vital functions in embryo development, ßFTZ-F1 is conspicuously expressed in the early stages of puparium formation (Yamada et al., 2000; Yu et al., 1997). Mutations in *ßFTZ-F1* disrupt the 20E signaling pathway at the onset of metamorphosis, resulting in prepupal lethality. Importantly, ßFTZ-F1 acts as a competence factor during prepupal development, promoting the expression of stage-specific genes like *E93* (Broadus et al., 1999; Woodard et al., 1994). In brief, the interplay among E75, HR3, and HR4 regulates the expression of *ßFTZ-F1* during the prepupal stage, thereby ensuring the appropriate sequence of programs needed for metamorphosis (see Ou and King-Jones, 2013).

The epistatic relationships between the components of the 20E pathway in *D. melanogaster* are relatively complex but generally conform to the Ashburner model. The ecdysone-receptor complex induces the expression of early genes, and the early proteins induce the expression of the late genes, while the ecdysone-receptor complex represses premature late gene induction (Fig. 7.4B). The interplay between these factors directs the proper timing of gene expression and sequence of developmental changes that define the early stages of metamorphosis.

Recent work in *D. melanogaster*, focusing on the larval-to-prepupal transition of developing wings and analyzing genome-wide DNA-binding profiles, has corroborated the Ashburner model in a more precise temporal dimension (Uyehara and McKay, 2019). The study revealed that EcR exhibits widespread binding across the genome, and that EcR binding is temporally dynamic, with thousands of binding sites changing over time. Importantly, EcR acts as both a temporal gate to block premature entry to the next developmental stage and as a temporal trigger to promote the subsequent program.

The 20-hydroxyecdysone signaling pathway in insects other than *Drosophila melanogaster*

Apart from *D. melanogaster*, the first nuclear receptors involved in 20E signaling were cloned in the 1990s from lepidopterans, like *Galleria mellonella* (Jindra and Riddiford, 1996), *M. sexta* (see Hiruma and Riddiford, 2010), and *B. mori* (see Swevers and Iatrou, 2003). Although they could not be functionally studied in vivo, the determination of their expression patterns along development, and experiments in cell lines, led to important information about their contributions to the regulation of metamorphosis. Regarding structural studies, the most significant results on the tertiary structure of EcR/USP have been obtained from species other than *D. melanogaster*. Specifically, the EcR LBD/USP LBD heterodimer structures have been determined by X-ray crystallography in the hemipteran *Bemisia tabaci*, in the coleopteran *T. castaneum*, and in the lepidopteran *Heliothis virescens* (see Hill et al., 2013).

In hemimetabolan species, the most complete studies have been carried out in the cockroach *Blattella germanica*. The following nuclear receptors related to 20E signaling have been characterized in this cockroach: EcR, USP (RXR), HR3, FTZ-F1, E75, and HR4 (see Mané-Padrós et al., 2012). Moreover, transcription factors of the 20E signaling pathway belonging to other protein families, such as BR-C (Huang et al., 2013; Piulachs et al., 2010) and E93 (Belles and Santos, 2014; Ureña et al., 2014), have also been studied in *B. germanica* (see below).

Hierarchically, the progression of the 20E signaling pathway in *B. germanica* is similar to that observed in *D. melanogaster*. It is initiated by EcR and USP, followed by the early gene *E75*, whereas *HR3* and *HR4* function as delayed early response genes. *FTZ-F1* is a late gene that is expressed when the ecdysone peak declines and has a differentially higher expression in the last nymphal instar, preluding metamorphosis, like in the prepupa of *D. melanogaster* (Fig. 7.4C). Studies in *B. germanica* have shown that FTZ-F1 plays a key role in the degeneration of the PG that occurs after the imaginal molt (Mané-Padrós et al., 2010) (see Chapter 9: Molting: the basis for growing and for changing the form). The epistatic relationships between the components of the pathway in *D. melanogaster* and *B. germanica* are also similar (Fig. 7.4D). An interesting difference lies in *E75*, which in *B. germanica* expresses at least four isoforms (while *D. melanogaster* has only three) that interact with each other in a complex way. Despite these differences of detail, the 20E signaling pathway appears essentially conserved at least from early-branching polyneopterans (cockroaches) to the most modified endopterygotes (flies).

The role of Broad-complex as promoter of the pupal stage

In *D. melanogaster*, *BR-C* null mutants remain in a prolonged larval state and do not pupate (Kiss et al., 1988). Subsequent studies in *D. melanogaster* and other holometabolan species like the lepidopteran *M. sexta* showed that BR-C determines the larval–pupal transformation (Karim et al., 1993; Zhou et al., 1998). An interesting methodological innovation was the use of a recombinant Sindbis virus expressing a *BR-C* antisense RNA fragment in the silkworm *B. mori*, which reduced endogenous *BR-C* mRNA levels in infected tissues, thereby preventing the insect from completing the larval–pupal transformation (Uhlirova et al., 2003). Later experiments depleting *BR-C* mRNA with RNAi were carried out in other holometabolan species, like the beetle *T. castaneum* (Konopová and Jindra, 2008; Parthasarathy et al., 2008) and the neuropteran *Chrysopa perla* (Konopová and Jindra, 2008), further confirming that BR-C is key for pupal differentiation (Fig. 7.5A–D).

RNAi depletion of BR-C has also been reported in hemimetabolan species, like the bugs *Oncopeltus fasciatus* (Erezyilmaz et al., 2006) and *Pyrrhocoris apterus* (Konopová and Jindra, 2008), and the cockroach *B. germanica* (Huang et al., 2013). Results indicate that the main role of BR-C in postembryonic development is regulating wing development during the nymphal period. In *O. fasciatus*, whose wings have a colored pattern, BR-C was shown to be necessary for nymphal heteromorphosis in pigmentation. Also in this bug, the destruction of the CA, thus interrupting JH production, downregulated BR-C expression (Erezyilmaz et al., 2006), thus suggesting that JH sustains the expression of BR-C during the nymphal period of hemimetabolan insects. In *B. germanica*, it has been additionally reported that BR-C depletion affects wing patterning, including venation (Fig. 7.5E and F) (Huang et al., 2013) although this BR-C action might be mediated by the microRNAs let-7, miR-100, and miR125 (Rubio and Belles, 2013). The *BR-C* gene expresses at least six isoforms in *B. germanica* (Piulachs et al., 2010), which contrasts with the four isoforms of the most modified *D. melanogaster* (DiBello et al., 1991).

The role of E93 as promoter of metamorphosis

E93 was discovered in the context of research into the degeneration of the salivary glands during *D. melanogaster* metamorphosis, as an ecdysone-induced prepupal-specific gene. A number of studies (Baehrecke and Thummel, 1995; Lee et al., 2000; Woodard et al., 1994) revealed that E93 is a key player in the degeneration process of these glands. More recently, it has been shown that E93 is also involved in the degeneration of the PG after the imaginal molt, at least in the cockroach *B. germanica* (Orathai Kamsoi and Xavier Belles, unpublished) (see Chapter 9: Molting: the basis for growing and for changing the form).

Regarding programmed cell death (PCD) processes, E93 plays also an important role as mediator of mushroom body (MB) neuroblasts autophagy,

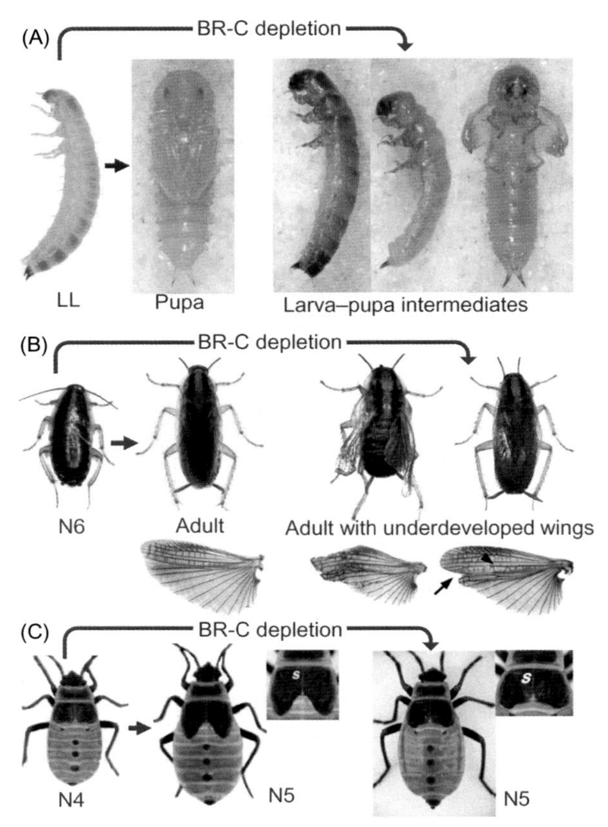

FIGURE 7.5 Experiments of Broad-complex (BR-C) depletion, showing its role as a pupal specifier in holometabolan species and as a promoter of wing development in hemimetabolans. (A) Depletion of BR-C in the larvae of the holometabolan *Tribolium castaneum*; the control group molted from last larval instar (LL, usually L7 or L8, depending on the strain and rearing conditions) to normal pupae; in the BR-C-depleted group, those experiencing a strong depletion effect molted to individuals with a general larval morphology (the two on the left), while those experiencing a mild depletion molted to individuals with a general pupal morphology, but showing short legs, partially developed gin traps, short wings, and urogomphi (like the individual on the right). (B) BR-C depletion in the last nymphal instar (N6) of the hemimetabolan *Blattella germanica*; the control group molted to normal adults, with correctly patterned wings; in the BR-C depleted group, 30% of the adults showed the wings not well extended, and 70% looked like normal adults, but in most cases, the membranous wings exhibited defects of shape (showing a short cup vein with the consequent formation of a notch (arrow) and/or vein patterning defects (arrowhead). (C) BR-C depletion in the penultimate nymphal instar (N4) of the hemimetabolan *Pyrrhocoris apterus*; the control group molted to normal N5, with correctly sized and shaped wing pads, whereas the BR-C depleted group molted to N5 with the wing pads similar to those of N4; *s*: scutellum. (A) Photos courtesy of Marek Jindra; (B and C) photos from Huang et al. (2013) and Konopová et al. (2011), respectively, with permission.

thus ending neurogenesis at the proper time during metamorphosis. In larvae of *D. melanogaster*, two neuroblast temporal factors, Imp (IGF-II, insulin growth factor, mRNA-binding protein) and Syp (Syncrip), contribute to regulate the timing of MB neurogenesis termination. If Imp is depleted, then MB neurogenesis terminates precociously, whereas if Syp is depleted, then MB neurogenesis extends into the adult stage (Yang et al., 2017). Moreover, decreased activity of PI3K reduces the growth and proliferation of MB neuroblasts, which are then primed for PCD (Siegrist et al., 2010). Importantly, ecdysteroid signaling is required for the Imp to Syp temporal switch, which is associated to the sequential expression of two late neuroblast factors, BR-C and E93 (Syed et al., 2017). Further work of Pahl et al. (2019) has revealed that E93 is a late-acting temporal factor, which downregulates PI3K levels, thus activating MB neuroblast autophagy and properly terminating MB neurogenesis in *D. melanogaster*.

The action of E93 in metamorphosis is not restricted to regulating PCD processes, as it also acts on morphogenesis. Mou et al. (2012) found that E93 is widely expressed in adult cells of the pupa of *D. melanogaster*, being required for patterning processes of morphogenesis. This suggested that E93 might play a general role in changing the responsiveness of target genes during metamorphosis. The approach followed by these authors was to focus on a simple E93-dependent process: the induction of the *Distal-less* (*Dll*) gene within bract cells of the pupal leg by epidermal growth factor (EGF) receptor signaling. Results indicated that E93 causes *Dll* to become responsive to EGF receptor signaling, showing that E93 is both necessary and sufficient for directing this switch (Mou et al., 2012).

The above results suggested that E93 controls the responsiveness of many other target genes because it is generally required for patterning during metamorphosis (Mou et al., 2012). Subsequent RNAi experiments showed that E93-depleted larvae of *D. melanogaster* were able to pupate but died at the end of the pupal stage (Ureña et al., 2014). RNAi experiments produced similar results in another holometabolan insect, the beetle *T. castaneum*, in which E93 depletion prevented the pupal–adult transition, resulting in the formation of a supernumerary second pupa (Fig. 7.6A). Similar results were obtained in a hemimetabolan species, the cockroach *B. germanica*, in which E93 depletion in nymphs prevented the nymphal–adult transition, resulting in reiterated supernumerary nymphal instars (Ureña et al., 2014) (Fig. 7.6B). At the same time, it was shown that the expression of *E93* in juvenile nymphs of *B. germanica* is inhibited by the transcription factor Krüppel homolog 1 (Kr-h1), the master transducer of the antimetamorphic action of JH, thus explaining the essential mechanism by which JH represses metamorphosis (Belles and Santos, 2014). The inhibitory action of Kr-h1 upon *E93* expression was later corroborated in the holometabolan *T. castaneum* (Chafino et al., 2019; Ureña et al., 2016), and studies in this beetle indicated that E93 also stimulates the prepupal peak of *BR-C* expression, which promotes the formation of the pupa (Chafino et al., 2019).

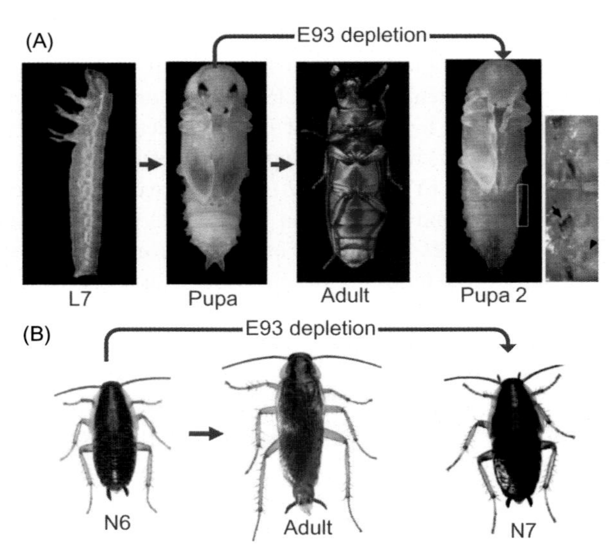

FIGURE 7.6 Experiments of E93 depletion, showing its role as an adult specifier in holometabolan and hemimetabolan species. (A) E93-depleted last larval instar (LL, usually L7 or L8, depending on the strain and rearing conditions) of the holometabolan *Tribolium castaneum* molt to normal pupa and then to a supernumerary pupa (pupa 2); the inset shows the gin traps (arrowhead) in the cuticle of the pupa 2 under the exuvia of the pupa, which still shows the original gin traps (arrow). (B) E93 depletion in the last nymphal instar (N6) of the hemimetabolan *Blattella germanica* and phenotype observed after the next molt; the control group molted to normal adults, whereas the E93-depleted nymphs molted into supernumerary nymphs (N7). Photos of the phenotypes from Ureña et al., 2014 (A) and Belles and Santos, 2014 (B), with permission.

TRANSDUCTION OF THE JUVENILE HORMONE SIGNAL

The elucidation of the JH receptor followed about 20 years after that of the 20E receptor. A remarkable attempt was made to consider the USP protein of *D. melanogaster* as a plausible candidate (Jones and Sharp, 1997). The idea made sense because USP is a homolog of the vertebrate retinoid X receptor (RXR), whose ligand, 9-*cis*-retinoic acid, is chemically similar to JH. The main drawback to considering USP as a good candidate, however, is the low-affinity binding of JH ($K_d \sim 4$ mM), which is not consistent with primary JH action, which occurs at nanomolar levels. Subsequently, the ligand binding properties of USP have been the object of further contributions maintaining the idea that USP binds to JH-related sesquiterpenoids (initially proposed to be JH III or JH III acid and more recently to be MF) (Jones et al., 2013a,b). Indeed, X-ray crystallographic studies have revealed that the LBD of USP can be relatively flexible and can configure a ligand binding pocket, as shown when using ligands like methylene lactam insecticides in the phtirapteran *Bovicola ovis* (Ren et al., 2014). However, studies carried out in the beetle *T. castaneum* have shown that USP is unable to bind JH in vitro and in vivo, and, in the context of JH and ecdysteroid action, USP simply acts as a constitutive structural partner for EcR (Iwema et al., 2007). In recent years, the transcription factor Methoprene tolerant (Met) has been confirmed as the intracellular JH receptor.

Met, the juvenile hormone intracellular receptor

Met was discovered in 1986 in *D. melanogaster* as a gene that confers resistance to Methoprene, an insecticide chemically similar to JH (Wilson and Fabian, 1986). It was later reported that the *Met* gene product is a protein belonging to the bHLH/PAS family of transcription factors (Ashok et al., 1998). However, the absence of a major visible phenotype clearly linking *Met* deficiency with metamorphosis, discouraged further inquiries on Met involvement in JH signaling. The situation changed when it was found that RNAi depletion of Met in young larvae of the beetle *T. castaneum* induces a precocious metamorphosis to pupa, which directly related Met with JH signaling (Konopová and Jindra, 2007) (Fig. 7.7A). The absence of developmental phenotypes in *Met* mutants of *D. melanogaster* was explained later, since in this species, *Met* has a paralog gene, *germ cell-expressed* (*gce*), with partially redundant functions with respect to *Met*, while *T. castaneum* has only one *Met* gene. Consistently, the simultaneous mutation of *Met* and *gce* in *D. melanogaster* was lethal during the larva–pupa transition, which is precisely the period in which a deficiency of JH is also lethal (Abdou et al., 2011). RNAi experiments demonstrated the role of Met as a transducer of the JH signal in hemimetabolan species, from cockroaches, like *B. germanica* (Lozano and Belles, 2014) (Fig. 7.7B), to bugs, like *P. apterus* (Konopová et al., 2011).

Studies conducted in vitro revealed that Met of *D. melanogaster* (Charles et al., 2011; Miura et al., 2005), *T. castaneum* (Charles et al., 2011), and *A. aegypti* (Li et al., 2014) binds JH III at nanomolar concentrations. Furthermore, if the PAS-B domain of *T. castaneum*, *A. aegypti*, or *D. melanogaster* Met/Gce is mutated, JH binding does not occur (Charles et al., 2011; Jindra et al., 2015b; Li et al., 2014). In *D. melanogaster*, it was also shown that Met (and Gce) mediates the bioactivity of MF (Bittova et al., 2019; Jindra et al., 2015b; Wen et al., 2015). A final piece of evidence of the role of Met (and Gce) as JH receptor was the demonstration that transgenic Met or Gce proteins restore the sensitivity to JH in Methoprene-tolerant mutants and rescue the lethality of *Met gce* double-mutant insects (Jindra et al., 2015b). Using the *D. melanogaster* Gce, Bittova et al. (2019) provided evidence of its ligand selectivity to natural JHs, even at the level of the carbon C11 optical isomers, as Gce preferentially bound the natural JH enantiomer. The data elegantly demonstrate that Gce indeed behaves as a specific JH receptor.

As in the case of 20E receptor, the JH receptor is not a single protein. Binding of JH stimulates Met or Gce to form a complex with another bHLH/PAS protein called Taiman (Tai), also known as FISC (ßFTZ-F1 interacting

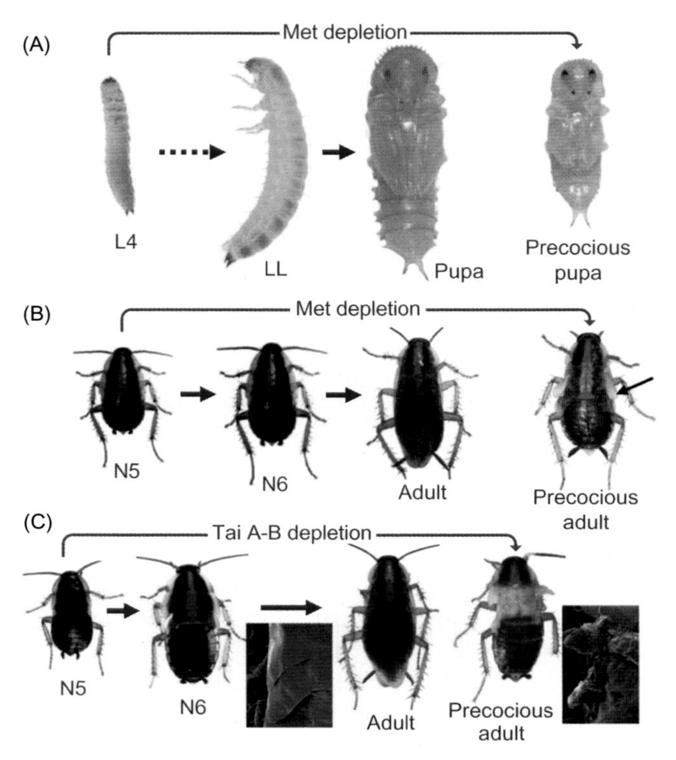

FIGURE 7.7 Experiments of Methoprene-tolerant (Met) and Taiman (Tai) depletion, showing their role as transducers of the antimetamorphic signal of JH in holometabolan and hemimetabolan insects. (A) Met depletion in the fourth larval instar (L4) of the holometabolan *Tribolium castaneum*; the control L4 molted to normal L5, successively to last larval instar (LL, usually L7 or L8, depending on the strain and rearing conditions) and then to pupa, whereas the Met-depleted L4 molted into precocious pupae. (B) Met depletion in the penultimate nymphal instar (N5) of the hemimetabolan *Blattella germanica*; the control group molted to normal last (N6) nymphal instar, whereas the Met-depleted nymphs molted into precocious adults (the arrow indicated the partially developed membranous wings). (C) Depletion of Tai isoforms A and B in the penultimate nymphal instar (N5) of *B. germanica*; the control group molted to normal last (N6) nymphal instar (with the wing pads encapsulated in the pteroteca, inset), whereas Tai-depleted nymphs molted into precocious adults (they were unable to ecdyse, but removal of the exuvia in the thoracic region allowed to observe the membranous wings partially developed, inset); Tai has four isoforms in *B. germanica*, but the depletion of all of them proved lethal. Photos of the phenotypes from Konopová and Jindra (2007) (A); Lozano and Belles (2014) (B); and Lozano et al. (2014) (C), with permission.

steroid receptor coactivator) or SRC (steroid receptor coactivator). The JH-Met + Tai complex binds to the JH response DNA motifs and activates the transcription of target genes (Charles et al., 2011; Kayukawa et al., 2012; Li et al., 2011, 2014; Zhang et al., 2011; Zou et al., 2013). In the cockroach *B. germanica*, RNAi experiments that have specifically depleted different Tai isoforms have demonstrated that Tai mediates the inhibitory effects of JH on metamorphosis (Lozano et al., 2014) (Fig. 7.7C).

Other studies have shown that Met of *D. melanogaster* interacts with the chaperone heat shock protein Hsp83, which facilitates their nuclear import, and the expression of genes induced by JH (He et al., 2014). Similarly, JH stimulates the nuclear translocation of Hsp90 and its phosphorylation, through the phospholipase C (PLC)/protein kinase C (PKC) pathway, in the cells of the lepidopteran *H. armigera* (Liu et al., 2013) (Fig. 7.8).

A membrane receptor for juvenile hormone

Several lines of evidence indicate that JH also uses a membrane receptor to transduce its signal into the cells. Data suggesting the existence of a JH membrane receptor were provided by experiments treating the *D. melanogaster* male accessory glands incubated in vitro with physiological doses of JH. This treatment mimics the mating-induced response of increased protein synthesis in glands from virgin flies. JH stimulation requires calcium in the medium and phorbol ester compounds, which activate PKC and increase protein synthesis in the accessory glands. In contrast, JH does not stimulate protein synthesis in mutants deficient in kinase C activity. The data suggest that a membrane protein mediates the effect of JH upon protein synthesis, involving calcium and kinase C (Yamamoto et al., 1988).

Much work devoted to characterize a possible JH membrane receptor has been carried out using the growing oocyte as an experimental model. During oocyte growth, the follicular cells contract, leaving intercellular spaces that allow the yolk protein precursors to access the oocyte membrane, a phenomenon that has been called patency (see Davey, 2000). Early studies in the kissing bug *R. prolixus* suggested that JH induces patency, and that PKC and Na^+/K^+ ATPase cascades are involved in this induction (Sevala and Davey, 1989, 1993). Work in the moth *H. virescens* suggested that the action of JH II and JH III on patency requires calcium stores and voltage-dependent calcium channels, whereas that of JH I appears to be largely calcium independent. Moreover, biochemical and pharmacological studies suggested that JH I would operate through a GPCR, involving cAMP-dependent signaling, whereas JH III and JH II would act through the inositol trisphosphate/diacylglycerol pathway (Pszczolkowski et al., 2005, 2008). Nevertheless, a work carried out in locusts (*S. gregaria* and *Locusta migratoria*) challenges the notion that patency is induced by JH. Different bioassays led to the conclusion that patency is induced by a "patency inducing factor" (PIF) produced by the lateral oviducts, which interacts with the basal follicle cells via the pedicel. Accordingly, JH would not be the direct inductor of patency although it would act enhancing the PIF action (Seidelmann et al., 2016).

In abdomens from newly emerged *A. aegypti* incubated in vitro, a hypothetical tyrosine kinase receptor followed by a PLC pathway is activated by

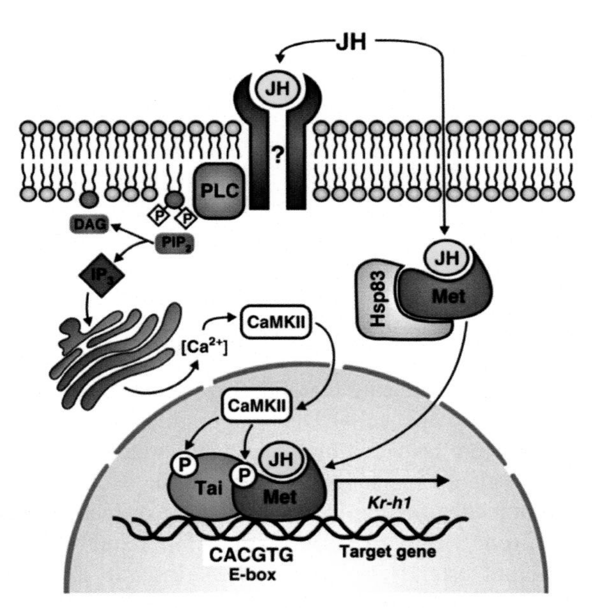

FIGURE 7.8 Interaction of juvenile hormone (JH) with a membrane receptor and a nuclear receptor. A still unidentified tyrosine kinase membrane receptor would activate phospholipase C (PLC)-dependent inositol trisphosphate (IP_3)/diacylglycerol (DAG) pathway, leading to Methoprene tolerant (Met) and Taiman (Tai) phosphorylation through a calcium/calmodulin-dependent kinase II (CaMKII). JH may also enter the cell by diffusion, then binding Met and stimulating the Hsp83-dependent nuclear import. The JH-Met + Tai complex would activate the downstream gene *Krüppel homolog 1* (*Kr-h1*), by binding to the response element containing the CACGTG E-box that is located in the promoter region of the gene. From Jindra et al. (2015a), with permission.

JH, leading to an increase of intracellular concentrations of diacylglycerol, inositol trisphosphate, and calcium. Remarkably, this pathway leads to the phosphorylation of Met and Tai through a calcium/calmodulin-dependent kinase II, which enhances the binding ability of the Met–Tai complex to the JH response elements of target genes (Liu et al., 2015). Thus the parallel membrane-based and intracellular branches of JH signaling converge at the Met–Tai receptor complex in the nucleus, leading to the previously known transcriptional activation (see Jindra et al., 2015a) (Fig. 7.8).

Krüppel homolog 1, a master transducer of the antimetamorphic action of juvenile hormone

Kr-h1 was discovered in *D. melanogaster* as a gene with structural similarity to the segmentation gene *Krüppel*, with which it shares the zinc-finger motifs and the amino acid spacers connecting them. In *D. melanogaster*, *Kr-h1* encodes two major isoforms, called α and β. The α isoform predominates in postembryonic stages and was initially reported as mediating ecdysone signaling in the larva–pupa transition (Beck et al., 2004; Pecasse et al., 2000), whereas the β isoform is abundantly expressed in embryonic neuronal cells (Beck et al., 2004). The first evidence connecting Kr-h1 and JH was also obtained in *D. melanogaster*. In this fly, the adult epidermis of the abdomen derives from larval histoblasts, which start proliferating after puparium formation. Administration of JH prior to the prepupal stage prevents the normal differentiation of the abdominal epidermis, and the bristles that should be formed in the adult are shorter or simply do not develop (Ashburner, 1970). Ectopically expressed *Kr-h1* in the abdominal epidermis during metamorphosis of *D. melanogaster* caused missing or short bristles, as when applying JH, thereby suggesting that Kr-h1 mediates the antimetamorphic action of JH (Minakuchi et al., 2008). In *T. castaneum*, RNAi depletion of Kr-h1 in young larvae caused a precocious larva–pupa transformation, which clearly demonstrated that Kr-h1 represses metamorphosis and that it works downstream of Met in the JH signaling pathway (Minakuchi et al., 2009). The antimetamorphic action of Kr-h1 was generalized to hemimetabolan species in two simultaneous works carried out, respectively, on the cockroach *B. germanica* (Lozano and Belles, 2011) and the bugs *P. apterus* and *R. prolixus* (Konopová et al., 2011). The RNAi experiments showed that depletion of Kr-h1 in nymphs in their penultimate stage induces the metamorphosis to a precocious adult in the next molt, while those treated in the antepenultimate nymphal stage require two molts to form the precocious adult, like in the case of Met depletion (Fig. 7.7). The key role of Kr-h1 as a transducer of the antimetamorphic action of JH has been reported in other species, hemimetabolans, like the bed bug *Cimex lectularius* (Gujar and Palli, 2016) and the planthopper *N. lugens* (Li et al., 2018), and holometabolans, like the silkworm *B. mori* (Kayukawa et al., 2014).

Regarding the mechanism by which JH stimulates the expression of *Kr-h1*, Kayukawa et al. (2012), using Kr-h1 from *B. mori* and reporter assays, identified a JH response element (kJHRE), comprising 141 nucleotides, located ~2 kb upstream from the *Kr-h1* gene transcription start site. Significantly, the core region of this kJHRE (GGCCTCCACGTG) contains a canonical E-box sequence to which Met and other bHLH/PAS proteins can bind. The JHREs previously described for other JH-dependent genes (see Riddiford, 2008) do not contain this E-box, but it is present in the promoter region of *Kr-h1* of the mosquito *A. aegypti* (Cui et al., 2014; Shin et al., 2012).

E93: A KEY TARGET OF KRÜPPEL HOMOLOG 1. THE MEKRE93 PATHWAY

The first evidence on the inhibitory effect of Kr-h1 upon *E93* was obtained in the cockroach *B. germanica*, on observing that the RNAi depletion of Kr-h1 resulted in a remarkable upregulation of *E93* expression (Belles and Santos, 2014). This led to the proposal of the MEKRE93 pathway (from

Met-Kr-h1-E93) as the central axis regulating metamorphosis. Accordingly, in the nymph–nymph transitions, the JH, acting through its receptor Met-Tai, induces the expression of *Kr-h1*, while Kr-h1 represses the expression of *E93*. In turn, the fall of JH production in the last juvenile instar interrupts the expression of *Kr-h1* and allows a strong induction of *E93* through ecdysone signaling, which triggers metamorphosis (Fig. 7.9A). RNAi experiments in *B. germanica* also showed that the depletion of E93 upregulated the expression of *Kr-h1*, which indicates that *Kr-h1* and *E93* are reciprocally repressed. The same interactions between Kr-h1 and E93 were subsequently reported in the beetle *T. castaneum* (Ureña et al., 2016), which extended the application of the MEKRE93 pathway to holometabolan species (Fig. 7.9A–C). The main regulatory difference between the hemimetabolan and holometabolan modes is the contribution of BR-C. In hemimetabolans, *BR-C* is mainly involved in promoting wing development, and its expression is enhanced by JH and Kr-h1 during juvenile stages (Huang et al., 2013) and repressed by E93 in the metamorphic transition (Ureña et al., 2014). RNAi studies in the cricket *G. bimaculatus*, also a hemimetabolan species, have confirmed the mentioned interactions and have revealed that depletion of BR-C results in a downregulation of *Kr-h1* and an upregulation of *E93* expression (Ishimaru et al., 2019). This might explain that BR-C depletion by RNAi in cricket late nymphs triggers the formation of precocious adults at the next molt. In holometabolans, BR-C triggers the formation of the pupa, and during larval instars, JH inhibits the expression of *BR-C*, although, intriguingly, artificially administered JH after pupal commitment stimulates it (Zhou et al., 1998). As shown in *T. castaneum*, E93 also contributes to trigger the pupal stage as it promotes the expression of *BR-C* in the prepupa (Chafino et al., 2019). Thus, the MEKRE93 pathway can be extended to the holometabolan metamorphosis by adding the interaction loop of E93 with BR-C, and corresponding interactions, to the ancestral (hemimetabolan) pathway (Belles, 2019) (Fig. 7.9A). As also shown in *T. castaneum*, *Kr-h1* activity maintains the larval status until the last larval instar, where a first and transient decrease of its expression, presumably derived from a transient decrease of JH production, is followed by a slight increase of *E93* expression and by a dramatic upregulation of *BR-C*, which triggers the larva-pupa transition. In the pupal stage, the expression of *Kr-h1* and *BR-C* vanishes, and that of *E93* becomes dramatically upregulated, which triggers the pupa-adult transition (Fig. 7.9B and C).

Research in holometabolan models showed that Kr-h1 prevents the larva–pupa transformation (Kayukawa et al., 2014; Minakuchi et al., 2008, 2009), and work in *B. mori* led to the identification of a Kr-h1-binding site (KBS) in the promoter region of the *BR-C* (Kayukawa et al., 2016). As *BR-C* is activated by 20E in the absence of JH, the authors conjectured that Kr-h1 might bind close to ecdysone response elements (EcREs) of *BR-C*, so as to prevent its activation by 20E. Following this idea,

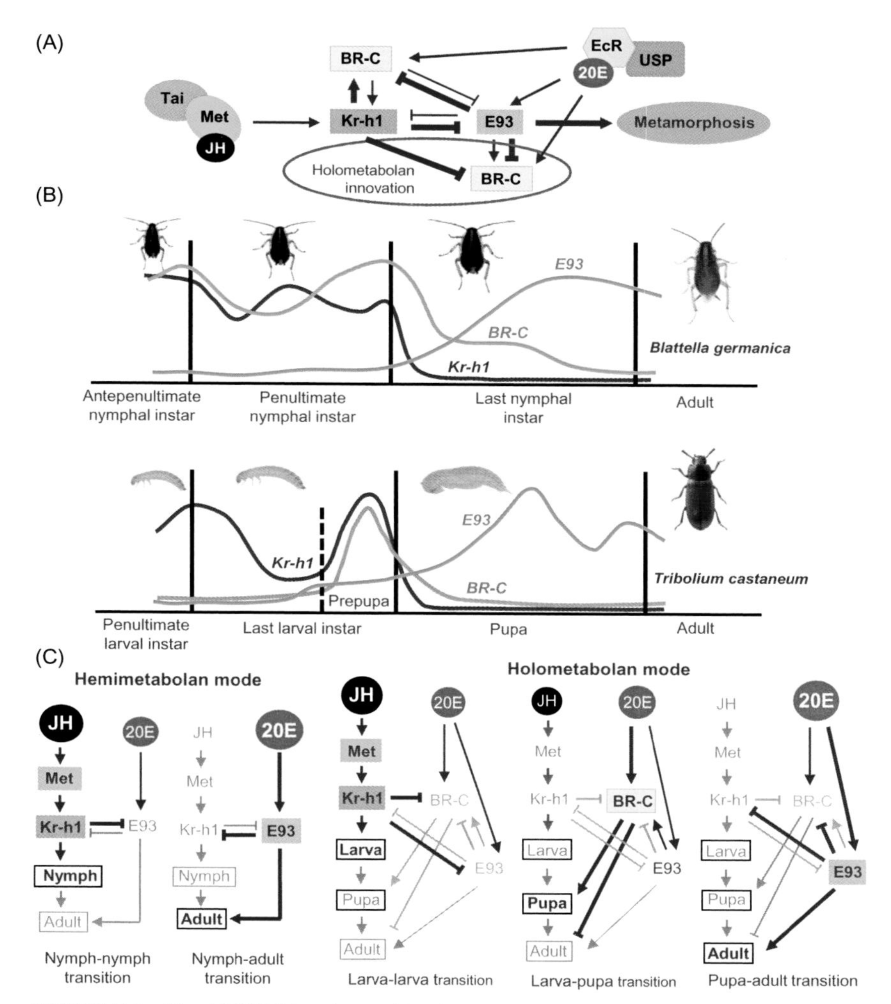

FIGURE 7.9 The MEKRE93 pathway. (A) The pathway in hemimetabolan and holometabolan metamorphosis. In the nymph–nymph transitions of hemimetabolans, juvenile hormone (JH) (through Methoprene-tolerant, Met, and Taiman, Tai) induces the expression of *Krüppel holomog* 1 (*Kr-h1*), whose gene product, in turn, represses the expression of *E93*. In contrast, the fall of JH production in the last juvenile stage interrupts the expression of *Kr-h1* and allows a strong induction of *E93* through ecdysone signaling, which triggers metamorphosis. The main difference between hemimetabolans and holometabolans is the contribution of *BR-C*, which in hemimetabolans is mainly involved in promoting wing development, whereas in holometabolans determines the formation of the pupa. (B) Expression of *Kr-h1*, *E93*, and *BR-C* in the hemimetabolan *Blattella germanica* and the holometabolan *Tribolium castaneum*; in both cases, the fall of JH production in the preadult stage interrupts the expression of *Kr-h1* and facilitates that of *E93*, although in *T. castaneum*, a first decrease of Kr-h1 in the last larval instar is needed to form the pupa. (C) Interactions among Kr-h1, BR-C, and E93 regulate the metamorphic transitions in hemimetabolan and holometabolan insects. Figures drawn with data from Belles and Santos (2014); Chafino et al. (2019); Huang et al. (2013); Ishimaru et al. (2019); Ureña et al. (2014 and 2016); Zhou et al. (1998).

Kayukawa et al. (2016) identified a 30-bp sequence, which was called KBS core region (GACCTACGCTAACGCTAAATAGAGTTCCGA), in the promoter region of *BR-C*, which is crucial for Kr-h1 binding. As conjectured, the KBS core region locates between two EcREs. This location, and further analysis of the 20E and JH regulation of the *BR-C* promoter, led to the proposal that *Kr-h1* expression is induced by JH via Met-Tai and that two molecules of Kr-h1 bind to the KBS core region, thereby preventing the 20E-induction of *BR-C* expression (Kayukawa et al., 2016).

These precedents led the same research group to presume that a similar mechanism could operate in the repression of *E93* by Kr-h1 (Kayukawa et al., 2017). Searching for sequences similar to the BR-C KBS core region in the promoter of *B. mori E93* led to finding a KBS candidate that, interestingly, was located near a putative EcRE. Using a *B. mori* cell line and reporter assays, it was shown that the *E93* reporter is strongly activated by 20E, whereas JH represses this activation. Mutations in the putative KBS region abolish this JHA-dependent repression, and deletion of the putative EcRE abolishes the 20E-induced expression. These data and additional RNAi experiments confirmed that *E93* is activated by 20E via the EcRE, and that the 20E-induction is repressed by JH and Kr-h1 via the KBS (Kayukawa et al., 2017).

The pathway regulating metamorphosis does not end with E93, as this factor activates the genes that contribute to the formation of the adult. An example of this is the enhancing effect of E93 upon *decapentaplegic* (*dpp*) expression in the wing of *D. melanogaster* (Wang et al., 2019b). Knockdown of *E93* in the wing and chromatin immunoprecipitation sequencing (ChIP-seq) analysis revealed that *dpp* is a downstream target of E93. ChIP-PCR analysis and dual luciferase reporter assay confirmed that E93 can bind to the *dpp* promoter, enhancing its activity. Moreover, *E93* overexpression in *Drosophila* S2 cells increases the expression of *dpp*, whereas *dpp* expression decreases after *E93* knockdown in the wing. These results indicate that E93 modulates the dpp signaling pathway, thus regulating wing development during *D. melanogaster* metamorphosis (Wang et al., 2019b).

REGULATION OF DEVELOPMENTAL GENE EXPRESSION AND CHROMATIN ACCESSIBILITY

We have seen a number of transcription factors acting on particular genes, activating or suppressing their expression, which leads to specify tissue and cell identity. However, the spatial expression of these factors must be coordinated with temporal signals to ensure the proper progression of development. The mechanisms based on the modulation of chromatin accessibility have contributed significantly to understanding that coordination.

Regarding metamorphosis, studies have focused on wing development, mainly using *D. melanogaster* as model. The seminal work of Uyehara et al.

(2017) revealed that temporal changes in gene expression correlate with genome-wide changes in chromatin accessibility at temporal-specific enhancers. Moreover, distinct combinations of transcription factors derived from the cascade of gene expression induced by a pulse of 20E characterize different times in wing formation. Importantly, E93 controls temporal identity by directly regulating chromatin accessibility across the genome. E93 controls enhancer activity through three different modes, including promoting accessibility of late-acting enhancers and decreasing accessibility of early-acting enhancers (Uyehara et al., 2017).

It has been also shown that modulation of chromatin accessibility ensures a robust cell cycle exit during terminal differentiation of *D. melanogaster* wings (Ma et al., 2019). The observations indicate that chromatin closes at a set of pupal wing enhancers for the rate-limiting cell cycle regulators *Cyclin E*, *E2F transcription factor 1*, and *string*. This closing coincides with wing cells entering a postmitotic state that is refractory to cell cycle reactivation. Unsurprisingly, the regions closed contain binding sites for mitogenic effectors. When cell cycle exit is genetically disrupted, chromatin accessibility at cell cycle genes results unaffected. This indicates that modulation of chromatin accessibility acts as a cell cycle-independent timer to limit the response to mitogenic signaling and abnormal cycling in terminally differentiating *D. melanogaster* wings (Ma et al., 2019).

Outside *D. melanogaster*, chromatin accessibility in relation to metamorphosis has been studied in the process of wing formation in the butterfly *Junonia coenia* (van der Burg et al., 2019). The work revealed that distinct sets of transcription factors are predictive of chromatin opening at different times in development. Moreover, data obtained suggest an important role for nuclear hormone receptors early in metamorphosis, whereas PAS-domain transcription factors appear strongly associated with later chromatin opening. ChIp-seq analyses revealed that spineless binding is a major predictor of opening chromatin. Intriguingly, binding of EcR, considered an accessibility factor in *D. melanogaster* (Shlyueva et al., 2014), is not predictive of opening but instead is largely associated with nondynamic, persistently accessible sites (van der Burg et al., 2019).

INTERACTIONS BETWEEN THE 20-HYDROXYECDYSONE AND THE JUVENILE HORMONE PATHWAYS

The regulation of *BR-C* and *E93*, two ecdysone-dependent genes, by JH through Kr-h1, is an example of interaction of the JH pathway with that of 20E, but there are other cases as well. For example, in the silk glands of the lepidopteran *G. mellonella*, the presence of JH increases the 20E-inducibility of *E75A* although JH alone does not affect that inducibility (Jindra and Riddiford, 1996). The same has been observed in the epidermis of another lepidopteran, *M. sexta*, where, in addition, JH modulates the 20E-regulated levels of expression of the *EcR* and *USP* genes (Riddiford et al., 2003). In *Plodia interpunctella*, another lepidopteran, experiments incubating wing precursor cells with 20E and JH showed that JH modulates the 20E-inducibility of *HR3*, *EcR-B1*, and *E75B* (Siaussat et al., 2004).

A number of studies have been carried out in *Drosophila* S2 cells, where JH activates the expression of *E75A* (Dubrovsky et al., 2004). Induction is rapid, the active concentration is low, and it does not require concurrent protein synthesis. The same group reported that JH activation of *E75A* in S2 cells requires gce (but not Met, which suggests that gce plays paralog-specific regulatory roles) as well as FTZ-F1 (Dubrovsky et al., 2011). Intriguingly, a FTZ-F1 response element (F1RE) confers JH induction, given that both Met and gce activate transcription of a reporter downstream of a DNA sequence containing F1REs. However, the precise role of FTZ-F1 in JH signaling requires further elucidation, as it is unclear whether the F1RE is sufficient to mediate endogenous JH-dependent transcription. While heterodimerization with FTZ-F1 is JH-dependent (Bernardo and Dubrovsky, 2012a), Met and gce mediate only weak JH-dependent transcriptional activation through the F1RE, suggesting that other factors may be involved in the JH response (Bernardo and Dubrovsky, 2012b).

Another example of interaction between the JH and 20E pathways affects the respective hormonal synthesis. As shown in *D. melanogaster*, JH biosynthesis can be prevented by 20E in the CA cells of the ring gland (Liu et al., 2018), whereas JH can inhibit ecdysone production in the PG cells or *D. melanogaster* and *B. mori*. The effect of JH is mediated by Kr-h1, which represses the expression of steroidogenic genes (Liu et al., 2018; Zhang et al., 2018b).

THE TGF-β SIGNALING PATHWAY

Several fundamental signaling pathways can affect the action of JH and 20E, but especially relevant in this context is the TGF-β pathway. In *D. melanogaster*, as in other animals, the TGF-β pathway operates through receptor-activated Smad proteins and can be classified into two branches, the TGF-β/Activin branch and the bone morphogenetic protein (BMP) branch. The TGF-β/Activin branch operates through the receptor babo and signals through a single R-Smad (Smox) and a single Co-Smad (Medea), whereas the BMP branch has two receptors: thick veins (tkv) and saxophone (sax) although it also operates through a single Smad (Mad) and the Co-Smad Medea (see Upadhyay et al., 2017).

In the absence of activin signaling in the PG, the third larval instar of *D. melanogaster* arrests and does not undergo metamorphosis, because activin signaling makes PG competent to respond to two other hormonal signals, PTTH and insulin. Moreover, without activin signaling, expression of the receptors of PTTH (*torso*) and insulin (*InR*) is downregulated in the PG

(Gibbens et al., 2011). Also in *D. melanogaster*, JH synthesis appears to be under the control of the BMP signaling through the ligand dpp. Using a genetic screen, Tkv and Mad were identified as positive regulators of JH signaling in the larva. JH is produced in the CA, and one of the key enzymes of its biosynthesis is JH acid methyltransferase (JHAMT). Glutamatergic signals from the brain induce *dpp* expression in the CA, and, in turn, dpp induces the expression of *jhamt*, which stimulates JH biosynthesis (Huang et al., 2011).

In contrast, in the cricket *G. bimaculatus*, myoglianin (myo), which is a ligand of the babo receptor of the activin branch of the TGF-β pathway, inhibits JH production in the CA through downregulating the expression of *jhamt* (Ishimaru et al., 2016). Therefore depletion of myo in the last nymphal instar of *G. bimaculatus* upregulates *jhamt* expression, hence JH production increases, and the crickets molt to supernumerary nymphs instead of adults (Ishimaru et al., 2016).

The inhibitory action of myo upon *jhamt* and the effects on metamorphosis have been also reported in the cockroach *B. germanica*, where myo depletion also upregulates *jhamt* expression and triggers the formation of supernumerary nymphs. In this cockroach, myo also plays significant roles in the PG, where a peak of *myo* expression in the penultimate nymphal instar stimulates the expression of ecdysteroidogenic genes, enhancing the production of the metamorphic ecdysone pulse in the last nymphal instar. The *myo* expression peak in penultimate nymphal instar also represses cell proliferation, which might enhance ecdysone production (Kamsoi and Belles, 2019). The inhibition of metamorphosis after depleting myo is explained by the MEKRE93 pathway. In the last nymphal instar, the *jhamt* upregulation induced by myo depletion increases JH production and *Kr-h1* expression, with a concomitant repression of *E93*, which prevents adult morphogenesis. These observations are consistent with a previous work reporting that the TGF-β/Activin signaling pathway contributes to the repression of *Kr-h1* expression and activation of *E93* expression during the metamorphosis of *B. germanica* (Santos et al., 2016). Depletion of Smox upregulates *Kr-h1*, which is consistent with the fact that myo signal is transduced by Smox, and in the end represses *jhamt* expression and JH production.

As discussed above, a group of corazonin-producing peptidergic neurons undergo apoptosis in response to ecdysone signaling via EcR and USP during early metamorphosis of *D. melanogaster* (Wang et al., 2019a). Interestingly, TGF-β signaling through the activin branch mediated by the ligand myo, type-I receptor Baboon and Smox, is also required for PCD of corazonin neurons. The studies of Wang et al. (2019a) have shown that ectopic expression of a constitutively active phosphomimetic form of Smox induces premature death of corazonin larval neurons, without affecting other neurons. This premature death is rescued by coexpressing a dominant negative form of EcR, which indicates that EcR is required for the proapoptotic function of Smox. Taken together, the data suggest that TGF-β and ecdysone signaling pathways act cooperatively, inducing PCD specifically in corazonin neurons (Wang et al., 2019a).

COMPETENCE TO METAMORPHOSE

While it is true that JH represses metamorphosis, this function does not seem to be met in very early juvenile stages. Classical works of suppression of the JH signal by means of allatectomy showed that precocious metamorphosis was only induced if the operation was carried out in relatively mature juvenile stages. Early experiments of CA removal showed that a precocious pupa of *B. mori* could not be obtained before the third larval molt (Bounhiol, 1938; Fukuda, 1944). As mentioned above, RNAi experiments on *B. germanica* (Lozano and Belles, 2011) and *P. apterus* (Konopová et al., 2011) showed that the depletion of Kr-h1 in nymphs in the antepenultimate stage required two molts to form the precocious adult. A number of examples using RNAi suppressing JH biosynthesis or signaling have confirmed that precocious metamorphosis cannot be induced in very young juvenile stages (Smykal et al., 2014). Supression of JH signaling through genome editing approaches has strongly supported this notion. Thus, knockout *B. mori* with null mutations in JH biosynthesis or JH receptor genes develop as wild-type silkworms during the first and second larval instars, and precocious pupal characters appear only after the second larval instar at the earliest (Daimon et al., 2015). Similar results have been obtained in the mosquito *A. aegypti*, where knockout of Met does not affect neither embryogenesis nor the development of the first two larval instars, L1 and L2, but triggers the appearance of metamorphic characters in the penultimate (L3) and last (L4) larval instars (Zhu et al., 2019). Taken together, the observations suggest that there should be a factor that confers competence for metamorphosis, which is not produced in very young nymphal or larval instars.

A first hint suggesting that the competence factor could be humoral was provided by the experiments reported in the 1930s and 1940s involving parabiosis or transplantation. Wigglesworth (1934), by connecting in parabiosis an individual of the kissing bug *R. prolixus* in the first nymphal instar with one in the fifth (last) nymphal instar, observed that the first nymphal instar molted to a precocious adult that showed a reduced genitalia, mid-developed wing pads, and adult-like cuticle. At the same time, Piepho (1938a,b) reported that fragments of epidermis of first larval instar of the lepidopteran *G. mellonella* produce pupal cuticles when transplanted into the body cavity of last larval instar. This happened at the time of pupal metamorphosis of the host (Piepho, 1938a,b). More recently, equivalent experiments of the transplantation of fragments of epidermis from first or second larval instars of *B. mori* into last instar host larvae produced similar results (Inui and Daimon, 2017). The search for the humoral factor that enables metamorphosis is ongoing.

References

Abdou, M.A., He, Q., Wen, D., Zyaan, O., Wang, Jing, Xu, J., et al., 2011. *Drosophila* Met and Gce are partially redundant in transducing juvenile hormone action. Insect Biochem. Mol. Biol. 41, 938–945. Available from: https://doi.org/10.1016/j.ibmb.2011.09.003.

Arakane, Y., Li, B., Muthukrishnan, S., Beeman, R.W., Kramer, K.J., Park, Y., 2008. Functional analysis of four neuropeptides, EH, ETH, CCAP and bursicon, and their receptors in adult ecdysis behavior of the red flour beetle, *Tribolium castaneum*. Mech. Dev. 125, 984–995.

Ashburner, M., 1970. Effects of juvenile hormone on adult differentiation of *Drosophila melanogaster*. Nature 227, 187–189.

Ashburner, M., 1974. Sequential gene activation by ecdysone in polytene chromosomes of *Drosophila melanogaster*. II. The effects of inhibitors of protein synthesis. Dev. Biol. 39, 141–157.

Ashburner, M., Chihara, C., Meltzer, P., Richards, G., 1974. Temporal control of puffing activity in polytene chromosomes. Cold Spring Harb. Symp. Quant. Biol. 38, 655–662.

Ashok, M., Turner, C., Wilson, T.G., 1998. Insect juvenile hormone resistance gene homology with the bHLH-PAS family of transcriptional regulators. Proc. Natl. Acad. Sci. U.S.A. 95, 2761–2766.

Audsley, N., Vandersmissen, H.P., Weaver, R., Dani, P., Matthews, J., Down, R., et al., 2013. Characterisation and tissue distribution of the PISCF allatostatin receptor in the red flour beetle, *Tribolium castaneum*. Insect Biochem. Mol. Biol. 43, 65–74. Available from: https://doi.org/10.1016/j.ibmb.2012.09.007.

Baehrecke, E.H., Thummel, C.S., 1995. The *Drosophila* E93 gene from the 93F early puff displays stage- and tissue-specific regulation by 20-hydroxyecdysone. Dev. Biol. 171, 85–97. Available from: https://doi.org/10.1006/dbio.1995.1262.

Bai, H., Palli, S.R., 2010. Functional characterization of bursicon receptor and genome-wide analysis for identification of genes affected by bursicon receptor RNAi. Dev. Biol. 344, 248–258. Available from: https://doi.org/10.1016/j.ydbio.2010.05.003.

Beck, Y., Pecasse, F., Richards, G., 2004. Krüppel-homolog is essential for the coordination of regulatory gene hierarchies in early *Drosophila* development. Dev. Biol. 268, 64–75. Available from: https://doi.org/10.1016/j.ydbio.2003.12.017.

Belles, X., 2019. Krüppel homolog 1 and E93: the doorkeeper and the key to insect metamorphosis. Arch. Insect Biochem. Physiol. e21609. Available from: https://doi.org/10.1002/arch.21609.

Belles, X., Santos, C.G., 2014. The MEKRE93 (methoprene tolerant-Krüppel homolog 1-E93) pathway in the regulation of insect metamorphosis, and the homology of the pupal stage. Insect Biochem. Mol. Biol. 52, 60–68. Available from: https://doi.org/10.1016/j.ibmb.2014.06.009.

Bender, M., Imam, F.B., Talbot, W.S., Ganetzky, B., Hogness, D.S., 1997. *Drosophila* ecdysone receptor mutations reveal functional differences among receptor isoforms. Cell 91, 777–788.

Ben-Shlomo, I., Yu Hsu, S., Rauch, R., Kowalski, H.W., Hsueh, A.J.W., 2003. Signaling receptome: a genomic and evolutionary perspective of plasma membrane receptors involved in signal transduction. Sci. Signal. 187, re9. Available from: https://doi.org/10.1126/stke.2003.187.re9.

Bernardo, T.J., Dubrovsky, E.B., 2012a. The *Drosophila* juvenile hormone receptor candidates methoprene-tolerant (MET) and germ cell-expressed (GCE) utilize a conserved LIXXL motif to bind the FTZ-F1 nuclear receptor. J. Biol. Chem. 287, 7821–7833. Available from: https://doi.org/10.1074/jbc.M111.327254.

Bernardo, T.J., Dubrovsky, E.B., 2012b. Molecular mechanisms of transcription activation by juvenile hormone: a critical role for bHLH-PAS and nuclear receptor proteins. Insects 3, 324–338. Available from: https://doi.org/10.3390/insects3010324.

Bialecki, M., Shilton, A., Fichtenberg, C., Segraves, W.A., Thummel, C.S., 2002. Loss of the ecdysteroid-inducible E75A orphan nuclear receptor uncouples molting from metamorphosis in *Drosophila*. Dev. Cell 3, 209–220.

Birgül, N., Weise, C., Kreienkamp, H.J., Richter, D., 1999. Reverse physiology in *Drosophila*: identification of a novel allatostatin-like neuropeptide and its cognate receptor structurally related to the mammalian somatostatin/galanin/opioid receptor family. EMBO J. 18, 5892–5900. Available from: https://doi.org/10.1093/emboj/18.21.5892.

Bittova, L., Jedlicka, P., Dracinsky, M., Kirubakaran, P., Vondrasek, J., Hanus, R., et al., 2019. Exquisite ligand stereoselectivity of a *Drosophila* juvenile hormone receptor contrasts with its broad agonist repertoire. J. Biol. Chem. 294, 410–423. Available from: https://doi.org/10.1074/jbc.RA118.005992.

Bounhiol, J., 1938. Recherches experimentales sur le determinisme de la metamorphose chez les Lépidoptères. Bull. Biol. Fr. Belg. 24, 1–199.

Broadus, J., McCabe, J.R., Endrizzi, B., Thummel, C.S., Woodard, C.T., 1999. The *Drosophila* beta FTZ-F1 orphan nuclear receptor provides competence for stage-specific responses to the steroid hormone ecdysone. Mol. Cell 3, 143–149.

Burtis, K.C., Thummel, C.S., Jones, C.W., Karim, F.D., Hogness, D.S., 1990. The *Drosophila* 74EF early puff contains E74, a complex ecdysone-inducible gene that encodes two ets-related proteins. Cell 61, 85–99.

Caceres, L., Necakov, A.S., Schwartz, C., Kimber, S., Roberts, I.J., Krause, H.M., 2011. Nitric oxide coordinates metabolism, growth, and development via the nuclear receptor E75. Genes Dev. 25, 1476–1485.

Caldwell, P.E., Walkiewicz, M., Stern, M., 2005. Ras activity in the *Drosophila* prothoracic gland regulates body size and developmental rate via ecdysone release. Curr. Biol. 15, 1785–1795. Available from: https://doi.org/10.1016/j.cub.2005.09.011.

Cazzamali, G., Saxild, N., Grimmelikhuijzen, C., 2002. Molecular cloning and functional expression of a *Drosophila* corazonin receptor. Biochem. Biophys. Res. Commun. 298, 31–36.

Chafino, S., Ureña, E., Casanova, J., Casacuberta, E., Franch-Marro, X., Martín, D., 2019. Upregulation of E93 gene expression acts as the trigger for metamorphosis independently of the threshold size in the beetle *Tribolium castaneum*. Cell Rep. 27, 1039–1049.e2. Available from: https://doi.org/10.1016/j.celrep.2019.03.094.

Chang, J.-C., Yang, R.-B., Adams, M.E., Lu, K.-H., 2009. Receptor guanylyl cyclases in Inka cells targeted by eclosion hormone. Proc. Natl. Acad. Sci. U. S. A. 106, 13371–13376.

Charles, J.-P., Iwema, T., Epa, V.C., Takaki, K., Rynes, J., Jindra, M., 2011. Ligand-binding properties of a juvenile hormone receptor, Methoprene-tolerant. Proc. Natl. Acad. Sci. U.S. A. 108, 21128–21133. Available from: https://doi.org/10.1073/pnas.1116123109.

Christ, P., Hill, S.R., Schachtner, J., Hauser, F., Ignell, R., 2018. Functional characterization of the dual allatostatin – a receptors in mosquitoes. Peptides 99, 44–55. Available from: https://doi.org/10.1016/j.peptides.2017.10.016.

Clever, U., Karlson, P., 1960. Induktion von Puff-Veränderungen in den Speicheldrüsenchromosomen von *Chironomus tentans* durch Ecdyson. Exp. Cell Res. 20, 623–626.

Coast, G.M., Schooley, D.A., 2011. Toward a consensus nomenclature for insect neuropeptides and peptide hormones. Peptides 32, 620–631. Available from: https://doi.org/10.1016/j.peptides.2010.11.006.

Colombani, J., Bianchini, L., Layalle, S., Pondeville, E., Dauphin-Villemant, C., Antoniewski, C., et al., 2005. Antagonistic actions of ecdysone and insulins determine final size in *Drosophila*. Science 310, 667–670. Available from: https://doi.org/10.1126/science.1119432.

Cui, Y., Sui, Y., Xu, J., Zhu, F., Palli, S.R., 2014. Juvenile hormone regulates *Aedes aegypti* Krüppel homolog 1 through a conserved E box motif. Insect Biochem. Mol. Biol. 52, 23–32. Available from: https://doi.org/10.1016/j.ibmb.2014.05.009.

Daimon, T., Uchibori, M., Nakao, H., Sezutsu, H., Shinoda, T., 2015. Knockout silkworms reveal a dispensable role for juvenile hormones in holometabolous life cycle. Proc. Natl. Acad. Sci. U. S. A. 112, E4226–E4235. Available from: https://doi.org/10.1073/pnas.1506645112.

Davey, K.G., 2000. The modes of action of juvenile hormones: some questions we ought to ask. Insect Biochem. Mol. Biol. 30, 663–669.

Davis, M.B., Carney, G.E., Robertson, A.E., Bender, M., 2005. Phenotypic analysis of EcR-A mutants suggests that EcR isoforms have unique functions during *Drosophila* development. Dev. Biol. 282, 385–396. Available from: https://doi.org/10.1016/j.ydbio.2005.03.019.

Davis, M.M., O'Keefe, S.L., Primrose, D.A., Hodgetts, R.B., 2007. A neuropeptide hormone cascade controls the precise onset of post-eclosion cuticular tanning in *Drosophila melanogaster*. Development 134, 4395–4404. Available from: https://doi.org/10.1242/dev.009902.

Diao, F., Mena, W., Shi, J., Park, D., Diao, F., Taghert, P., et al., 2016. The splice isoforms of the *Drosophila* ecdysis triggering hormone receptor have developmentally distinct roles. Genetics 202, 175–189.

DiBello, P.R., Withers, D.A., Bayer, C.A., Fristrom, J.W., Guild, G.M., 1991. The *Drosophila* broad-complex encodes a family of related proteins containing zinc fingers. Genetics 129, 385–397.

Dubrovsky, E.B., Dubrovskaya, V.A., Berger, E.M., 2004. Hormonal regulation and functional role of *Drosophila* E75A orphan nuclear receptor in the juvenile hormone signaling pathway. Dev. Biol. 268, 258–270. Available from: https://doi.org/10.1016/j.ydbio.2004.01.009.

Dubrovsky, E.B., Dubrovskaya, V.A., Bernardo, T., Otte, V., DiFilippo, R., Bryan, H., 2011. The Drosophila FTZ-F1 nuclear receptor mediates juvenile hormone activation of E75A gene expression through an intracellular pathway. J. Biol. Chem. 286, 33689–33700. Available from: https://doi.org/10.1074/jbc.M111.273458.

Eben Massari, M., Murre, C., 2000. Helix-loop-helix proteins: regulators of transcription in eucaryotic organisms. Mol. Cell. Biol. 20, 429–440.

Erezyilmaz, D.F., Riddiford, L.M., Truman, J.W., 2006. The pupal specifier broad directs progressive morphogenesis in a direct-developing insect. Proc. Natl. Acad. Sci. U.S.A. 103, 6925–6930.

Evans, R.M., Mangelsdorf, D.J., 2014. Nuclear receptors, RXR, and the Big Bang. Cell 157, 255–266. Available from: https://doi.org/10.1016/j.cell.2014.03.012.

Ewer, J., Gammie, S.C., Truman, J.W., 1997. Control of insect ecdysis by a positive-feedback endocrine system: roles of eclosion hormone and ecdysis triggering hormone. J. Exp. Biol. 200, 869–881.

Fletcher, J.C., Burtis, K.C., Hogness, D.S., Thummel, C.S., 1995. The *Drosophila* E74 gene is required for metamorphosis and plays a role in the polytene chromosome puffing response to ecdysone. Development 121, 1455–1465.

Fukuda, S., 1944. The hormonal mechanism of larval molting and metamorphosis in the silkworm. J. Fac. Sci. Tokyo Univ. Sect. IV 6, 477–532.

Gibbens, Y.Y., Warren, J.T., Gilbert, L.I., O'Connor, M.B., 2011. Neuroendocrine regulation of *Drosophila* metamorphosis requires TGFbeta/Activin signaling. Development 138, 2693–2703. Available from: https://doi.org/10.1242/dev.063412.

Gilman, A.G., 1987. G proteins: transducers of receptor-generated signals. Annu. Rev. Biochem. 56, 615–649. Available from: https://doi.org/10.1146/annurev.bi.56.070187.003151.

Gujar, H., Palli, S.R., 2016. Krüppel homolog 1 and E93 mediate Juvenile hormone regulation of metamorphosis in the common bed bug, *Cimex lectularius*. Sci. Rep. 6, 26092. Available from: https://doi.org/10.1038/srep26092.

Hall, B.L., Thummel, C.S., 1998. The RXR homolog ultraspiracle is an essential component of the *Drosophila* ecdysone receptor. Development 125, 4709–4717.

Hamoudi, Z., Lange, A.B., Orchard, I., 2016. Identification and characterization of the corazonin receptor and possible physiological roles of the corazonin-signaling pathway in *Rhodnius prolixus*. Front. Neurosci. 10, 357.

He, Q., Wen, D., Jia, Q., Cui, C., Wang, J., Palli, S.R., et al., 2014. Heat shock protein 83 (Hsp83) facilitates Methoprene-tolerant (Met) nuclear import to modulate juvenile hormone signaling. J. Biol. Chem. 289, 27874–27885. Available from: https://doi.org/10.1074/jbc.M114.582825.

He, X., Zang, J., Yang, H., Huang, H., Shi, Y., Zhu, C., et al., 2015. *Bombyx mori* prothoracicostatic peptide receptor is allosterically activated via a Gα(i/o)-protein-biased signalling cascade by *Drosophila* sex peptide. Biochem. J. 466, 391–400.

Hill, R.J., Billas, I.M.L., Bonneton, F., Graham, L.D., Lawrence, M.C., 2013. Ecdysone receptors: from the Ashburner model to structural biology. Annu. Rev. Entomol. 58, 251–271. Available from: https://doi.org/10.1146/annurev-ento-120811-153610.

Hiruma, K., Riddiford, L.M., 2010. Developmental expression of mRNAs for epidermal and fat body proteins and hormonally regulated transcription factors in the tobacco hornworm, *Manduca sexta*. J. Insect Physiol. 56, 1390–1395. Available from: https://doi.org/10.1016/j.jinsphys.2010.03.029.

Honegger, H.-W., Dewey, E.M., Ewer, J., 2008. Bursicon, the tanning hormone of insects: recent advances following the discovery of its molecular identity. J. Comp. Physiol. A. Neuroethol. Sens. Neural. Behav. Physiol. 194, 989–1005. Available from: https://doi.org/10.1007/s00359-008-0386-3.

Hua, Y.J., Tanaka, Y., Nakamura, K., Sakakibara, M., Nagata, S., Kataoka, H., 1999. Identification of a prothoracicostatic peptide in the larval brain of the silkworm, *Bombyx mori*. J. Biol. Chem. 274, 31169–31173.

Huang, J., Tian, L., Peng, C., Abdou, M., Wen, D., Wang, Y., et al., 2011. DPP-mediated TGFbeta signaling regulates juvenile hormone biosynthesis by activating the expression of juvenile hormone acid methyltransferase. Development 138, 2283–2291. Available from: https://doi.org/10.1242/dev.057687.

Huang, J.-H., Lozano, J., Belles, X., 2013. Broad-complex functions in postembryonic development of the cockroach *Blattella germanica* shed new light on the evolution of insect metamorphosis. Biochim. Biophys. Acta Gen. Subj. 1830, 2178–2187. Available from: https://doi.org/10.1016/j.bbagen.2012.09.025.

Huang, J., Marchal, E., Hult, E.F., Zels, S., Vanden Broeck, J., Tobe, S.S., 2014. Mode of action of allatostatins in the regulation of juvenile hormone biosynthesis in the cockroach, *Diploptera punctata*. Insect Biochem. Mol. Biol. 54, 61–68. Available from: https://doi.org/10.1016/j.ibmb.2014.09.001.

Iga, M., Nakaoka, T., Suzuki, Y., Kataoka, H., 2014. Pigment dispersing factor regulates ecdysone biosynthesis via *Bombyx* neuropeptide G protein coupled receptor-B2 in the prothoracic glands of *Bombyx mori*. PLoS One 9, e103239. Available from: https://doi.org/10.1371/journal.pone.0103239.

Inui, T., Daimon, T., 2017. Implantation assays using the integument of early stage *Bombyx* larvae: insights into the mechanisms underlying the acquisition of competence for metamorphosis. J. Insect Physiol. 100, 35–42. Available from: https://doi.org/10.1016/j.jinsphys.2017.05.002.

Ishimaru, Y., Tomonari, S., Matsuoka, Y., Watanabe, T., Miyawaki, K., Bando, T., et al., 2016. TGF-β signaling in insects regulates metamorphosis via juvenile hormone biosynthesis. Proc. Natl. Acad. Sci. U.S.A. 113, 5634–5639. Available from: https://doi.org/10.1073/pnas.1600612113.

Ishimaru, Y., Tomonari, S., Watanabe, T., Noji, S., Mito, T., 2019. Regulatory mechanisms underlying the specification of the pupal-homologous stage in a hemimetabolous insect. Philos. Trans. R. Soc. Lond. B. Biol. Sci. 374, 20190225. Available from: https://doi.org/10.1098/rstb.2019.0225.

Iwema, T., Billas, I.M., Beck, Y., Bonneton, F., Nierengarten, H., Chaumot, A., et al., 2007. Structural and functional characterization of a novel type of ligand-independent RXR-USP receptor. EMBO J. 26, 3770–3782. Available from: https://doi.org/10.1038/sj.emboj.7601810.

Jindra, M., Riddiford, L.M., 1996. Expression of ecdysteroid-regulated transcripts in the silk gland of the wax moth, *Galleria mellonella*. Dev. Genes Evol. 206, 305–314. Available from: https://doi.org/10.1007/s004270050057.

Jindra, M., Belles, X., Shinoda, T., 2015a. Molecular basis of juvenile hormone signaling. Curr. Opin. Insect Sci. 11, 39–46. Available from: https://doi.org/10.1016/j.cois.2015.08.004.

Jindra, M., Uhlirova, M., Charles, J.-P., Smykal, V., Hill, R.J., 2015b. Genetic evidence for function of the bHLH-PAS protein Gce/Met as a juvenile hormone receptor. PLOS Genet. 11, e1005394. Available from: https://doi.org/10.1371/journal.pgen.1005394.

Jones, G., Sharp, P.A., 1997. Ultraspiracle: an invertebrate nuclear receptor for juvenile hormones. Proc. Natl. Acad. Sci. U.S.A. 94, 13499–13503.

Jones, D., Jones, G., Teal, P.E.A., 2013a. Sesquiterpene action, and morphogenetic signaling through the ortholog of retinoid X receptor, in higher Diptera. Gen. Comp. Endocrinol. 194, 326–335. Available from: https://doi.org/10.1016/j.ygcen.2013.09.021.

Jones, G., Teal, P., Henrich, V.C., Krzywonos, A., Sapa, A., Wozniak, M., et al., 2013b. Ligand binding pocket function of *Drosophila* USP is necessary for metamorphosis. Gen. Comp. Endocrinol. 182, 73–82. Available from: https://doi.org/10.1016/j.ygcen.2012.11.009.

Kamsoi, O., Belles, X., 2019. Myoglianin triggers the premetamorphosis stage in hemimetabolan insects. FASEB J. 33, 3659–3669. Available from: https://doi.org/10.1096/fj.201801511R.

Karim, F.D., Guild, G.M., Thummel, C.S., 1993. The *Drosophila* Broad-Complex plays a key role in controlling ecdysone-regulated gene expression at the onset of metamorphosis. Development 118, 977–988.

Kayukawa, T., Minakuchi, C., Namiki, T., Togawa, T., Yoshiyama, M., Kamimura, M., et al., 2012. Transcriptional regulation of juvenile hormone-mediated induction of Krüppel homolog 1, a repressor of insect metamorphosis. Proc. Natl. Acad. Sci. U.S.A. 109, 11729–11734. Available from: https://doi.org/10.1073/pnas.1204951109.

Kayukawa, T., Murata, M., Kobayashi, I., Muramatsu, D., Okada, C., Uchino, K., et al., 2014. Hormonal regulation and developmental role of Krüppel homolog 1, a repressor of metamorphosis, in the silkworm *Bombyx mori*. Dev. Biol. 388, 48–56. Available from: https://doi.org/10.1016/j.ydbio.2014.01.022.

Kayukawa, T., Nagamine, K., Ito, Y., Nishita, Y., Ishikawa, Y., Shinoda, T., 2016. Krüppel homolog 1 inhibits insect metamorphosis via direct transcriptional repression of Broad-Complex, a pupal specifier gene. J. Biol. Chem. 291, 1751–1762. Available from: https://doi.org/10.1074/jbc.M115.686121.

Kayukawa, T., Jouraku, A., Ito, Y., Shinoda, T., 2017. Molecular mechanism underlying juvenile hormone-mediated repression of precocious larval-adult metamorphosis. Proc. Natl. Acad. Sci. U.S.A. 114, 1057–1062. Available from: https://doi.org/10.1073/pnas.1615423114.

Kim, Y.-J., Spalovska-Valachova, I., Cho, K.-H., Zitnanova, I., Park, Y., Adams, M.E., et al., 2004. Corazonin receptor signaling in ecdysis initiation. Proc. Natl. Acad. Sci. U.S.A. 101, 6704–6709. Available from: https://doi.org/10.1073/pnas.0305291101.

Kim, Y.-J., Bartalska, K., Audsley, N., Yamanaka, N., Yapici, N., Lee, J.-Y., et al., 2010. MIPs are ancestral ligands for the sex peptide receptor. Proc. Natl. Acad. Sci. U.S.A. 107, 6520–6525. Available from: https://doi.org/10.1073/pnas.0914764107.

King-Jones, K., Thummel, C.S., 2005. Nuclear receptors – a perspective from *Drosophila*. Nat. Rev. Genet. 6, 311–323. Available from: https://doi.org/10.1038/nrg1581.

King-Jones, K., Charles, J.P., Lam, G., Thummel, C.S., 2005. The ecdysone-induced DHR4 orphan nuclear receptor coordinates growth and maturation in *Drosophila*. Cell 121, 773–784.

Kiss, I., Beaton, A.H., Tardiff, J., Fristrom, D., Fristrom, J.W., 1988. Interactions and developmental effects of mutations in the Broad-Complex of *Drosophila melanogaster*. Genetics 118, 247–259.

Koelle, M.R., Talbot, W.S., Segraves, W.A., Bender, M.T., Cherbas, P., Hogness, D.S., 1991. The *Drosophila* EcR gene encodes an ecdysone receptor, a new member of the steroid receptor superfamily. Cell 67, 59–77.

Komura-Kawa, T., Hirota, K., Shimada-Niwa, Y., Yamauchi, R., Shimell, M.J., Shinoda, T., et al., 2015. The Drosophila zinc finger transcription factor Ouija board controls ecdysteroid biosynthesis through specific regulation of spookier. PLoS Genet. 11, e1005712. Available from: https://doi.org/10.1371/journal.pgen.1005712.

Konopová, B., Jindra, M., 2007. Juvenile hormone resistance gene Methoprene-tolerant controls entry into metamorphosis in the beetle *Tribolium castaneum*. Proc. Natl. Acad. Sci. U.S.A. 104, 10488–10493. Available from: https://doi.org/10.1073/pnas.0703719104.

Konopová, B., Jindra, M., 2008. Broad-Complex acts downstream of Met in juvenile hormone signaling to coordinate primitive holometabolan metamorphosis. Development 135, 559–568. Available from: https://doi.org/10.1242/dev.016097.

Konopová, B., Smykal, V., Jindra, M., 2011. Common and distinct roles of juvenile hormone signaling genes in metamorphosis of holometabolous and hemimetabolous insects. PLoS One 6, e28728. Available from: https://doi.org/10.1371/journal.pone.0028728.

Kozlova, T., Thummel, C.S., 2003. Essential roles for ecdysone signaling during *Drosophila* mid-embryonic development. Science 301, 1911–1914.

Kreienkamp, H.-J., Larusson, H.J., Witte, I., Roeder, T., Birgul, N., Honck, H.-H., et al., 2002. Functional annotation of two orphan G-protein-coupled receptors, Drostar1 and -2, from *Drosophila melanogaster* and their ligands by reverse pharmacology. J. Biol. Chem. 277, 39937–39943. Available from: https://doi.org/10.1074/jbc.M206931200.

Kremer, L.P.M., Korb, J., Bornberg-Bauer, E., 2018. Reconstructed evolution of insulin receptors in insects reveals duplications in early insects and cockroaches. J. Exp. Zool. B: Mol. Dev. Evol. 330, 305–311.

Lam, G.T., Jiang, C., Thummel, C.S., 1997. Coordination of larval and prepupal gene expression by the DHR3 orphan receptor during *Drosophila* metamorphosis. Development 124, 1757–1769.

Larsen, M.J., Burton, K.J., Zantello, M.R., Smith, V.G., Lowery, D.L., Kubiak, T.M., 2001. Type A allatostatins from *Drosophila melanogaster* and *Diploptera punctata* activate two *Drosophila* allatostatin receptors, DAR-1 and DAR-2, expressed in CHO cells. Biochem. Biophys. Res. Commun. 286, 895–901. Available from: https://doi.org/10.1006/bbrc.2001.5476.

Lavorgna, G., Ueda, H., Clos, J., Wu, C., 1991. FTZ-F1, a steroid hormone receptor-like protein implicated in the activation of fushi tarazu. Science 252, 848–851.

Lee, C.Y., Wendel, D.P., Reid, P., Lam, G., Thummel, C.S., Baehrecke, E.H., 2000. E93 directs steroid-triggered programmed cell death in *Drosophila*. Mol. Cell 6, 433–443.

Lee, B., Beeman, R.W., Park, Y., 2011. Functions of duplicated genes encoding CCAP receptors in the red flour beetle, *Tribolium castaneum*. J. Insect Physiol. 57, 1190–1197.

Lee, D., Orchard, I., Lange, A.B., 2013. Evidence for a conserved CCAP-signaling pathway controlling ecdysis in a hemimetabolous insect, *Rhodnius prolixus*. Front. Neurosci. 7, 207.

Lee, G., Sehgal, R., Wang, Z., Park, J.H., 2019. Ultraspiracle-independent anti-apoptotic function of ecdysone receptors is required for the survival of larval peptidergic neurons via suppression of grim expression in *Drosophila melanogaster*. Apoptosis 24, 256–268. Available from: https://doi.org/10.1007/s10495-019-01514-2.

Li, M., Mead, E.A., Zhu, J., 2011. Heterodimer of two bHLH-PAS proteins mediates juvenile hormone-induced gene expression. Proc. Natl. Acad. Sci. U. S. A. 108, 638–643. Available from: https://doi.org/10.1073/pnas.1013914108.

Li, M., Liu, P., Wiley, J.D., Ojani, R., Bevan, D.R., Li, J., et al., 2014. A steroid receptor coactivator acts as the DNA-binding partner of the methoprene-tolerant protein in regulating juvenile hormone response genes. Mol. Cell. Endocrinol. 394, 47–58. Available from: https://doi.org/10.1016/j.mce.2014.06.021.

Li, K.L., Yuan, S.Y., Nanda, S., Wang, W.X., Lai, F.X., Fu, Q., et al., 2018. The roles of E93 and Kr-h1 in metamorphosis of *Nilaparvata lugens*. Front. Physiol. 9, 1677. Available from: https://doi.org/10.3389/fphys.2018.01677.

Liu, W., Zhang, F.-X., Cai, M.-J., Zhao, W.-L., Li, X.-R., Wang, J.-X., et al., 2013. The hormone-dependent function of Hsp90 in the crosstalk between 20-hydroxyecdysone and juvenile hormone signaling pathways in insects is determined by differential phosphorylation and protein interactions. Biochim. Biophys. Acta Gen. Subj. 1830, 5184–5192. Available from: https://doi.org/10.1016/j.bbagen.2013.06.037.

Liu, P., Peng, H.-J., Zhu, J., 2015. Juvenile hormone-activated phospholipase C pathway enhances transcriptional activation by the methoprene-tolerant protein. Proc. Natl. Acad. Sci. U.S.A. 112, E1871–E1879. Available from: https://doi.org/10.1073/pnas.1423204112.

Liu, S., Li, K., Gao, Y., Liu, X., Chen, W., Ge, W., et al., 2018. Antagonistic actions of juvenile hormone and 20-hydroxyecdysone within the ring gland determine developmental transitions in *Drosophila*. Proc. Natl. Acad. Sci. U.S.A. 115, 139–144. Available from: https://doi.org/10.1073/pnas.1716897115.

Lozano, J., Belles, X., 2011. Conserved repressive function of Krüppel homolog 1 on insect metamorphosis in hemimetabolous and holometabolous species. Sci. Rep. 1, 163. Available from: https://doi.org/10.1038/srep00163.

Lozano, J., Belles, X., 2014. Role of methoprene-tolerant (Met) in adult morphogenesis and in adult ecdysis of *Blattella germanica*. PLoS One 9, e103614. Available from: https://doi.org/10.1371/journal.pone.0103614.

Lozano, J., Kayukawa, T., Shinoda, T., Belles, X., 2014. A role for Taiman in insect metamorphosis. PLoS Genet. 10, e1004769. Available from: https://doi.org/10.1371/journal.pgen.1004769.

Lungchukiet, P., Donly, B.C., Zhang, J., Tobe, S.S., Bendena, W.G., 2008. Molecular cloning and characterization of an allatostatin-like receptor in the cockroach *Diploptera punctata*. Peptides 29, 276–285. Available from: https://doi.org/10.1016/j.peptides.2007.10.029.

Luo, C.-W., Dewey, E.M., Sudo, S., Ewer, J., Hsu, S.Y., Honegger, H.-W., et al., 2005. Bursicon, the insect cuticle-hardening hormone, is a heterodimeric cystine knot protein that activates G protein-coupled receptor LGR2. Proc. Natl. Acad. Sci. U.S.A. 102, 2820–2825. Available from: https://doi.org/10.1073/pnas.0409916102.

Ma, Y., McKay, D.J., Buttitta, L., 2019. Changes in chromatin accessibility ensure robust cell cycle exit in terminally differentiated cells. PLoS Biol 17, e3000378. Available from: https://doi.org/10.1371/journal.pbio.3000378.

Mané-Padrós, D., Cruz, J., Vilaplana, L., Nieva, C., Ureña, E., Belles, X., et al., 2010. The hormonal pathway controlling cell death during metamorphosis in a hemimetabolous insect. Dev. Biol. 346, 150–160. Available from: https://doi.org/10.1016/j.ydbio.2010.07.012.

Mané-Padrós, D., Borràs-Castells, F., Belles, X., Martín, D., 2012. Nuclear receptor HR4 plays an essential role in the ecdysteroid-triggered gene cascade in the development of the hemimetabolous insect *Blattella germanica*. Mol. Cell. Endocrinol. 348, 322–330. Available from: https://doi.org/10.1016/j.mce.2011.09.025.

Mayoral, J.G., Nouzova, M., Brockhoff, A., Goodwin, M., Hernandez-Martinez, S., Richter, D., et al., 2010. Allatostatin-C receptors in mosquitoes. Peptides 31, 442–450. Available from: https://doi.org/10.1016/j.peptides.2009.04.013.

Mena, W., Diegelmann, S., Wegener, C., Ewer, J., 2016. Stereotyped responses of *Drosophila* peptidergic neuronal ensemble depend on downstream neuromodulators. Elife 5, e19686. Available from: https://doi.org/10.7554/eLife.19686.

Mendive, F.M., Van Loy, T., Claeysen, S., Poels, J., Williamson, M., Hauser, F., et al., 2005. *Drosophila* molting neurohormone bursicon is a heterodimer and the natural agonist of the orphan receptor DLGR2. FEBS Lett. 579, 2171–2176. Available from: https://doi.org/10.1016/j.febslet.2005.03.006.

Minakuchi, C., Zhou, X., Riddiford, L.M., 2008. Krüppel homolog 1 (Kr-h1) mediates juvenile hormone action during metamorphosis of *Drosophila melanogaster*. Mech. Dev. 125, 91–105. Available from: https://doi.org/10.1016/j.mod.2007.10.002.

Minakuchi, C., Namiki, T., Shinoda, T., 2009. Krüppel homolog 1, an early juvenile hormone-response gene downstream of Methoprene-tolerant, mediates its anti-metamorphic action in the red flour beetle *Tribolium castaneum*. Dev. Biol. 325, 341–350. Available from: https://doi.org/10.1016/j.ydbio.2008.10.016.

Mirth, C., Truman, J.W., Riddiford, L.M., 2005. The role of the prothoracic gland in determining critical weight for metamorphosis in *Drosophila melanogaster*. Curr. Biol. 15, 1796–1807. Available from: https://doi.org/10.1016/j.cub.2005.09.017.

Miura, K., Oda, M., Makita, S., Chinzei, Y., 2005. Characterization of the *Drosophila* Methoprene-tolerant gene product. FEBS J. 272, 1169–1178. Available from: https://doi.org/10.1111/j.1742-4658.2005.04552.x.

Mou, X., Duncan, D.M., Baehrecke, E.H., Duncan, I., 2012. Control of target gene specificity during metamorphosis by the steroid response gene E93. Proc. Natl. Acad. Sci. U.S.A. 109, 2949–2954. Available from: https://doi.org/10.1073/pnas.1117559109.

Niwa, Y.S., Niwa, R., 2016. Transcriptional regulation of insect steroid hormone biosynthesis and its role in controlling timing of molting and metamorphosis. Dev. Growth Differ. 58, 94–105. Available from: https://doi.org/10.1111/dgd.12248.

Ohhara, Y., Shimada-Niwa, Y., Niwa, R., Kayashima, Y., Hayashi, Y., Akagi, K., et al., 2015. Autocrine regulation of ecdysone synthesis by β3-octopamine receptor in the prothoracic gland is essential for *Drosophila* metamorphosis. Proc. Natl. Acad. Sci. U.S.A. 112, 1452–1457. Available from: https://doi.org/10.1073/pnas.1414966112.

Okamoto, N., Viswanatha, R., Bittar, R., Li, Z., Haga-Yamanaka, S., Perrimon, N., et al., 2018. A membrane transporter is required for steroid hormone uptake in *Drosophila*. Dev. Cell 47, 294–305.e7. Available from: https://doi.org/10.1016/j.devcel.2018.09.012.

Oro, A.E., McKeown, M., Evans, R.M., 1992. The *Drosophila* retinoid X receptor homolog ultraspiracle functions in both female reproduction and eye morphogenesis. Development 115, 449–462.

Ou, Q., King-Jones, K., 2013. What goes up must come down: transcription factors have their say in making ecdysone pulses. Curr. Top. Dev. Biol. 103, 35–71. Available from: https://doi.org/10.1016/B978-0-12-385979-2.00002-2.

Pahl, M.C., Doyle, S.E., Siegrist, S.E., 2019. E93 integrates neuroblast intrinsic state with developmental time to terminate MB neurogenesis via autophagy. Curr. Biol 29, 750–762. Available from: https://doi.org/10.1016/j.cub.2019.01.039.

Park, Y., Kim, Y.J., Adams, M.E., 2002. Identification of G protein-coupled receptors for *Drosophila* PRXamide peptides, CCAP, corazonin, and AKH supports a theory of ligand-receptor coevolution. Proc. Natl. Acad. Sci. U.S.A. 99, 11423–11428.

Park, Y., Kim, Y., Dupriez, V., Adams, M.E., 2003. Two subtypes of ecdysis-triggering hormone receptor in *Drosophila melanogaster*. J. Biol. Chem. 278, 17710–17715.

Parthasarathy, R., Tan, A., Bai, H., Palli, S.R., 2008. Transcription factor broad suppresses precocious development of adult structures during larval-pupal metamorphosis in the red flour beetle, *Tribolium castaneum*. Mech. Dev. 125, 299–313.

Pecasse, F., Beck, Y., Ruiz, C., Richards, G., 2000. Krüppel-homolog, a stage-specific modulator of the prepupal ecdysone response, is essential for *Drosophila* metamorphosis. Dev. Biol. 221, 53–67. Available from: https://doi.org/10.1006/dbio.2000.9687.

Piepho, H., 1938a. Wachstum und totale metamorphose an hautimplantaten bei der wachsmotte *Galleria mellonella* L. Biol. Zbl. 58, 356–366.

Piepho, H., 1938b. Über die auslösung der raupenhautung, verpuppung und imaginalentwicklung an hautimplantaten von schmetterlingen. Biol. Zbl. 58, 481–495.

Piulachs, M.-D., Pagone, V., Belles, X., 2010. Key roles of the Broad-Complex gene in insect embryogenesis. Insect Biochem. Mol. Biol. 40, 468–475. Available from: https://doi.org/10.1016/j.ibmb.2010.04.006.

Poels, J., Van Loy, T., Vandersmissen, H.P., Van Hiel, B., Van Soest, S., Nachman, R.J., et al., 2010. Myoinhibiting peptides are the ancestral ligands of the promiscuous *Drosophila* sex peptide receptor. Cell. Mol. Life Sci. 67, 3511–3522. Available from: https://doi.org/10.1007/s00018-010-0393-8.

Pszczolkowski, M.A., Peterson, A., Srinivasan, A., Ramaswamy, S.B., 2005. Pharmacological analysis of ovarial patency in *Heliothis virescens*. J. Insect Physiol. 51, 445–453. Available from: https://doi.org/10.1016/j.jinsphys.2005.01.008.

Pszczolkowski, M.A., Olson, E., Rhine, C., Ramaswamy, S.B., 2008. Role for calcium in the development of ovarial patency in *Heliothis virescens*. J. Insect Physiol. 54, 358–366. Available from: https://doi.org/10.1016/j.jinsphys.2007.10.005.

Ren, B., Peat, T.S., Streltsov, V.A., Pollard, M., Fernley, R., Grusovin, J., et al., 2014. Unprecedented conformational flexibility revealed in the ligand-binding domains of the *Bovicola ovis* ecdysone receptor (EcR) and ultraspiracle (USP) subunits. Acta Crystallogr. D: Biol. Crystallogr. 70, 1954–1964. Available from: https://doi.org/10.1107/S1399004714009626.

Rewitz, K.F., Yamanaka, N., Gilbert, L.I., O'Connor, M.B., 2009. The insect neuropeptide PTTH activates receptor tyrosine kinase torso to initiate metamorphosis. Science 326, 1403–1405. Available from: https://doi.org/10.1126/science.1176450.

Riddiford, L.M., 2008. Juvenile hormone action: a 2007 perspective. J. Insect Physiol. 54, 895–901. Available from: https://doi.org/10.1016/j.jinsphys.2008.01.014.

Riddiford, L.M., Hiruma, K., Zhou, X., Nelson, C.A., 2003. Insights into the molecular basis of the hormonal control of molting and metamorphosis from *Manduca sexta* and *Drosophila melanogaster*. Insect Biochem. Mol. Biol. 33, 1327–1338.

Roller, L., Zitnanova, I., Dai, L., Simo, L., Park, Y., Satake, H., et al., 2010. Ecdysis triggering hormone signaling in arthropods. Peptides 31, 429–441.

Rubio, M., Belles, X., 2013. Subtle roles of microRNAs let-7, miR-100 and miR-125 on wing morphogenesis in hemimetabolan metamorphosis. J. Insect Physiol. 59, 1089–1094. Available from: https://doi.org/10.1016/j.jinsphys.2013.09.003.

Santos, C.G., Fernandez-Nicolas, A., Belles, X., 2016. Smads and insect hemimetabolan metamorphosis. Dev. Biol. 417, 104–113. Available from: https://doi.org/10.1016/j.ydbio.2016.07.006.

Schubiger, M., Truman, J.W., 2000. The RXR ortholog USP suppresses early metamorphic processes in *Drosophila* in the absence of ecdysteroids. Development 127, 1151–1159.

Schubiger, M., Wade, A.A., Carney, G.E., Truman, J.W., Bender, M., 1998. *Drosophila* EcR-B ecdysone receptor isoforms are required for larval molting and for neuron remodeling during metamorphosis. Development 125, 2053–2062.

Segraves, W.A., Hogness, D.S., 1990. The E75 ecdysone-inducible gene responsible for the 75B early puff in *Drosophila* encodes two new members of the steroid receptor superfamily. Genes Dev. 4, 204–219.

Seidelmann, K., Helbing, C., Göbeler, N., Weinert, H., 2016. Sequential oogenesis is controlled by an oviduct factor in the locusts *Locusta migratoria* and *Schistocerca gregaria*: overcoming the doctrine that patency in follicle cells is induced by juvenile hormone. J. Insect Physiol. 90, 1–7. Available from: https://doi.org/10.1016/j.jinsphys.2016.03.008.

Sevala, V.L., Davey, K.G., 1989. Action of juvenile hormone on the follicle cells of *Rhodnius prolixus*: evidence for a novel regulatory mechanism involving protein kinase C. Experientia 45, 355–356. Available from: https://doi.org/10.1007/BF01957476.

Sevala, V.L., Davey, K.G., 1993. Juvenile hormone dependent phosphorylation of a 100 kDa polypeptide is mediated by protein kinase C in the follicle cells of *Rhodnius prolixus*. Invertebr. Reprod. Dev. 23, 189–193. Available from: https://doi.org/10.1080/07924259.1993.9672314.

Shi, Y., Jiang, H.-B., Gui, S.-H., Liu, X.-Q., Pei, Y.-X., Xu, L., et al., 2017. Ecdysis triggering hormone signaling (ETH/ETHR-A) is required for the larva-larva ecdysis in *Bactrocera dorsalis* (Diptera: Tephritidae). Front. Physiol. 8, 587. Available from: https://doi.org/10.3389/fphys.2017.00587.

Shin, S.W., Zou, Z., Saha, T.T., Raikhel, A.S., 2012. bHLH-PAS heterodimer of methoprene-tolerant and Cycle mediates circadian expression of juvenile hormone-induced mosquito genes. Proc. Natl. Acad. Sci. U.S.A. 109, 16576–16581. Available from: https://doi.org/10.1073/pnas.1214209109.

Shlyueva, D., Stelzer, C., Gerlach, D., Yáñez-Cuna, J.O., Rath, M., Boryń, M., et al., 2014. Hormone-responsive enhancer-activity maps reveal predictive motifs, indirect repression, and targeting of closed chromatin. Mol. Cell 54, 180–192. Available from: https://doi.org/10.1016/j.molcel.2014.02.026.

Siaussat, D., Bozzolan, F., Queguiner, I., Porcheron, P., Debernard, S., 2004. Effects of juvenile hormone on 20-hydroxyecdysone-inducible EcR, HR3, E75 gene expression in imaginal wing cells of *Plodia interpunctella* lepidoptera. Eur. J. Biochem. 271, 3017–3027. Available from: https://doi.org/10.1111/j.1432-1033.2004.04233.x.

Siegrist, S.E., Haque, N.S., Chen, C.-H., Hay, B.A., Hariharan, I.K., 2010. Inactivation of both Foxo and reaper promotes long-term adult neurogenesis in *Drosophila*. Curr. Biol. 20, 643–648. Available from: https://doi.org/10.1016/j.cub.2010.01.060.

Smykal, V., Daimon, T., Kayukawa, T., Takaki, K., Shinoda, T., Jindra, M., 2014. Importance of juvenile hormone signaling arises with competence of insect larvae to metamorphose. Dev. Biol. 390, 221–230. Available from: https://doi.org/10.1016/j.ydbio.2014.03.006.

Sopko, R., Perrimon, N., 2013. Receptor tyrosine kinases in *Drosophila* development. Cold Spring Harb. Perspect. Biol. 5, pii: a009050. Available from: https://doi.org/10.1101/cshperspect.a009050a009050.

Sullivan, A.A., Thummel, C.S., 2003. Temporal profiles of nuclear receptor gene expression reveal coordinate transcriptional responses during *Drosophila* development. Mol. Endocrinol. 17, 2125–2137. Available from: https://doi.org/10.1210/me.2002-0430.

Sutherland, J.D., Kozlova, T., Tzertzinis, G., Kafatos, F.C., 1995. *Drosophila* hormone receptor 38: a second partner for *Drosophila* USP suggests an unexpected role for nuclear receptors of the nerve growth factor-induced protein B type. Proc. Natl. Acad. Sci. U.S.A. 92, 7966–7970. Available from: https://doi.org/10.1073/pnas.92.17.7966.

Swevers, L., Iatrou, K., 2003. The ecdysone regulatory cascade and ovarian development in lepidopteran insects: insights from the silkmoth paradigm. Insect Biochem. Mol. Biol. 33, 1285–1297.

Syed, M.H., Mark, B., Doe, C.Q., 2017. Steroid hormone induction of temporal gene expression in Drosophila brain neuroblasts generates neuronal and glial diversity. Elife 6, pii: e26287. Available from: https://doi.org/10.7554/eLife.26287.

Talbot, W.S., Swyryd, E.A., Hogness, D.S., 1993. *Drosophila* tissues with different metamorphic responses to ecdysone express different ecdysone receptor isoforms. Cell 73, 1323–1337.

Tanaka, Y., 2011. Recent topics on the regulatory mechanism of ecdysteroidogenesis by the prothoracic glands in insects. Front. Endocrinol. 2, 107. Available from: https://doi.org/10.3389/fendo.2011.00107.

Thomas, H.E., Stunnenberg, H.G., Stewart, A.F., 1993. Heterodimerization of the *Drosophila* ecdysone receptor with retinoid X receptor and ultraspiracle. Nature 362, 471–475.

Tsukamoto, Y., Nagata, S., 2016. Newly identified allatostatin Bs and their receptor in the two-spotted cricket, *Gryllus bimaculatus*. Peptides 80, 25–31. Available from: https://doi.org/10.1016/j.peptides.2016.03.015.

Ueda, H., Sonoda, S., Brown, J.L., Scott, M.P., Wu, C., 1990. A sequence-specific DNA-binding protein that activates fushi tarazu segmentation gene expression. Genes Dev. 4, 624–635.

Uhlirova, M., Foy, B.D., Beaty, B.J., Olson, K.E., Riddiford, L.M., Jindra, M., 2003. Use of Sindbis virus-mediated RNA interference to demonstrate a conserved role of Broad-Complex in insect metamorphosis. Proc. Natl. Acad. Sci. U.S.A. 100, 15607–15612.

Upadhyay, A., Moss-Taylor, L., Kim, M.-J., Ghosh, A.C., O'Connor, M.B., 2017. TGF-β family signaling in *Drosophila*. Cold Spring Harb. Perspect. Biol. 9, pii: a022152. Available from: https://doi.org/10.1101/cshperspect.a022152.

Ureña, E., Manjón, C., Franch-Marro, X., Martín, D., 2014. Transcription factor E93 specifies adult metamorphosis in hemimetabolous and holometabolous insects. Proc. Natl. Acad. Sci. U.S.A. 111, 7024–7029. Available from: https://doi.org/10.1073/pnas.1401478111.

Ureña, E., Chafino, S., Manjón, C., Franch-Marro, X., Martín, D., 2016. The occurrence of the holometabolous pupal stage requires the interaction between E93, Krüppel-homolog 1 and Broad-Complex. PLoS Genet. 12, e1006020. Available from: https://doi.org/10.1371/journal.pgen.1006020.

Uryu, O., Ou, Q., Komura-Kawa, T., Kamiyama, T., Iga, M., Syrzycka, M., et al., 2018. Cooperative control of ecdysone biosynthesis in *Drosophila* by transcription factors Séance, Ouija board, and Molting defective. Genetics 208, 605–622. Available from: https://doi.org/10.1534/genetics.117.300268.

Uyehara, C.M., McKay, D.J., 2019. Direct and widespread role for the nuclear receptor EcR in mediating the response to ecdysone in *Drosophila*. Proc. Natl. Acad. Sci. U. S. A 116, 9893–9902. Available from: https://doi.org/10.1073/pnas.1900343116.

Uyehara, C.M., Nystrom, S.L., Niederhuber, M.J., Leatham-Jensen, M., Ma, Y., Buttitta, L.A., et al., 2017. Hormone-dependent control of developmental timing through regulation of chromatin accessibility. Genes Dev. 31, 862–875. Available from: https://doi.org/10.1101/gad.298182.117.

van der Burg, K.R.L., Lewis, J.J., Martin, A., Nijhout, H.F., Danko, C.G., Reed, R.D., 2019. Contrasting roles of transcription factors spineless and EcR in the highly dynamic chromatin landscape of butterfly wing metamorphosis. Cell Rep 27, 1027–1038. Available from: https://doi.org/10.1016/j.celrep.2019.03.092.

Verlinden, H., Gijbels, M., Lismont, E., Lenaerts, C., Vanden Broeck, J., Marchal, E., 2015. The pleiotropic allatoregulatory neuropeptides and their receptors: a mini-review. J. Insect Physiol. 80, 2–14. Available from: https://doi.org/10.1016/j.jinsphys.2015.04.004.

von Kalm, L., Crossgrove, K., Von Seggern, D., Guild, G.M., Beckendorf, S.K., 1994. The Broad-Complex directly controls a tissue-specific response to the steroid hormone ecdysone at the onset of *Drosophila* metamorphosis. EMBO J. 13, 3505–3516.

Walkiewicz, M.A., Stern, M., 2009. Increased insulin/insulin growth factor signaling advances the onset of metamorphosis in *Drosophila*. PLoS One 4, e5072. Available from: https://doi.org/10.1371/journal.pone.0005072.

Wang, C., Zhang, J., Tobe, S.S., Bendena, W.G., 2012. Defining the contribution of select neuropeptides and their receptors in regulating sesquiterpenoid biosynthesis by *Drosophila melanogaster* ring gland/corpus allatum through RNAi analysis. Gen. Comp. Endocrinol. 176, 347–353. Available from: https://doi.org/10.1016/j.ygcen.2011.12.039.

Wang, Z., Lee, G., Vuong, R., Park, J.H., 2019a. Two-factor specification of apoptosis: TGF-β signaling acts cooperatively with ecdysone signaling to induce cell- and stage-specific apoptosis of larval neurons during metamorphosis in *Drosophila melanogaster*. Apoptosis 24, 972–989. Available from: https://doi.org/10.1007/s10495-019-01574-4.

Wang, W., Peng, J., Li, Z., Wang, P., Guo, M., Zhang, T., et al., 2019b. Transcription factor E93 regulates wing development by directly promoting Dpp signaling in *Drosophila*. Biochem. Biophys. Res. Commun. 513, 280–286. Available from: https://doi.org/10.1016/j.bbrc.2019.03.100.

Wen, D., Rivera-Perez, C., Abdou, M., Jia, Q., He, Q., Liu, X., et al., 2015. Methyl farnesoate plays a dual role in regulating *Drosophila* metamorphosis. PLoS Genet. 11, e1005038. Available from: https://doi.org/10.1371/journal.pgen.1005038.

White, K.P., Hurban, P., Watanabe, T., Hogness, D.S., 1997. Coordination of *Drosophila* metamorphosis by two ecdysone-induced nuclear receptors. Science 276, 114–117.

Wigglesworth, V.B., 1934. The physiology of ecdysis in *Rhodnius prolixus* (Hemiptera). II. Factors controlling moulting and 'metamorphosis.'. Quart. J. Microsc. Sci. 77, 191–222.

Wilson, T.G., Fabian, J., 1986. A *Drosophila melanogaster* mutant resistant to a chemical analog of juvenile hormone. Dev. Biol. 118, 190–201.

Woodard, C.T., Baehrecke, E.H., Thummel, C.S., 1994. A molecular mechanism for the stage specificity of the *Drosophila* prepupal genetic response to ecdysone. Cell 79, 607–615.

Xu, H.-J., Zhang, C.-X., 2017. Insulin receptors and wing dimorphism in rice planthoppers. Philos. Trans. R. Soc. Lond. B. Biol. Sci. 372, pii: 20150489. Available from: https://doi.org/10.1098/rstb.2015.0489.

Yamada, M., Murata, T., Hirose, S., Lavorgna, G., Suzuki, E., Ueda, H., 2000. Temporally restricted expression of transcription factor betaFTZ-F1: significance for embryogenesis, molting and metamorphosis in *Drosophila melanogaster*. Development 127, 5083–5092.

Yamamoto, K., Chadarevian, A., Pellegrini, M., 1988. Juvenile hormone action mediated in male accessory glands of *Drosophila* by calcium and kinase C. Science 239, 916–919.

Yamanaka, N., Hua, Y.-J., Mizoguchi, A., Watanabe, K., Niwa, R., Tanaka, Y., et al., 2005. Identification of a novel prothoracicostatic hormone and its receptor in the silkworm *Bombyx mori*. J. Biol. Chem. 280, 14684–14690. Available from: https://doi.org/10.1074/jbc.M500308200.

Yamanaka, N., Zitnan, D., Kim, Y.-J., Adams, M.E., Hua, Y.-J., Suzuki, Y., et al., 2006. Regulation of insect steroid hormone biosynthesis by innervating peptidergic neurons. Proc. Natl. Acad. Sci. U.S.A. 103, 8622–8627. Available from: https://doi.org/10.1073/pnas.0511196103.

Yamanaka, N., Yamamoto, S., Zitnan, D., Watanabe, K., Kawada, T., Satake, H., et al., 2008. Neuropeptide receptor transcriptome reveals unidentified neuroendocrine pathways. PLoS One 3, e3048.

Yamanaka, N., Hua, Y.-J., Roller, L., Spalovská-Valachová, I., Mizoguchi, A., Kataoka, H., et al., 2010. *Bombyx* prothoracicostatic peptides activate the sex peptide receptor to regulate ecdysteroid biosynthesis. Proc. Natl. Acad. Sci. U.S.A. 107, 2060–2065. Available from: https://doi.org/10.1073/pnas.0907471107.

Yang, C.-P., Samuels, T.J., Huang, Y., Yang, L., Ish-Horowicz, D., Davis, I., Lee, T., 2017. Imp and Syp RNA-binding proteins govern decommissioning of *Drosophila* neural stem cells. Development 144, 3454–3464. Available from: https://doi.org/10.1242/dev.149500.

Yao, T.P., Segraves, W.A., Oro, A.E., McKeown, M., Evans, R.M., 1992. *Drosophila* ultraspiracle modulates ecdysone receptor function via heterodimer formation. Cell 71, 63–72.

Yao, T.P., Forman, B.M., Jiang, Z., Cherbas, L., Chen, J.D., McKeown, M., et al., 1993. Functional ecdysone receptor is the product of EcR and Ultraspiracle genes. Nature 366, 476–479.

Yu, Y., Li, W., Su, K., Yussa, M., Han, W., Perrimon, N., et al., 1997. The nuclear hormone receptor Ftz-F1 is a cofactor for the *Drosophila* homeodomain protein Ftz. Nature 385, 552–555.

Zelhof, A.C., Yao, T.P., Chen, J.D., Evans, R.M., McKeown, M., 1995. Seven-up inhibits ultraspiracle-based signaling pathways in vitro and in vivo. Mol. Cell. Biol. 15, 6736–6745. Available from: https://doi.org/10.1128/MCB.15.12.6736.

Zhang, Z., Xu, J., Sheng, Z., Sui, Y., Palli, S.R., 2011. Steroid receptor co-activator is required for juvenile hormone signal transduction through a bHLH-PAS transcription factor, methoprene tolerant. J. Biol. Chem. 286, 8437–8447. Available from: https://doi.org/10.1074/jbc.M110.191684.

Zhang, F., Wang, J., Thakur, K., Hu, F., Zhang, J.G., Jiang, X.F., et al., 2018. Isolation functional characterization of allatotropin receptor from the cotton bollworm, *Helicoverpa armigera*. Peptides 122, 169874.

Zhang, T., Song, W., Li, Z., Qian, W., Wei, L., Yang, Y., et al., 2018b. Krüppel homolog 1 represses insect ecdysone biosynthesis by directly inhibiting the transcription of steroidogenic enzymes. Proc. Natl. Acad. Sci. U.S.A. 115, 3960–3965. Available from: https://doi.org/10.1073/pnas.1800435115.

Zhou, B., Hiruma, K., Shinoda, T., Riddiford, L.M., 1998. Juvenile hormone prevents ecdysteroid-induced expression of broad complex RNAs in the epidermis of the tobacco hornworm, *Manduca sexta*. Dev. Biol. 203, 233–244.

Zhu, T.-T., Meng, Q.-W., Guo, W.-C., Li, G.-Q., 2015. RNA interference suppression of the receptor tyrosine kinase Torso gene impaired pupation and adult emergence in *Leptinotarsa decemlineata*. J. Insect Physiol. 83, 53–64. Available from: https://doi.org/10.1016/j.jinsphys.2015.10.005.

Zitnan, D., Kingan, T.G., Hermesman, J.L., Adams, M.E., 1996. Identification of ecdysis-triggering hormone from an epitracheal endocrine system. Science 271, 88–91.

Zhu, G.-H., Jiao, Y., Chereddy, S.C.R.R., Noh, M.Y., Palli, S.R., 2019. Knockout of juvenile hormone receptor, Methoprene-tolerant, induces black larval phenotype in the yellow fever mosquito, *Aedes aegypti*. Proc. Natl. Acad. Sci. U. S. A 116, 21501–21507. Available from: https://doi.org/10.1073/pnas.1905729116.

Zou, Z., Saha, T.T., Roy, S., Shin, S.W., Backman, T.W.H., Girke, T., et al., 2013. Juvenile hormone and its receptor, methoprene-tolerant, control the dynamics of mosquito gene expression. Proc. Natl. Acad. Sci. U. S. A. 110, E2173–2181. Available from: https://doi.org/10.1073/pnas.1305293110.

Chapter 8

Epigenetic-related mechanisms

The term "epigenetics" was coined by Waddington in 1942 to refer to changes in phenotype without changes in genotype (see Waddington, 1957), when explaining aspects of development for which there was little mechanistic understanding. In recent times, the emphasis has been placed on molecular mechanisms that directly affect, alter, or interact with chromatin, which have consequences on gene expression and on their heritability. Epigenetic mechanisms work in addition to the DNA template to stabilize gene expression programs and thereby canalize cell-type identities (Allis and Jenuwein, 2016). Within individuals, epigenetic information would be transmitted through mitotic cell division, whereas meiotic cell division would lead to intergenerational epigenetic inheritance (Glastad et al., 2019). As regard to epigenetic mechanisms, the most prominent would be the methylation of cytosine in DNA, the modifications of histone proteins, and the nucleosome positioning and regulation by noncoding RNAs (ncRNAs). However, these epigenetic mechanisms have been rarely considered in the studies of insect metamorphosis, particularly with respect to the inheritance aspects, so there is very little information available in this sense.

As a general warning regarding the following sections, it is worth noting that the observations are generally based on depleting methylating or acetylating enzymes, considering that the phenotype observed is directly due to the corresponding histone modification. However, work based on mutations of the residues of all histones has shown that the phenotypes observed can be inconsistent with those obtained when mutating the enzyme genes (McKay et al., 2015). This suggests that the role of the enzymes extends beyond the canonical catalytic activity, and that the epigenetic processes related to histone modifications are more complex than previously thought. Another note of caution to point out is that the reported effects on epigenetic marks mostly derive from simple correlations, which can lead to unjustified cause—effect conclusions. Indeed, it has been shown that genes can be active in mutants with no epigenetic marks (Hödl and Basler, 2012) and also that there are key developmental genes that are also perfectly active without having any epigenetic mark (Pérez-Lluch et al., 2015).

Insect Metamorphosis. DOI: https://doi.org/10.1016/B978-0-12-813020-9.00008-9

DNA METHYLATION

DNA methylation in animals has been implicated in several biological processes including developmental progression and regulation. Mechanistically, DNA methylation is a covalent addition of a methyl group to DNA. In animals methylation exclusively affects cytosine bases and is largely confined to CpG dinucleotides (Glastad et al., 2019). DNA methylation is catalyzed by enzymes known as DNA methyltransferases (DNMTs), and studies in mammals indicate that they can be separated into "de novo" and "maintenance" methyltransferases. De novo methyltransferases establish new methylation patterns in the genome and are represented by the DNMT3 family. In contrast, maintenance methyltransferases, represented by the DNMT1, preferentially methylate hemimethylated DNA substrates, thus maintaining methylation patterns across cell generations (Lyko, 2018).

In contrast to vertebrates, in which DNA methylation occurs throughout the genome, in insects it is relatively sparse. Moreover, in most groups (with the possible exception of termites) DNA methylation occurs predominantly over transcribed genes, specifically exons, whereas in mammals it occurs globally, and promoter methylation represses gene transcription. Currently, there is no robust evidence for DNA methylation enrichment in gene promoters of insect genomes (Glastad et al., 2019). DNA methylation has been detected in all major insect subclasses although the levels of DNA methylation and the types of DNMTs possessed vary in different groups. For example, the hemipteran *Acyrthosiphon pisum* has two isoforms of both DNMT1 and DNMT3, whereas the phthirapteran *Pediculus humanus* does not have DNMT3. Most studies have focused on endopterygotes where DNMT1 and DNM3 are present, for instance, in the hymenopterans *Nasonia vitripennis* and *Apis mellifera*, where de novo and maintenance DNMTs in insects were first fully reported (Wang et al., 2006). In contrast, the lepidopteran *Bombyx mori* and the coleopteran *Tribolium castaneum* have lost DNMT3. However, the most dramatic loss of DNA methylation has been found in dipterans, where genome sequencing projects have not detected either DNMT1 or DNMT3 (Glastad et al., 2019). Interestingly, hemimetabolan insects show much higher levels of genomic DNA methylation (which concentrates in protein-coding sequences) than holometabolans (Bewick et al., 2016; Provataris et al., 2018). This suggests that the evolution of holometaboly was accompanied by a reduction in DNA methylation levels. The functional sense of these relationships, however, remains a mystery. There are not many studies that relate DNA methylation to insect development and metamorphosis. Some of them examine the role of DNA methylation in embryogenesis, both in hemimetabolan and holometabolan species, whereas a few focus on postembryonic development.

DNA methylation and embryogenesis

The sequencing of the genome of the hymenopteran *N. vitripennis* revealed the occurrence of three *DNMT1* (*DNMT1a*, *DNMT1b*, and *DNMT1c*) genes and one *DNMT3*, and quantitative real-time PCR analysis showed that *DNMT1a*, *DNMT1c*, and *DNMT3* mRNAs are maternally provided to the embryo. Maternal RNA interference (RNAi) of *DNMT1a* results in embryonic lethality at the onset of gastrulation (Zwier et al., 2012). In hemimetabolan species, maternal RNAi targeting of *DNMT1* in the bug *Oncopeltus fasciatus* dramatically reduced egg production. The oocytes of the DNMT1-depleted females showed structural defects in the nucleus of the follicular epithelium, and the few eggs oviposited were inviable (Bewick et al., 2019). Intriguingly, maternal knockdown of *DNMT1* in *T. castaneum* caused a developmental arrest in offspring embryos. However, the function of *DNMT1* appears to be unrelated to CpG DNA methylation in this beetle, since no evidence of that modification was found, which suggests an alternative role of this protein (Schulz et al., 2018).

DNA methylation and postembryonic development

A study that directly relates DNA methylation and metamorphosis has been carried out in the context of the inhibitory action of juvenile hormone (JH) on ecdysone production. In *D. melanogaster*, JH represses ecdysone synthesis through its main transducer, Krüppel homolog 1 (Kr-h1), which inhibits the transcription of steroidogenic genes. Using the steroidogenic gene *spookier* (*spok*) as a case study, bisulfite sequencing PCR (BSP) results indicate that cytosines located in the ~200-bp region upstream of the proximal Kr-h1-binding site (KBS) in the *spok* promoter are methylated following Kr-h1 overexpression. Methylation of these cytosines was prevented with the DNA methylation inhibitor 5-aza-2,9-deoxycytidine (Aza). Furthermore, luciferase reporter assays confirmed that Kr-h1-associated inhibition on transcriptional activity of *spok* promoter was abolished by Aza. The results suggest that direct repression of Kr-h1 on the transcription of *spok* is most likely caused by Kr-h1−induced DNA methylation of the *spok* promoter following its direct binding to the KBS (Zhang et al., 2018). It remains to be elucidated what the enzyme that catalyzes that methylation would be, since *D. melanogaster* lacks DNMT1 and DNMT3 (Glastad et al., 2019).

HISTONE MODIFICATIONS

The nucleosome, the basic unit of DNA packaging in eukaryotes, consists of a segment of DNA wound in sequence around eight histone protein cores. In turn the nucleosome core particle consists of approximately 146 base pairs of DNA wrapped around a protein complex composed of two copies each of the histone proteins H2A, H2B, H3, and H4 (Talbert and Henikoff, 2010). Histones can experience a number of modifications, which can affect gene transcription. The most common modifications are acetylation, methylation or ubiquitination of lysine, methylation of arginine and lysine, and phosphorylation of serine (Talbert and Henikoff, 2010). Histone modifications can influence transcription, especially if they affect the promoter and/or the enhancer regions of the gene. Indeed epigenomic profiling has helped to better identify gene enhancers and promoters in some cases (Allis and Jenuwein, 2016). A number of histone modifications seem to correlate with gene silencing, whereas others appear to correlate with gene activation. In general, modifications that lower the charge of the globular histone core (like acetylation or phosphorylation) are predicted to "loosen" core-DNA association (Fenley et al., 2010). The information available related to the regulation of insect metamorphosis is scarce and refers to acetylation of lysine and methylation of lysine or arginine residues.

Histone acetylation at lysine residues

Histone acetylation is catalyzed by histone acetyltransferase (HAT) enzymes. In *D. melanogaster*, a thoroughly studied HAT is Gcn5, which can acetylate lysine 9 in histone 3 (H3K9) and H3K14, as well as H4K5 and H4K12, depending on the complex where Gcn5 is integrated. Integration of HATs into multisubunit complexes facilitates their enzymatic activity and substrate specificity. In addition to HATs, important components of these complexes are Ada proteins, transcriptional adapters that enhance the enzymatic activity. The best known HAT complexes in insects are Ada2a-containing complex (ATAC) and pt-Ada-Gcn5 acetyltransferase (SAGA) complex, both containing Gcn5 but appearing to be specific for the acetylation of histones H4 and H3, respectively. ATAC contains Ada2a, which is required for the correct acetylation of H4K5 and H4K12, while SAGA contains Ada2b, which is required for the acetylation of H3K9 and H3K14. Both HAT complexes share Ada3, and mutants of this adaptor protein show deficient acetylation of H3K9, H3K14, and H4K12, but not H4K5, which suggests that the role of Ada3 may be different in each HAT complex (Pankotai et al., 2010). A pioneering study of HATs performed in *D. melanogaster* on Gcn5 revealed that null *Gcn5* alleles cannot initiate metamorphosis, while hypomorphic *Gcn5* alleles are unable to form adult appendages and cuticle. Strong cell proliferation defects were observed in imaginal tissues of *Gcn5* mutants, while, remarkably, loss of acetylation of H3K9 and H3K14 in these mutants did not affect larval tissues (Carré et al., 2005).

A number of works have examined the effects of histone acetylation on ecdysteroid production and signaling in *D. melanogaster*. Thus *D. melanogaster* with mutated Ada2a (present in the ATAC complex) and Ada3 (common to ATAC and SAGA complexes) die at the onset of metamorphosis, while ecdysteroid titers and ecdysone receptor (EcR) levels are reduced. Moreover, the expression of the steroidogenic genes *spok*, *phantom*, *disembodied*, and *shadow* is downregulated, while that of *shade*, which converts ecdysone into 20-hydroxyecdysone (20E) in peripheral tissues, is upregulated in these ATAC subunit mutants. Driven expression of Ada3 at the prothoracic gland cells partially rescues Ada3 mutants. In contrast, loss-of-function mutations in Ada2b do not affect the above features, thus indicating that the steroidogenic genes are regulated by the ATAC complex, but not by the SAGA complex (Pankotai et al., 2010).

Also in *D. melanogaster*, acetylation of H3K23 in the promoter region of the ecdysteroid signaling genes *E74* and *E75* correlates with gene activation induced by ecdysone/20E. In contrast, acetylation of H3K9, H4K8, H4K12, and H4K16 shows insignificant correlation with ecdysteroid-induced transcriptional activity. RNAi of *nejire*, whose gene product, a CREB-binding protein (CBP), acetylates H3K23, leads to reduced *E74* and *E75* expression (Bodai et al., 2012). The results suggest that acetylation of specific histone lysine residues contributes to the expression of genes involved in the ecdysteroid signaling. It has been also reported that the ATAC complex might play a dual regulatory role in *D. melanogaster* ecdysteroid synthesis and signaling through the acetylation of Fushi tarazu-factor 1 (FTZ-F1) protein and the regulation of the H4K5 acetylation at the promoters of steroidogenic genes (Borsos et al., 2015).

The CBP expressed by *nejire* has been studied in the German cockroach, *Blattella germanica*. RNAi depletion of CBP in freshly emerged sixth (last) nymphal instar delays the imaginal molt and impairs the ecdysis, as CBP-depleted insects cannot shed off the exuviae and fully extend the wings (Fig. 8.1A). The mRNA levels of the ecdysteroid signal transducers HR3A, E75A, and E75B are lower in the CBP-depleted insects than in controls (Fig. 8.1B), which can explain the molting delay. Moreover, CBP depletion affects the expression of *Kr-h1* and *E93*, which repress and trigger metamorphosis, respectively (Belles and Santos, 2014). In mid-last nymphal instar, when the expression of *Kr-h1* should be low and that of *E93* high, the mRNA levels are high and low, respectively, in CBP-depleted nymphs (Fernandez-Nicolas and Belles, 2016). This suggests that CBP-depletion causes a delay in the fall of *Kr-h1* expression that occurs in the last nymphal instar as a consequence of the concomitant drop of with JH production. Simultaneous treatments with dsRNA targeting *nejire* transcripts and with JH (which stimulate *Kr-h1* expression) showed that control specimens treated with JH increase the expression of *Kr-h1* fourfold, whereas CBP depletion

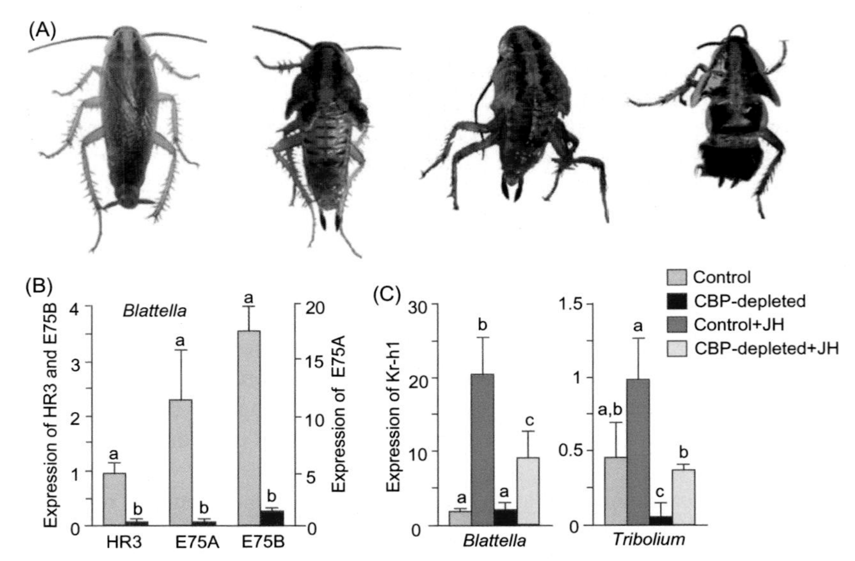

FIGURE 8.1 Effect of CREB-binding protein (CBP) depletion on molting and metamorphosis in *Blattella germanica* and *Tribolium castaneum*. (A) Phenotypes obtained in the adult after depleting CBP in last nymphal instar of *B. germanica*, a normal adult is shown on the left, and different levels of ecdysis inhibition and wing extension are shown on the right. (B) Reduction of the expression of *HR3*, *E75A*, and *E75B* ecdysteroid response genes in CBP-depleted last nymphal instar of *B. germanica*. (C) Effect of juvenile hormone (JH) in CBP-depleted insects and in controls of last nymphal instar of *B. germanica* or last larval instar of *T. castaneum*. Different letters at the top of the columns in (B) and (C) indicate statistically significant differences (t-test, $p < 0.05$). Note that JH treatment induces a lower upregulation of *Kr-h1* expression in CBP-depleted insects than in controls. (A) From Fernandez-Nicolas and Belles (2016), with permission. (B and C) Drawn with data from the same authors and Roy et al. (2017), respectively.

reduces by 50% such a JH-driven increase (Fig. 8.1C) (Fernandez-Nicolas and Belles, 2016). The data suggest that CBP contributes to regulation of the correct fall of *Kr-h1* expression in the last nymphal instar, perhaps acting on the regulation of JH, and that CBP enhances the expression of *Kr-h1* when stimulated by JH. The enhancing effect of CBP on JH-induced expression of *Kr-h1* has been also reported in *T. castaneum*. In this beetle, CBP was knockdown by RNAi treatment in freshly ecdysed last larval instar, and the JH analog hydroprene was applied to these insects 72 hours later. Results showed that JH enhances the expression of *Kr-h1*, whereas CBP depletion reduces the JH-driven increase of *Kr-h1* expression (Roy et al., 2017), as in *B. germanica*.

Histone acetylation marks can be eliminated by the action of histone deacetylases (HDACs). A recent paper on *T. castaneum* of the same group (George et al., 2019) provides mechanistic explanations to the epigenetic component of the stimulatory effect of JH on the expression of *Kr-h1*.

HDAC1 RNAi experiments and JH treatments, combined with chromatin immunoprecipitation (ChIP) assays and luciferase tests of promoter activity, showed that JH treatments suppressed HDAC1 expression, whereas HDAC1 depletion upregulated *Kr-h1*. Moreover, HDAC1 depletion or JH treatment increased the acetylation levels of core histones in the promoter of *Kr-h1*. Finally, overexpression or knockdown of HDAC1 decreased or increased *Kr-h1* expression and *Kr-h1* promoter activity, respectively (George et al., 2019). The results as a whole indicate that (1) acetylation of core histones in the *Kr-h1* promoter increases promoter activity and *Kr-h1* expression, (2) HDAC1 activity triggers the reverse situation, and (3) JH suppresses HDAC1 expression. In Chapter 7 (Molecular mechanisms regulating hormone production and action), we have seen that the JH bound to the Methoprene tolerant Taiman receptor complex binds to the promoter of *Kr-h1* inducing the expression of the gene. The results of George et al. (2019) suggest that JH also contributes to the expression of *Kr-h1* by suppressing HDAC1 expression, thus keeping high the acetylation levels in the promoter of *Kr-h1* and thus the expression of *Kr-h1*.

Experimental manipulation of HDACs in the beetle *Gnathocerus cornutus* revealed that acetylation of histones affects the morphogenesis of the male hypertrophied mandibles. The size of the mandibles of *G. cornutus* is influenced by nutritional conditions, and RNAi experiments targeting HDACs showed that HDAC1 depletion in larvae causes specific curtailment of mandibles in adults, whereas HDAC3 depletion increases their size. Intriguingly, these depletions result in the opposite effect on wing size, but little effect on the size of the core body and genital modules (Ozawa et al., 2016). These results suggest that histone acetylation mediates the nutritional effects on the size of mandibles, and that plastic development of exaggerated traits is controlled in a module-specific manner by HDACs.

Histone methylation at lysine and arginine residues

In general, the occurrence of lysine or arginine methylation in H3 or H4 has consequences in many biological processes, including heterochromatin formation, X-chromosome inactivation, and transcriptional regulation. Three arginine-methylation sites (R2, R17, and R26 in H3) and six lysine-methylation sites (K4, K9, K27, K36, and K79 in H3, and K20 in H4) have been identified to date. In addition, the amino acid residue can be mono-, di-, or trimethylated, and this differential methylation provides further functional diversity to the site (Cheung and Lau, 2005). Importantly, methylation of histone residues can have different outcomes depending upon the biochemical context. For example, H3K9me3 usually correlates with heterochromatin repression, while H3K4me3 usually correlates with activated transcription (Glastad et al., 2019).

Methylation of arginine residues

In *D. melanogaster*, two *Drosophila* protein arginine methyltransferases (DART), DART1 and DART4, have proven arginine methyltransferase activity (Boulanger et al., 2004). DART4, also known as CARMER, mediates the dimethylation of H3R17 and has been studied in the context of ecdysteroid-mediated apoptosis during metamorphosis. CARMER expression is low in larval stages before being upregulated at the early prepupal stage, coinciding with the histolysis of larval tissues. Moreover, suppression of CARMER in ecdysteroid-responsive cultured cells blocks ecdysteroid-induced cell death. CARMER associates with the EcR-Ultraspiracle (USP) receptor complex and modulates the ecdysteroid-induced transcription of a number of apoptotic genes. Thus the ecdysone-induced expression of the initiator caspase Dronc and the downstream caspases DrICE and dcp-1 is abolished or significantly reduced in CARMER-depleted cells. Similar effects are observed when examining Dark, an adaptor protein required for *Dronc* activation, and the upstream proapoptotic genes *Reaper* and *Hid*, which are upregulated by ecdysteroids during salivary gland and midgut cell death (Cakouros et al., 2004). The data suggest that CARMER coordinates the EcR/USP-mediated regulation of several core cell death molecules. Of interest in this context is the cofactor *Drosophila* lysine ketoglutarate reductase/saccharopine dehydrogenase (dLKR/SDH), which binds histones H3 and H4 and suppresses ecdsyteroid-mediated transcription of cell death genes by inhibiting the histone H3R17me2 mediated by CARMER. This suggests that the dynamic recruitment of dLKR/SDH to ecdsyteroid-regulated gene promoters controls the timing of hormone-induced gene expression (Cakouros et al., 2008). DART1, which dimethylates the H4R3, has been also studied in *D. melanogaster* in relation to EcR. Disruption of DART1 results in high mortality concentrated in the pupal stage. Moreover, experiments in vitro show that ligand-bound EcR physically associates with DART1, while RNAi experiments and reporter assays with DART1, EcR, and the ecdysone analogue Muristerone A indicate that DART1 corepresses *EcR* expression (Kimura et al., 2008).

Methylation of lysine residues

Of importance in the methylation of histone lysine residues are the Trithorax and Polycomb complexes of chromatin proteins. In general, the Trithorax complex is required to maintain the active state of a gene, whereas the Polycomb complex is necessary for maintaining a repression state. The opposing activities of the Trithorax and Polycomb complexes contribute to maintain epigenetic memory and allow dynamic switches in gene expression during development (Schuettengruber et al., 2017).

An important component of the Trithorax complex is Trithorax-related (Trr), a Su(var)3-9, Enhancer-of-zeste and Trithorax (SET) domain histone methyltransferase that trimethylates H3K4. In *D. melanogaster*, the *hedgehog* (*hh*) gene is expressed in the latest stages of eye disc development posteriorly to the morphogenetic furrow, and Trr is required for compound eye development, acting upstream of *hh*. During compound eye development, Trr interacts with EcR. Moreover, Trr, EcR, and trimethylated H3K4 are detected at the ecdysteroid-inducible promoters of *hh* and *Broad-complex* (*BR-C*) in cultured S2 cells, and H3K4 trimethylation at these promoters decreases in embryos lacking a functional copy of *trr* (Sedkov et al., 2003). The data suggest that Trr functions as a coactivator of *EcR* by altering the chromatin structure at ecdysteroid-responsive promoters. Another important component of the Trithorax complex is the coactivator Ash2. In *D. melanogaster*, *ash2* mutants show severe defects in pupariation due to a lack of activation of ecdysteroid-responsive genes. This phenotype results from the absence of the H3K4me3 marks set by Trr in these genes. Importantly, Ash2 interacts with Trr and is required for its stabilization. The data suggest that Ash2 functions together with Trr as an ecdysone receptor coactivator by stabilizing Trr (Carbonell et al., 2013).

An extensive work of Rickels et al. (2017) has unveiled the influence of the level of methylation in H3K4 on *D. melanogaster* development. Flies from *trr* RNAi die during pupation, whereas embryos expressing catalytically deficient Trr eclose and develop to reproductive adults although these mutants show subtle developmental abnormalities when subjected to the temperature stress of 29°C. Flies with a *trr* allele that decrease H3K4 methylation show an additional wing cross-vein between L3 and L4 veins (Fig. 8.2A) compared with controls. In contrast, mutants with a *trr* allele that increases H3K4 methylation indicate that conversion of H3K4me1 to H3K4me2 and H3K4me3 is compatible with life, but the adults show darker pigmentation on the seventh abdominal segment (Fig. 8.2B) and a higher occurrence of supernumerary thoracic bristles compared with controls. Remarkably, this phenotype is also observed in mutants of the H3K4me3 demethylase Kdm5 (also known as *little imaginal discs* or *lid*, see below), which have increased levels of H3K4me3. This suggests that the phenotype

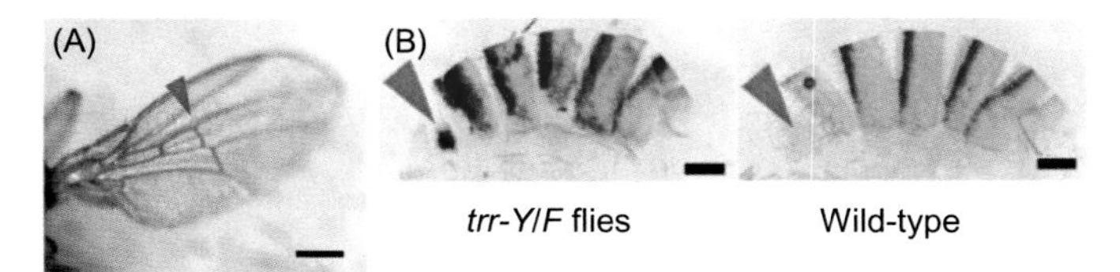

FIGURE 8.2 Subtle phenotypes of lines expressing *trithorax-related* (*trr*) catalytic mutants in *Drosophila melanogaster*. (A) Flies with a *trr* allele that decrease H3K4 methylation and maintained at 29°C showing an additional wing cross-vein between L3 and L4 veins (arrowhead). (B) Flies with a *trr* allele that increases H3K4 methylation (*trr-Y/F* flies) showing darker pigmentation on the seventh abdominal segment (arrowheads) as compared to wild-type flies. Scale bar, 0.25 mm. From Rickels et al. (2017), with permission.

of thoracic bristles results from increased H3K4me3 (Rickels et al., 2017). The data point to a wide tolerance for differential H3K4me1 levels at developmental enhancers in *D. melanogaster* but also that this modification is important for fine-tuning enhancer activity, especially under temperature stress and in the context of metamorphosis.

The polycomb repressive complexes (PRCs) are divided into two main classes: PRC1 and PRC2, which are able to monoubiquitylate and di- and trimethylate specific lysine residues on H2A and H3, respectively. *D. melanogaster* PRC1 is composed of the core components such as Polycomb (Pc), Polyhomeotic (Ph), Posterior sex combs (Psc), and Sex combs extra (Sce, also known as dRing1). Pc can bind the H3K27me3 histone modification through its chromodomain, and this is thought to be important for anchoring the complex to chromatin. The PRC2 core complex is composed of enhancer of zeste (E(z)), suppressor of zeste 12 (Su(z)), extra sex combs (Esc), and p55 (Nurf55 or Caf1). E(z) contains a SET domain and is the PRC2 subunit responsible for the deposition of H3K27 (Schuettengruber et al., 2017).

In a study on diapause regulation by environmental signals in the moth *Heliothis armigera*, immunostaining, ChIP, and RNAi experiments showed that the *prothoracicotropic hormone* (*PTTH*) gene may be an Esc target (Lu et al., 2013). The observations suggest that high levels of H3K27me3 in the brain of nondiapausing pupae, which depend on PRC2, can promote insect development and metamorphosis by activating *PTTH*, whereas low levels of H3K27me3 can induce diapause by reducing *PTTH* expression. In brief, the PRC2 protein Esc appears to activate *PTTH* expression by mediating H3K27me3, which regulates the insect developmental timing (Lu et al., 2013).

An extensive study in *D. melanogaster* larvae, which combined information on PRC1 and H3K27me3 state, unveiled subtle mechanisms contributing to the progression of the morphogenetic furrow in the formation of the compound eye (Loubière et al., 2016). The study shows that *ph*-null mutant cells elicit extensive overproliferation in the posterior part of the eye disc, whereas *E*(*z*)-null mutant cells hypoproliferated. Moreover, quantitative ChIP at the *Notch* (*N*) locus indicates that Pc and Ph bind to *N* in eye discs in the absence of H3K27me3. A large set of genes that are bound by cPRC1 proteins in larval tissues in the absence of H3K27me3 was uncovered, including 894 genes in eye discs. The comparison of Pc binding and H3K27me3 in *ph*- and in *E*(*z*)-null eye disc tissue reveals that H3K27me3 at PRC1–PRC2 target genes is lower in *E*(*z*)-null mutants. As expected, Pc binding also decreases in both *ph*- and in *E*(*z*)-null mutants. H3K27me3 also decreases upon null mutation of *ph* at PRC1–PRC2 target genes, suggesting that PRC1 may have an effect in stabilizing PRC2 function (Loubière et al., 2016). The results point to an important role of PRC1 on the development of the compound eye during holometabolan metamorphosis, especially around the formation of the morphogenetic furrow.

Also important is the action of histone demethylases that eliminate methylation marks. Regarding metamorphosis, the H3K27me3 demethylase Utx is required for hormone-mediated transcriptional regulation of apoptosis and autophagy genes during ecdysteroid-regulated degeneration of salivary glands in *D. melanogaster*. Utx binds to the complex EcR/USP and is recruited to the promoters of key cell death genes. Salivary gland degeneration is delayed in *Utx* mutants, with reduced caspase activity and autophagy, coincident with reduced expression of cell death genes (Denton et al., 2013). The data show how a particular demethylation activity regulates a complex hormone-dependent cell death process in the context of metamorphosis.

Another histone demethylase involved in metamorphosis is Kdm4, which demethylates H3K9. *D. melanogaster* possess two *Kdm4* genes, *Kdm4A* and *Kdm4B*, and the loss-of-function of both triggers a lethal phenotype associated with molting problems (Tsurumi et al., 2013). Most of the double mutant insects die as early second larval instar, showing a double set of mouth hooks and posterior spiracles. This indicates that at least the ecdysis process was prevented, which suggests deficiencies in the ecdysteroid signaling. Indeed, the ecdysteroid-dependent genes are either downregulated (like *EcR*, *E75B*, *BR-C*, and *HR3*) or upregulated (E74) in *Kdm4A–Kdm4B* double mutants. Studies in *Drosophila* S2 cells show that ecdysteroid-dependent stimulation of *BR-C* and *HR3* is impaired in Kdm4A + Kdm4B-depleted cells, while other experiments indicate that Kdm4A and Kdm4B demethylate H3K9me3 at the *BR-C* promoter in response to ecdysteroids, and that Kdm4A physically interacts with EcR (Tsurumi et al., 2013). The data suggest that Kdm4 demethylases may function as transcriptional cofactors required for transcriptional activation in the ecdysteroid signaling pathway, through removing the repressive histone mark H3K9me2,3 from the respective promoters.

Also in *D. melanogaster*, Drelon et al. (2018) have shown that the histone demethylase Kdm5, which eliminates H3K4me3, plays crucial roles in postembryonic development. Insects with a null allele of *kdm5* (or *lid*) show a delayed larval development, which coincides with decreased proliferation and increased cell death in wing discs. The demethylase Kdm5 contains the enzyme activity domain Jumonji C, but, intriguingly, the developmental delay observed in kdm5 mutants is independent of this domain. Transcriptome analyses of wing discs from *kdm5* null mutants revealed that genes involved in several cellular processes were dysregulated (Drelon et al., 2018).

NONCODING RNAS

ncRNAs form an heterogeneous class of RNAs that are not translated into proteins although regulatory functions have been reported for a number of them (Cech and Steitz, 2014). Four types of ncRNAs have been associated with epigenetic mechanisms: PIWI-interacting RNAs (piRNAs), microRNAs (miRNAs), small interfering RNAs (siRNAs), and long noncoding RNAs (lncRNAs) (Glastad et al., 2019). Of these, piRNAs and lncRNAs are those with more clear evidence regarding epigenetic effects in insects although there is practically no data describing strictly epigenetic functions of ncRNAs in the context of metamorphosis. In contrast, a series of regulatory functions of miRNAs have been described in the metamorphosis of hemimetabolan and holometabolan species (Belles, 2017), and a handful of studies suggest that lncRAs can also play roles related to metamorphosis.

miRNAs and hemimetabolan metamorphosis

In insects, the endonuclease dicer-1 transforms the miRNA precursor into a mature miRNA, and dicer-1-depleted last instar nymphs of the cockroach *B. germanica* molt to supernumerary nymphs instead of adults (Gomez-Orte and Belles, 2009). The supernumerary nymphs are identical to those obtained after a JH treatment in last nymphal instar, which triggers a dramatic upregulation of *Kr-h1* expression (Lozano and Belles, 2011). Subsequent work revealed that dicer-1 depletion increases *Kr-h1* mRNA levels, that reduction of *Kr-h1* expression in dicer-1-depleted nymphs rescues metamorphosis, and that the 3′-UTR region of *Kr-h1* mRNA contains a functional binding site for the miRNA miR-2 (Lozano et al., 2015). A treatment of the last nymphal instar with a miR-2 inhibitor impaired metamorphosis, while a treatment of dicer-1–depleted nymphs with a miR-2 mimic restored it (Fig. 8.3). The data indicate that in *B. germanica* miR-2 scavenge *Kr-h1* transcripts just after the molt to the last nymphal instar, which, together with the decrease in production of JH, allows a strong increase of *E93* expression and, hence, metamorphosis according to the MEKRE93 pathway (Belles and Santos, 2014).

Intriguingly, in the phylogenetically close species *Locusta migratoria* *Kr-h1* is regulated by the miRNAs let-7 and miR-278. Both miRNAs bind to the *Kr-h1* mRNA, and administration of let-7 and miR-278 mimics in the penultimate nymphal instar reduces the *Kr-h1* transcript levels and triggers the formation of precocious adult features at the next molt (Song et al., 2018). No miR-2 binding sites were predicted in the mRNA of *L. migratoria Kr-h1*, which suggests that different miRNA systems have evolved in cockroaches and locusts to regulate the same target within the same physiological process.

Underlining again the species specificity of miRNA functions, let-7 and miR-100 have been found to be involved in wing patterning in *B. germanica*. Depletion of miR-100 in the last nymphal instar with a specific anti-miR-100 triggers the formation of adults with the wings slightly reduced in size and with vein anomalies. Equivalent experiments with let-7 elicited the same vein anomalies in the adult wings (Rubio and Belles, 2013). This wing phenotype is similar to that obtained after depleting *BR-C* expression (Huang et al., 2013), which suggests that BR-C directly or indirectly induces the

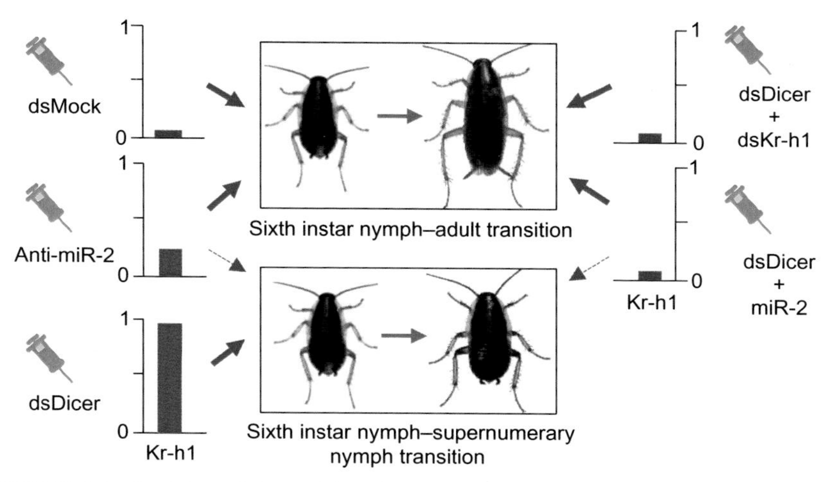

FIGURE 8.3 Depletion of *Krüppel homolog 1* (*Kr-h1*) transcripts by miR-2 during the last nymphal instar controls metamorphosis in *Blattella germanica*. RNAi depletion of dicer-1 (dsDicer treatment) in the sixth (last) nymphal instar inhibits metamorphosis, triggering the formation of a supernumerary seventh nymphal instar; the treatment impairs the decrease in *Kr-h1* transcripts observed in controls (dsMock treatment). Depletion of Kr-h1 (dsKr-h1 treatment) in dsDicer-1−treated animals rescues metamorphosis. Administration of a miR-2 inhibitor (anti-miR-2 treatment) impairs metamorphosis. Administration of a miR-2 mimic (miR-2 treatment) in dsDicer-1−treated insects rescues metamorphosis. The expression levels of *Kr-h1* are indicated in relation to the maximum value (=1) obtained in dsDicer-treated experiments. Figure drawn with data from Lozano et al. (2015).

expression of miR-100 and let-7, as occurs in *D. melanogaster* (Sempere et al., 2003).

miRNAs and holometabolan metamorphosis

Several miRNAs regulate gene transcripts involved in different aspects of holometabolan metamorphosis, from ecdysteroid signaling to cell death processes. In *D. melanogaster*, ecdysteroid signaling involves a positive autoregulatory loop that increases EcR levels, and miR-14 modulates this loop by limiting the expression of EcR, whereas ecdysteroid signaling through EcR downregulates miR-14 (Varghese and Cohen, 2007). This mechanism may be crucial due to the intrinsic lability of the positive EcR autoregulatory loop. In the silkworm *B. mori*, miR-14 plays important roles in ecdysteroid-regulated development, as ubiquitous transgenic overexpression of this miRNA results in delayed larval development and smaller body size of larva and pupa, with a concomitant decrease in ecdysteroid titers. Complementarily, miR-14 disruption triggers a precocious wandering stage associated to an increase in ecdysteroid titers. The ecdysteroid signaling transducers EcR-B and E75 have been identified as miR-14 targets

(Liu et al., 2018). The whole data suggest that miR-14 contributes to maintaining correct ecdysteroid signaling during larval development and metamorphosis in *B. mori*.

Regarding wing formation, the loss of let-7 and miR-125 delays the cell cycle exit in wing cells of *D. melanogaster*. The gene *abrupt* was shown to be a target in vivo of let-7 in pupal wing discs, where a specific temporal blocking of *abrupt* transcript by let-7 is required for correct wing development (Caygill and Johnston, 2008). Another miRNA, miR-7, modulates cell growth and cell cycle progression during *D. melanogaster* wing development, by targeting the cell cycle regulator Dacapo and the Notch signaling pathway (Aparicio et al., 2015). The *D. melanogaster LIM-only* gene (*dLMO*) encodes a transcription cofactor that represses the expression of *apterous*, a crucial gene for determining wing dorsal identity. During wing development, miR-9 regulates *dLMO* mRNA through a specific target site, and thus mutants without this site produce high levels of dLMO and exhibit a wing phenotype characterized by a lack of margins (Biryukova et al., 2009). Also in *D. melanogaster*, the Hox gene *Ultrabithorax* (*Ubx*) determines the formation of halteres in the metathorax instead of flying wings, whereas ectopic expression of the miRNA iab-4 triggers a haltere-to-wing homeotic transformation (Ronshaugen et al., 2005) (Fig. 8.4). This effect is mediated by iab-4 binding to the *Ubx* mRNA. The *iab-4* locus is part of the *Bithorax* complex that contains up to nine homeotic genes, including *Ubx* (Ronshaugen et al., 2005). Notably, antisense transcription of this *iab-4* locus generates a symmetric hairpin containing a miRNA (named iab-8 or iab-4AS), which is almost identical to iab-4 and thus also targets the *Ubx* mRNA and plays the same role as iab-4 (Ronshaugen et al., 2005). An important aspect of wing morphogenesis is the

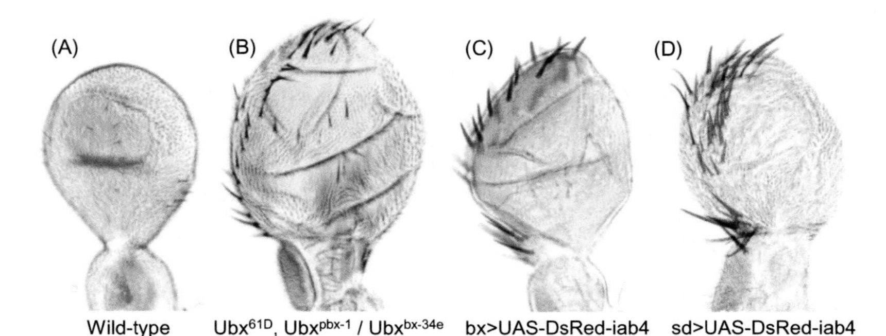

FIGURE 8.4 Haltere-to-wing homeotic transformation triggered by directed expression of *iab-4* in *Drosophila melanogaster*. (A) Wild-type haltere, which contains small lightly pigmented sensilla but lacks the triple row of sensory bristles seen in wings. (B) Haltere-to-wing transformation in a mild *Ubx* loss-of-function mutant background. Misexpression of *iab-4* miRNA hairpin in (C) *bx-Gal4/Y, UAS-DsRed-iab-4* and (D) *sd-Gal4, UAS-DsRed-iab-4* animals induces a similar haltere-to-wing transformation. Adapted from Ronshaugen et al. (2005), with permission.

subdivision of proliferating tissues into adjacent compartments. Cell interactions mediated by the Notch receptor are involved in the specification of wing compartment boundaries in *D. melanogaster*. Notch mediates boundary formation in the wing partly through repression of the miRNA bantam, which triggers cell proliferation through blocking the transcripts of *enabled*, an actin regulator. Thus proliferation rates are modulated by enabled and the miRNA bantam, which contributes to maintaining the dorsal—ventral affinity boundary (Becam et al., 2011).

In the beetle *T. castaneum*, miR-8 is highly expressed in the cuticle of late larval instars and its inhibition causes metamorphosis defects in wings, eyes, and legs. The mRNA of a number of genes related to metamorphosis, including *Wingless*, *Eyegone* (*Eyg*), *Farnesyl pyrophosphate synthase*, and *Semaphorin-1A*, contains putative miR-8-binding sites, and luciferase reporter assays and loss-of-function experiments indicate that *Eyg*, a key gene of eye development, is targeted by miR-8 (Wu et al., 2019). This suggests that miR-8 is essential, at least for the correct formation of the compound eye, during *T. castaneum* metamorphosis.

The adult abdomen of *D. melanogaster* is formed through a process of replacement of larval epidermal cells by adult cells. The cells that will form the adult epidermis, the abdominal histoblasts, are specified during embryogenesis and remain quiescent during larval development, without being wrapped in imaginal discs (see Chapter 5: The holometabolan development). During pupal development, when the abdominal histoblasts replace the larval abdominal cells, miR-965 controls histoblast proliferation and migration by regulating the genes *string* and *wingless*. Interestingly, ecdysteroid signaling downregulates miR-965 at the beginning of pupariation, linking activation of the histoblast cells to the endocrine control of metamorphosis (Verma and Cohen, 2015).

A process associated with holometabolan metamorphosis is the dramatic destruction of most tissues and organs and the formation of new ones that take place during the pupal stage. The first miRNA studied in this context was bantam, which was identified as a key modulator of apoptosis through the control of the proapoptotic gene *hid* in *D. melanogaster* (Brennecke et al., 2003). Moreover, bantam mediates the interaction between the epidermal growth factor receptor and Hippo growth control pathways in different tissues of *D. melanogaster*, notably in wing and eye development. The Hippo signaling pathway acts through the transcriptional coactivator Yorkie and p53, thus controlling the expression of the proapoptotic gene *reaper*. Yorkie further modulates *reaper* transcripts posttranscriptionally by regulating miR-2, which prevents apoptosis during metamorphosis (Zhang and Cohen, 2013). In the process of histolysis of larval salivary glands during the formation of the pupa of *D. melanogaster*, loss of miR-14 prevents the induction of autophagy during the gland cell death, whereas misexpression of miR-14 prematurely induces autophagy specifically in these glands.

miR-14 regulates this context-specific autophagy through its target, inositol 1,4,5-trisphosphate kinase 2, thereby affecting inositol 1,4,5-trisphosphate signaling and calcium levels during salivary gland cell death (Nelson et al., 2014).

Finally, miRNAs are also involved in the processes of neuronal reorganization that occur during metamorphosis. Thus *D. melanogaster* mutants devoid of let-7 and miR-125 expression display defects in the maturation of neuromuscular junctions of adult abdominal muscles (Caygill and Johnston, 2008). Also let-7 regulates the maturation of abdominal neuromuscular junctions during metamorphosis by targeting the transcripts of *abrupt*, which encodes a nuclear protein expressed in muscle cells (Caygill and Johnston, 2008). Moreover, let-7 knockout adults of *D. melanogaster* look morphologically normal but have juvenile neuromuscular features, which causes flight and motility deficiencies (Sokol et al., 2008). Regarding neurogenesis and neuromaturation, let-7 and miR-125 act on the transcription factor Chinmo, thus regulating temporal cell fate in the mushroom body lineage. Significantly, let-7 is activated in postmitotic neurons formed during pupal morphogenesis of *D. melanogaster*, when transitions among three mushroom body neuron subtypes take place (Wu et al., 2012).

Long noncoding RNAs

lncRNAs represent a broad category of ncRNAs that are over 200 nucleotides long, which can affect the function of genes through different mechanisms (Fatica and Bozzoni, 2014). A number of lncRNAs have been described as facilitators of epigenetic regulation. They can bind to particular targets and recruit chromatin-modifying enzymes, which initiates the formation of a silent or active chromatin state (Fatica and Bozzoni, 2014; Meller et al., 2015). The studies that relate lncRNAs and metamorphosis are few and only one provide functional data. However, several of them report correlations between the expression profiles of lncRNAs and progress of metamorphosis that are suggestive of cause–effect relationships.

Recent studies in *D. melanogaster* report the identification of lncRNAs and the quantification of their abundance in RNA libraries covering the entire development. Results reveal that the expression patterns of lncRNAs are very dynamic, but, interestingly, some of them concentrate their expression at metamorphosis (Chen et al., 2016). Similar correlations between high expression of given lncRNAs and metamorphosis have also been reported in the lepidopterans *Plutella xylostella* (Liu et al., 2017) and *B. mori* (Zhou et al., 2018). The latter study analyzes the differences of lncRNAs in silk glands between domesticated (*B. mori*) and wild (*Bombyx mandarina*) silkworms. As expected, many of the detected lncRNAs are expressed preferentially in the last larval instar, when the larva makes the cocoon (e.g., the lncRNA dw4sg_0040). However, some have their peak expression in the transition from pupa to adult (e.g., the lncRNA dw4sg_0178) (Zhou et al., 2018), which suggests that it may contribute to the regulation of adult morphogenesis.

A more recent study has shown that knockdown of the *dFIG4* gene specifically in the eye imaginal disc of *D. melanogaster* results in abnormal morphology of the adult compound eye, a phenotype known as the rough eye. A genetic screening led to identification of the gene *CR18854*, which suppresses the rough eye phenotype. The *CR18854* gene encodes a lncRNA of about 2566 bases that form a hairpin structure (Muraoka et al., 2018). Similarly, knockdown of the gene *Cabeza* in the eye imaginal disc of *D. melanogaster* triggers the rough eye phenotype, whereas normal eye development is restored by suppressing the activity of the gene hsrω, which codes for another lncRNA (Lo Piccolo and Yamaguchi, 2017). The mechanism by which CR18854 and hsrω interact with *dFIG4* and *Cabeza*, respectively, is enigmatic. In any case, the evidence suggests that lncRNAs such as CR18854 and hsrω are involved in a common pathway in the context of eye formation during holometabolan metamorphosis.

REFERENCES

Allis, C.D., Jenuwein, T., 2016. The molecular hallmarks of epigenetic control. Nat. Rev. Genet. 17, 487–500. Available from: https://doi.org/10.1038/nrg.2016.59.

Aparicio, R., Simoes Da Silva, C.J., Busturia, A., 2015. MicroRNA *miR-7* contributes to the control of *Drosophila* wing growth. Dev. Dyn. 244, 21–30. Available from: https://doi.org/10.1002/dvdy.24214.

Becam, I., Rafel, N., Hong, X., Cohen, S.M., Milan, M., 2011. Notch-mediated repression of bantam miRNA contributes to boundary formation in the *Drosophila* wing. Development 138, 3781–3789. Available from: https://doi.org/10.1242/dev.064774.

Belles, X., 2017. MicroRNAs and the evolution of insect metamorphosis. Annu. Rev. Entomol. 62, 111–125. Available from: https://doi.org/10.1146/annurev-ento-031616-034925.

Belles, X., Santos, C.G., 2014. The MEKRE93 (Methoprene tolerant-Küppel homolog 1-E93) pathway in the regulation of insect metamorphosis, and the homology of the pupal stage. Insect Biochem. Mol. Biol. 52, 6068. Available from: https://doi.org/10.1016/j.ibmb.2014.06.009.

Bewick, A.J., Sanchez, Z., Mckinney, E.C., Moore, A.J., Moore, P.J., Schmitz, R.J., 2019. Dnmt1 is essential for egg production and embryo viability in the large milkweed bug, *Oncopeltus fasciatus*. Epigenet. Chromatin 12, 6. Available from: https://doi.org/10.1186/s13072-018-0246-5.

Bewick, A.J., Vogel, K.J., Moore, A.J., Schmitz, R.J., 2016. Evolution of DNA methylation across insects. Mol. Biol. Evol 34, 654–665. Available from: https://doi.org/10.1093/molbev/msw264.

Biryukova, I., Asmar, J., Abdesselem, H., Heitzler, P., 2009. *Drosophila* mir-9a regulates wing development via fine-tuning expression of the LIM only factor, dLMO. Dev. Biol. 327, 487–496. Available from: https://doi.org/10.1016/j.ydbio.2008.12.036.

Bodai, L., Zsindely, N., Gáspár, R., Kristó, I., Komonyi, O., Boros, I.M., 2012. Ecdysone induced gene expression is associated with acetylation of histone H3 lysine 23 in *Drosophila melanogaster*. PLoS One 7, e40565. Available from: https://doi.org/10.1371/journal.pone.0040565.

Borsos, B.N., Pankotai, T., Kovács, D., Popescu, C., Páhi, Z., Boros, I.M., 2015. Acetylations of Ftz-F1 and histone H4K5 are required for the fine-tuning of ecdysone biosynthesis during *Drosophila* metamorphosis. Dev. Biol. 404, 80–87. Available from: https://doi.org/10.1016/j.ydbio.2015.04.020.

Boulanger, M.-C., Miranda, T.B., Clarke, S., di Fruscio, M., Suter, B., Lasko, P., et al., 2004. Characterization of the *Drosophila* protein arginine methyltransferases DART1 and DART4. Biochem. J. 379, 283–289. Available from: https://doi.org/10.1042/bj20031176.

Brennecke, J., Hipfner, D.R., Stark, A., Russell, R.B., Cohen, S.M., 2003. Bantam encodes a developmentally regulated microRNA that controls cell proliferation and regulates the proapoptotic gene *hid* in *Drosophila*. Cell 113, 25–36.

Cakouros, D., Daish, T.J., Mills, K., Kumar, S., 2004. An arginine-histone methyltransferase, CARMER, coordinates ecdysone-mediated apoptosis in *Drosophila* cells. J. Biol. Chem. 279, 18467–18471. Available from: https://doi.org/10.1074/jbc.M400972200.

Cakouros, D., Mills, K., Denton, D., Paterson, A., Daish, T., Kumar, S., 2008. dLKR/SDH regulates hormone-mediated histone arginine methylation and transcription of cell death genes. J. Cell Biol. 182, 481–495. Available from: https://doi.org/10.1083/jcb.200712169.

Carbonell, A., Mazo, A., Serras, F., Corominas, M., 2013. Ash2 acts as an ecdysone receptor coactivator by stabilizing the histone methyltransferase Trr. Mol. Biol. Cell 24, 361–372. Available from: https://doi.org/10.1091/mbc.e12-04-0267.

Carré, C., Szymczak, D., Pidoux, J., Antoniewski, C., 2005. The histone H3 acetylase dGcn5 is a key player in *Drosophila melanogaster* metamorphosis. Mol. Cell. Biol. 25, 8228–8238. Available from: https://doi.org/10.1128/MCB.25.18.8228-8238.2005.

Caygill, E.E., Johnston, L.A., 2008. Temporal regulation of metamorphic processes in *Drosophila* by the let-7 and miR-125 heterochronic microRNAs. Curr. Biol. 18, 943–950. Available from: https://doi.org/10.1016/j.cub.2008.06.020.

Cech, T.R., Steitz, J.A., 2014. The noncoding RNA revolution-trashing old rules to forge new ones. Cell 157, 77–94. Available from: https://doi.org/10.1016/j.cell.2014.03.008.

Chen, B., Zhang, Y., Zhang, X., Jia, S., Chen, S., Kang, L., 2016. Genome-wide identification and developmental expression profiling of long noncoding RNAs during *Drosophila* metamorphosis. Sci. Rep. 6, 23330. Available from: https://doi.org/10.1038/srep23330.

Cheung, P., Lau, P., 2005. Epigenetic regulation by histone methylation and histone variants. Mol. Endocrinol. 19, 563–573. Available from: https://doi.org/10.1210/me.2004-0496.

Denton, D., Aung-Htut, M.T., Lorensuhewa, N., Nicolson, S., Zhu, W., Mills, K., et al., 2013. UTX coordinates steroid hormone-mediated autophagy and cell death. Nat. Commun. 4, 2916. Available from: https://doi.org/10.1038/ncomms3916.

Drelon, C., Belalcazar, H.M., Secombe, J., 2018. The histone demethylase KDM5 is essential for larval growth in *Drosophila*. Genetics 209, 773–787. Available from: https://doi.org/10.1534/genetics.118.301004.

Fatica, A., Bozzoni, I., 2014. Long non-coding RNAs: new players in cell differentiation and development. Nat. Rev. Genet. 15, 7–21. Available from: https://doi.org/10.1038/nrg3606.

Fenley, A.T., Adams, D.A., Onufriev, A.V., 2010. Charge state of the globular histone core controls stability of the nucleosome. Biophys. J. 99, 1577–1585. Available from: https://doi.org/10.1016/j.bpj.2010.06.046.

Fernandez-Nicolas, A., Belles, X., 2016. CREB-binding protein contributes to the regulation of endocrine and developmental pathways in insect hemimetabolan pre-metamorphosis. Biochim. Biophys. Acta Gen. Subj. 1860, 508–515. Available from: https://doi.org/10.1016/j.bbagen.2015.12.008.

George, S., Gaddelapati, S.C., Palli, S.R., 2019. Histone deacetylase 1 suppresses Krüppel homolog 1 gene expression and influences juvenile hormone action in *Tribolium castaneum*. Proc. Natl. Acad. Sci. U.S.A. 116, 17759–17764. Available from: https://doi.org/10.1073/pnas.1909554116.

Glastad, K.M., Hunt, B.G., Goodisman, M.A.D., 2019. Epigenetics in insects: genome regulation and the generation of phenotypic diversity. Annu. Rev. Entomol. 64, 185–203. Available from: https://doi.org/10.1146/annurev-ento-011118-111914.

Gomez-Orte, E., Belles, X., 2009. MicroRNA-dependent metamorphosis in hemimetabolan insects. Proc. Natl. Acad. Sci. U.S.A. 106, 21678–21682. Available from: https://doi.org/10.1073/pnas.0907391106.

Hödl, M., Basler, K., 2012. Transcription in the absence of histone H3.2 and H3K4 methylation. Curr. Biol. 22, 2253–2257. Available from: https://doi.org/10.1016/j.cub.2012.10.008.

Huang, J.-H., Lozano, J., Belles, X., 2013. Broad-complex functions in postembryonic development of the cockroach *Blattella germanica* shed new light on the evolution of insect metamorphosis. Biochim. Biophys. Acta Gen. Subj. 1830, 2178–2187. Available from: https://doi.org/10.1016/j.bbagen.2012.09.025.

Kimura, S., Sawatsubashi, S., Ito, S., Kouzmenko, A., Suzuki, E., Zhao, Y., et al., 2008. *Drosophila* arginine methyltransferase 1 (DART1) is an ecdysone receptor co-repressor. Biochem. Biophys. Res. Commun. 371, 889–893. Available from: https://doi.org/10.1016/j.bbrc.2008.05.003.

Liu, F., Guo, D., Yuan, Z., Chen, C., Xiao, H., 2017. Genome-wide identification of long non-coding RNA genes and their association with insecticide resistance and metamorphosis in diamondback moth, *Plutella xylostella*. Sci. Rep. 7, 15870. Available from: https://doi.org/10.1038/s41598-017-16057-2.

Liu, Z., Ling, L., Xu, J., Zeng, B., Huang, Y., Shang, P., et al., 2018. MicroRNA-14 regulates larval development time in *Bombyx mori*. Insect Biochem. Mol. Biol. 93, 57–65. Available from: https://doi.org/10.1016/j.ibmb.2017.12.009.

Lo Piccolo, L., Yamaguchi, M., 2017. RNAi of arcRNA hsrω affects sub-cellular localization of *Drosophila* FUS to drive neurodiseases. Exp. Neurol. 292, 125–134. Available from: https://doi.org/10.1016/j.expneurol.2017.03.011.

Loubière, V., Delest, A., Thomas, A., Bonev, B., Schuettengruber, B., Sati, S., et al., 2016. Coordinate redeployment of PRC1 proteins suppresses tumor formation during *Drosophila* development. Nat. Genet. 48, 1436–1442. Available from: https://doi.org/10.1038/ng.3671.

Lozano, J., Belles, X., 2011. Conserved repressive function of Krüppel homolog 1 on insect metamorphosis in hemimetabolous and holometabolous species. Sci. Rep. 1, 163. Available from: https://doi.org/10.1038/srep00163.

Lozano, J., Montañez, R., Belles, X., 2015. MiR-2 family regulates insect metamorphosis by controlling the juvenile hormone signaling pathway. Proc. Natl. Acad. Sci. U.S.A. 112, 3740–3745. Available from: https://doi.org/10.1073/pnas.1418522112.

Lu, Y.-X., Denlinger, D.L., Xu, W.-H., 2013. Polycomb repressive complex 2 (PRC2) protein ESC regulates insect developmental timing by mediating H3K27me3 and activating prothoracicotropic hormone gene expression. J. Biol. Chem. 288, 23554–23564. Available from: https://doi.org/10.1074/jbc.M113.482497.

Lyko, F., 2018. The DNA methyltransferase family: a versatile toolkit for epigenetic regulation. Nat. Rev. Genet. 19, 81–92. Available from: https://doi.org/10.1038/nrg.2017.80.

Lyko, F., Ramsahoye, B.H., Jaenisch, R., 2000. DNA methylation in *Drosophila melanogaster*. Nature 408, 538–540. Available from: https://doi.org/10.1038/35046205.

McKay, D.J., Klusza, S., Penke, T.J.R., Meers, M.P., Curry, K.P., McDaniel, S.L., et al., 2015. Interrogating the function of metazoan histones using engineered gene clusters. Dev. Cell 32, 373–386. Available from: https://doi.org/10.1016/j.devcel.2014.12.025.

Meller, V.H., Joshi, S.S., Deshpande, N., 2015. Modulation of chromatin by noncoding RNA. Annu. Rev. Genet. 49, 673–695. Available from: https://doi.org/10.1146/annurev-genet-112414-055205.

Muraoka, Y., Nakamura, A., Tanaka, R., Suda, K., Azuma, Y., Kushimura, Y., et al., 2018. Genetic screening of the genes interacting with *Drosophila* FIG4 identified a novel link between CMT-causing gene and long noncoding RNAs. Exp. Neurol. 310, 1–13. Available from: https://doi.org/10.1016/j.expneurol.2018.08.009.

Nelson, C., Ambros, V., Baehrecke, E.H., 2014. miR-14 regulates autophagy during developmental cell death by targeting ip3-kinase 2. Mol. Cell 56, 376–388. Available from: https://doi.org/10.1016/j.molcel.2014.09.011.

Ozawa, T., Mizuhara, T., Arata, M., Shimada, M., Niimi, T., Okada, K., et al., 2016. Histone deacetylases control module-specific phenotypic plasticity in beetle weapons. Proc. Natl. Acad. Sci. U.S.A. 113, 15042–15047. Available from: https://doi.org/10.1073/pnas.1615688114.

Pankotai, T., Popescu, C., Martín, D., Grau, B., Zsindely, N., Bodai, L., et al., 2010. Genes of the ecdysone biosynthesis pathway are regulated by the dATAC histone acetyltransferase complex in *Drosophila*. Mol. Cell. Biol. 30, 4254–4266. Available from: https://doi.org/10.1128/MCB.00142-10.

Pérez-Lluch, S., Blanco, E., Tilgner, H., Curado, J., Ruiz-Romero, M., Corominas, M., et al., 2015. Absence of canonical marks of active chromatin in developmentally regulated genes. Nat. Genet. 47, 1158–1167. Available from: https://doi.org/10.1038/ng.3381.

Provataris, P., Meusemann, K., Niehuis, O., Grath, S., Misof, B., 2018. Signatures of DNA methylation across insects suggest reduced DNA methylation levels in holometabola. Genome Biol. Evol 10, 1185–1197. Available from: https://doi.org/10.1093/gbe/evy066.

Rickels, R., Herz, H.-M., Sze, C.C., Cao, K., Morgan, M.A., Collings, C.K., et al., 2017. Histone H3K4 monomethylation catalyzed by Trr and mammalian COMPASS-like proteins at enhancers is dispensable for development and viability. Nat. Genet. 49, 1647–1653. Available from: https://doi.org/10.1038/ng.3965.

Ronshaugen, M., Biemar, F., Piel, J., Levine, M., Lai, E.C., 2005. The *Drosophila* microRNA iab-4 causes a dominant homeotic transformation of halteres to wings. Genes Dev. 19, 2947–2952. Available from: https://doi.org/10.1101/gad.1372505.

Roy, A., George, S., Palli, S.R., 2017. Multiple functions of CREB-binding protein during post-embryonic development: identification of target genes. BMC Genom. 18, 996. Available from: https://doi.org/10.1186/s12864-017-4373-3.

Rubio, M., Belles, X., 2013. Subtle roles of microRNAs let-7, miR-100 and miR-125 on wing morphogenesis in hemimetabolan metamorphosis. J. Insect Physiol. 59, 1089–1094. Available from: https://doi.org/10.1016/j.jinsphys.2013.09.003.

Schuettengruber, B., Bourbon, H.-M., Di Croce, L., Cavalli, G., 2017. Genome regulation by Polycomb and Trithorax: 70 years and counting. Cell 171, 34–57. Available from: https://doi.org/10.1016/j.cell.2017.08.002.

Schulz, N.K.E., Wagner, C.I., Ebeling, J., Raddatz, G., Diddens-de Buhr, M.F., Lyko, F., et al., 2018. Dnmt1 has an essential function despite the absence of CpG DNA methylation in the red flour beetle *Tribolium castaneum*. Sci. Rep. 8, 16462. Available from: https://doi.org/10.1038/s41598-018-34701-3.

Sedkov, Y., Cho, E., Petruk, S., Cherbas, L., Smith, S.T., Jones, R.S., et al., 2003. Methylation at lysine 4 of histone H3 in ecdysone-dependent development of *Drosophila*. Nature 426, 78–83. Available from: https://doi.org/10.1038/nature02080.

Sempere, L.F., Sokol, N.S., Dubrovsky, E.B., Berger, E.M., Ambros, V., 2003. Temporal regulation of microRNA expression in *Drosophila melanogaster* mediated by hormonal signals and Broad-Complex gene activity. Dev. Biol. 259, 9–18.

Sokol, N.S., Xu, P., Jan, Y.-N., Ambros, V., 2008. *Drosophila* let-7 microRNA is required for remodeling of the neuromusculature during metamorphosis. Genes Dev. 22, 1591–1596. Available from: https://doi.org/10.1101/gad.1671708.

Song, J., Li, W., Zhao, H., Gao, L., Fan, Y., Zhou, S., 2018. The microRNAs let-7 and miR-278 regulate insect metamorphosis and oogenesis by targeting the juvenile hormone early-response gene *Krüppel-homolog 1*. Development 145, dev170670. Available from: https://doi.org/10.1242/dev.170670.

Talbert, P.B., Henikoff, S., 2010. Histone variants – ancient wrap artists of the epigenome. Nat. Rev. Mol. Cell Biol. 11, 264–275. Available from: https://doi.org/10.1038/nrm2861.

Tsurumi, A., Dutta, P., Shang, R., Yan, S.-J., Sheng, R., Li, W.X., 2013. *Drosophila* Kdm4 demethylases in histone H3 lysine 9 demethylation and ecdysteroid signaling. Sci. Rep. 3, 2894. Available from: https://doi.org/10.1038/srep02894.

Varghese, J., Cohen, S.M., 2007. microRNA miR-14 acts to modulate a positive autoregulatory loop controlling steroid hormone signaling in *Drosophila*. Genes Dev. 21, 2277–2282. Available from: https://doi.org/10.1101/gad.439807.

Verma, P., Cohen, S.M., 2015. miR-965 controls cell proliferation and migration during tissue morphogenesis in the *Drosophila* abdomen. Elife 4, e07389. Available from: https://doi.org/10.7554/eLife.07389.

Waddington, C.H., 1957. The Strategy of the Genes. Allen & Unwin, London.

Wang, Y., Jorda, M., Jones, P.L., Maleszka, R., Ling, X., Robertson, H.M., et al., 2006. Functional CpG methylation system in a social insect. Science 314, 645–647.

Wu, Y.-C., Chen, C.-H., Mercer, A., Sokol, N.S., 2012. Let-7-complex microRNAs regulate the temporal identity of *Drosophila* mushroom body neurons via chinmo. Dev. Cell 23, 202–209. Available from: https://doi.org/10.1016/j.devcel.2012.05.013.

Wu, W., Zhai, M., Li, C., Yu, X., Song, X., Gao, S., et al., 2019. Multiple functions of miR-8-3p in the development and metamorphosis of the red flour beetle, *Tribolium castaneum*. Insect Mol. Biol. 28, 208–221. Available from: https://doi.org/10.1111/imb.12539.

Zhang, W., Cohen, S.M., 2013. The Hippo pathway acts via p53 and microRNAs to control proliferation and proapoptotic gene expression during tissue growth. Biol. Open 2, 822–828. Available from: https://doi.org/10.1242/bio.20134317.

Zhang, T., Song, W., Li, Z., Qian, W., Wei, L., Yang, Y., et al., 2018. Krüppel homolog 1 represses insect ecdysone biosynthesis by directly inhibiting the transcription of steroidogenic enzymes. Proc. Natl. Acad. Sci. U.S.A. 115, 3960–3965. Available from: https://doi.org/10.1073/pnas.1800435115.

Zhou, Q.-Z., Fang, S.-M., Zhang, Q., Yu, Q.-Y., Zhang, Z., 2018. Identification and comparison of long non-coding RNAs in the silk gland between domestic and wild silkworms. Insect Sci. 25, 604–616. Available from: https://doi.org/10.1111/1744-7917.12443.

Zwier, M.V., Verhulst, E.C., Zwahlen, R.D., Beukeboom, L.W., van de Zande, L., 2012. DNA methylation plays a crucial role during early *Nasonia* development. Insect Mol. Biol. 21, 129–138. Available from: https://doi.org/10.1111/j.1365-2583.2011.01121.x.

Chapter 9

Molting: the basis for growing and for changing the form

The chitin-based cuticle that forms the rigid exoskeleton of insects gives them many advantages, as it supports muscle attachment for locomotion, including flight, prevents physical and chemical damage, and protects against water loss and infectious diseases. These advantages largely explain the extraordinary evolutionary success of insects. Chitin is also an important layer in internal structures, including the tracheal system, inner cuticular linings of the alimentary canal, the ducts of the genital apparatus, and those of dermal glands. A rigid exoskeleton, however, poses a problem for growth, and insects have solved that problem by means of molting, by which the old cuticle that has become too small to allow further growth is replaced by a new, more spacious one. Moreover, molting is not only a mechanism that solves the problem of growth, but it also serves to allow changes in form. Metamorphosis, thus, requires molting. We will see in this chapter that the cuticle is formed with materials produced by the subjacent epidermal cells through a sequence of complex and perfectly coordinated steps, thanks to the action of hormones that regulate the whole process. As discussed in Chapter 6 (Hormones involved in the regulation of metamorphosis), the most important effector of the molting process is the ecdysone produced in the prothoracic gland (PG) and transformed into the biologically more active derivative 20-hydroxyecdysone (20E) in peripheral tissues.

THE INSECT CUTICLE

The cuticle is composed of chitin, which is a polysaccharide, together with diverse cuticular proteins, lipids, catecholamines, and minerals. The cuticle is distributed in well-differentiated horizontal layers, which rest on the apical plasma membrane of the epidermal cells (Moussian, 2010) (Fig. 9.1). Cuticular proteins (and potentially chitin) are cross-linked by *o*-quinones, which are oxidized catecholamines derived from the amino acid tyrosine through the action of different enzymes, including laccase-2 (Zhu et al., 2016). Considering the physiological properties and biochemical composition, two main cuticular layers have been distinguished: the inner chitinous

Insect Metamorphosis. DOI: https://doi.org/10.1016/B978-0-12-813020-9.00009-0

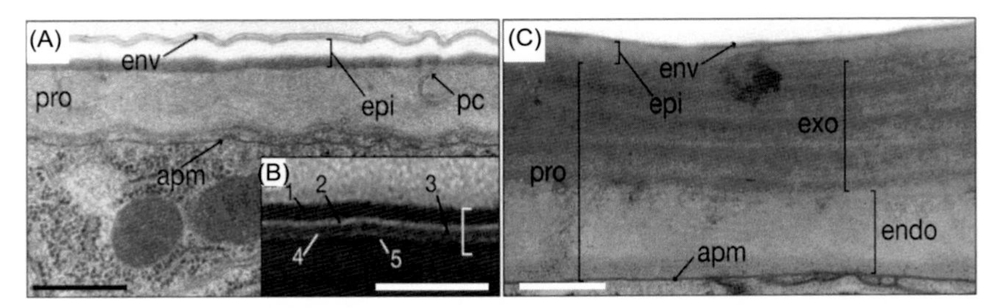

FIGURE 9.1 The ultrastructure of the cuticle in the fruit fly *Drosophila melanogaster*. (A) Larval cuticle produced during embryogenesis showing the three layers defined by their mode of formation. The procuticle (pro) is the inner chitin–protein matrix attached to the apical plasma membrane (apm) of the epidermal cells. It is overlain by the protein network of the epicuticle (epi). The outermost envelope (env) with five alternating electron-dense and electron-lucid sheets shown with a bracket in the inset (B). (C) The three layers of adult cuticle built during metamorphosis. The procuticle, which is thicker than in the larva, is subdivided into the exocuticle (exo) and the endocuticle (endo). Scale bars: (A) and (C) 0.5 μm and (B) 0.1 μm. From Moussian (2010), with permission.

procuticle and the outer chitin-free epicuticle (Neville, 1975). On the basis of ultrastructural features, the procuticle is subdivided into a lower endocuticle and an upper exocuticle. In turn, the epicuticle is divided into an inner and an outer epicuticle, the latter is also known as the cuticulin layer. Moreover, the surface of the cuticle is usually covered by wax and cement layers. Considering the mode of formation, the sublayers of the epicuticle may be considered as separate entities; this has favored a three-layered cuticle nomenclature: the envelope (outer epicuticle or cuticulin layer), the epicuticle (the inner epicuticle), and the procuticle (Fig. 9.1).

DETERMINATION OF MOLTING TIME: BASIC MECHANISMS

Due to the rigid cuticle, the increase of body volume within an instar is limited, and a subsequent molt is triggered when the insect reaches a given size. When this is attained, the insect usually stops feeding, enters into a wandering stage, and then molts. However, the mechanisms by which size is internally sensed are not well understood. Two main triggering mechanisms have been proposed. One is based on cuticle stretch reception, and the other invokes a decoupling of whole-body growth from the growth of the oxygen supply system.

The influence of stretch reception has been experimentally tested with injections of saline solutions, which trigger a premature molt, as shown, for example, in the milkweed bug *Oncopeltus fasciatus* (Nijhout, 1994). The fact that a saline injection only stretches the abdominal wall suggests that molting is triggered by stimulation of abdominal stretch receptors.

The mechanism may be similar to that operating in bloodsucking bugs, like *Rhodnius prolixus* and *Dipetalogaster maximus*, which sense a blood meal through abdominal stretch receptors, resulting in molting. However, in other insects, like in the hawk moth *Manduca sexta*, artificial stretch of the body wall does not trigger a molt, which suggests that abdominal stretch is not the mechanism used by caterpillars for body size sensing (Nijhout, 1994).

The proponents of the mechanism based on a decoupling of body growth and capacity for oxygen supply argue that as body mass increases within an instar, the demand for oxygen also increases, but the fixed tracheal system does not allow a corresponding increase in oxygen supply (Callier and Nijhout, 2011; Greenlee and Harrison, 2004) (Fig. 9.2). In insects, the atmospheric air enters through spiracles in the body wall and passes into tracheal tubes that develop from invaginations of the exoskeletal integument. The tracheal system is flexible enough to respond to varying O_2 demands, thus small tracheoles not sclerotized can grow and increase branching to improve O_2 delivery during the intermolt period. However, larger sclerotized tracheal tubes can only grow when the insect molts (Wigglesworth, 1983). Therefore at certain critical points in development, the O_2 delivery capacity is unable to meet O_2 demands. During the intermolt periods, when insect body mass can double, O_2 demand surpasses the almost fixed tracheal system supply. For example, during the fifth nymphal instar, juvenile *Schistocerca americana* locusts increase their body mass by 90%, but mass-specific metabolic rate drops by 15%, which suggests a mismatch between O_2 demand and delivery (Greenlee and Harrison, 2004). It has been proposed that an increase in body mass during the intermolt period compresses the air-filled tracheal

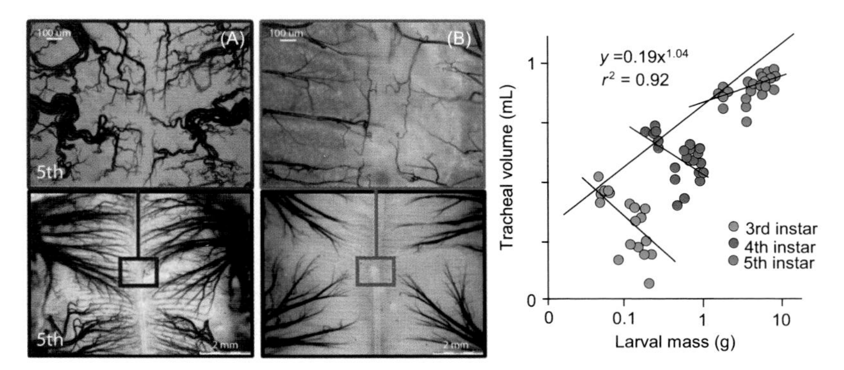

FIGURE 9.2 Tracheal system and body growth in the hawk moth *Manduca sexta*. (A) Tracheal network along the midline of the midgut at the beginning and end of the fifth larval instar. The comparison of both situations shows that the tracheal system does not grow within each instar, but the body does. (B) Relationships between the larval mass and the tracheal volume in three larval instars. They show that the tracheal system does not grow within each instar but increases in size discretely at each molt; there is an apparent slight decrease in tracheal volume during the third and fourth larval instars. From Callier and Nijhout (2011), with permission.

system, reducing O_2 delivery in late-stage insects (Greenlee and Harrison, 2004). This leads to the idea that specific threshold levels of hypoxia could trigger a molting process. In other words, in each instar, the insect has an oxygen delivery ceiling imposed by the constant dimension of the tracheal system, and the mechanism that senses this ceiling defines the critical size for molting (Callier and Nijhout, 2011).

THE APOLYSIS AND THE SECRETION OF A NEW CUTICLE

When the insect is going to molt, it stops feeding and becomes practically inactive. The first molting step that occurs is the apolysis, that is, the separation of the old cuticle from the epidermis and the secretion of a molting fluid from certain epidermal cells and dermal glands. The molting fluid fills the space between the old and new cuticles during the molting process. Chitinase, β-acetylglucosaminidase, and peptidases have been identified in molting fluids. Chitinases bind to the old cuticle to degrade chitin, and peptidases transform cuticle proteins into polypeptide fragments and free amino acids for recycling. Additional proteins have been recently characterized through proteomic and functional analyses, a number of them being associated with immune response (Zhang et al., 2014).

As apolysis begins, the epidermis starts a cycle of mitosis and cell divisions, subsequently forming the cuticulin layer of the epicuticle. This cuticulin layer is constituted by lipoproteins that combine with other proteins so that the ensemble becomes sclerotized. In this way, the epicuticle acquires resistance against chemical and enzymatic aggressions (Moussian, 2010; Neville, 1975). Importantly, as the new cuticle has to have a greater extension, while it is being formed it folds in the space available (Fig. 9.3). Once the new epicuticle has been laid, the molting fluid becomes active and starts to digest the old cuticle, beginning with the endocuticle basal layer. As the epicuticle is not affected by the enzymes of the molting fluid, the new cuticle and epidermal cells are protected from it. The products of the digestion of the old cuticle, mainly *N*-acetylglucosamine (derived from chitin), short peptides, and amino acids (Zhang et al., 2014), are continuously absorbed by the cuticle that is being formed. The old exocuticle and epicuticle cannot be digested, but these layers represent between 20% and 30%, respectively, of the whole cuticle. Thus most of the proteins and carbohydrates of the old cuticle are reutilized in the new one.

Epidermal cells communicate with the cuticle that is being formed through a system of fine ducts. Depending on the species, there are between 50 and 200 of these ducts that arise from each epidermal cell. Once the old endocuticle has been digested, the molting fluid is absorbed and the space it occupied becomes practically dry. Immediately, the epidermal cells secrete a complex mixture of carbohydrates and resins that flows through the fine

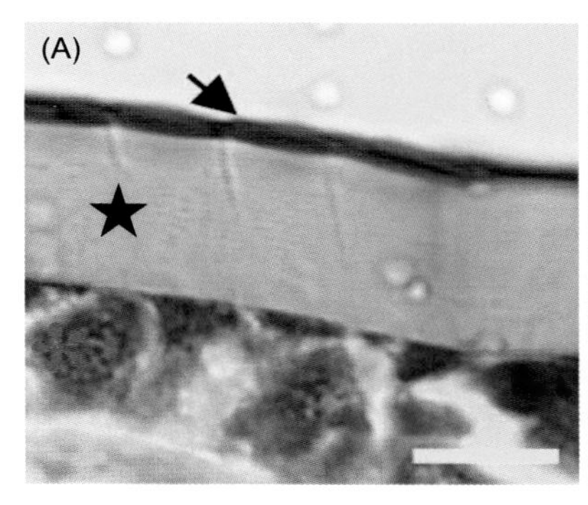

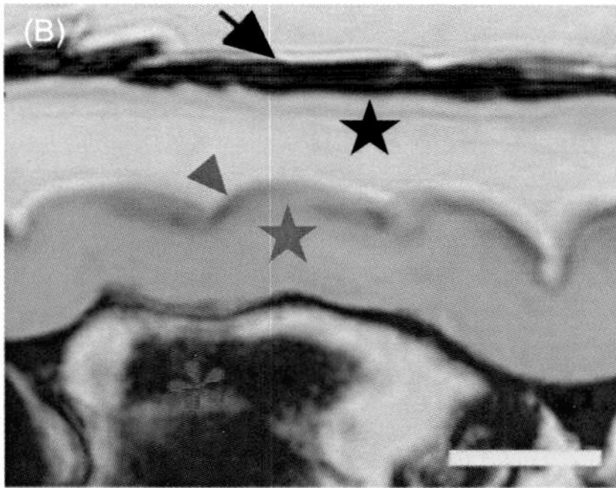

FIGURE 9.3 The apolysis and production of a new cuticle in the cockroach *Blattella germanica*. (A) The main cuticle layers in the last nymphal instar: epicuticle (arrow) and procuticle (star) and the epidermal cells (asterisk). (B) Secretion of new cuticle during molting to the adult stage, with the old epicuticle (arrow) and procuticle (black star), and the new procuticle (red arrowhead) and epicuticle (red star) being secreted by the epidermal cells (asterisk). Note that the new cuticle folds while it is formed. Scale bar: 10 μm. From Mané-Padrós et al. (2005), with permission.

ducts to the outside and is distributed over the new epicuticle to form an impermeable layer of waxes, which protects it from the external water and prevents the loss of internal water (Neville, 1975). At that moment, the insect is ready to detach itself from the undigested remains of the former exocuticle and epicuticle (which is called exuvia) through the process known as ecdysis.

THE ECDYSIS

The ecdysis, which essentially consists in the detachment of the remains of the old cuticle or exuvia. While the new cuticle is produced and the old one is digested, the insect has consolidated the musculature, particularly the musculature of the abdomen, which will play a fundamental role in the ecdysis. The abdominal muscles form bands that go from one segment to another and are structured in such a way that they can cause notable increases in hemolymph pressure. When the new cuticle is formed and the exuvia has dried, the abdominal muscles contract and push the hemolymph toward the front of the body, while the insect drinks liquid or takes in air, which further increases the pressure. This pressure is what causes the old cuticle to break, so that the insect, under its new cuticle, leaves the remains of the old one (Ewer and Reynolds, 2002; Truman, 2005).

The ecdysis process is composed of two behavioral main steps such as preecdysis and ecdysis, which are characterized by distinct contractions of skeletal muscles. In the case of the adult ecdysis (which is also called adult emergence or eclosion), these two steps can be separated by a quiescent period of approximately 20–30 minutes. In the preecdysis step, the behavior consists of various contractions of the head, thorax, abdomen, and their appendages. These contractions can last between 1 and 2 hours and are required for the loosening and splitting of the old cuticle. Subsequently, the ecdysis step is characterized by peristaltic movements of the abdomen and various contractions of legs and/or prolegs that last between 10 and 15 minutes. These abdominal peristaltic movements are similar in most insects and are required for the complete shedding of the exuvia (Zitnan et al., 2007).

A spectacular example of the effects of hemolymph pressure is the ptilinum. In flies, the pupal stage takes place inside the puparium, which is the hardened barrel-shaped cover that derives from the tegument of the last larval instar. The adult fly has a sort of longitudinal groove in the front part of the head, between the eyes, which is nothing more than an inward fold of the cuticle and is called the ptilinum. At the beginning of the ecdysis, when hemolymph pressure increases by the contraction of the abdominal muscles, the fold projects forward and swells like a balloon. By means of expansions and retraction, the ptilinum is used to break the puparium, so that the adult fly can emerge from it (Gibson et al., 2014) (Fig. 9.4).

The exuvia contains a significant proportion of sclerotins, proteins that give hardness to the cuticle and that are not digested by the molting fluid. Therefore the large increase in hemolymph pressure achieved by the insect would not be enough to break the exuvia, especially in the harder parts, such as the cephalic capsule or the thorax, were it not for the existence of rupture lines. They are not sclerotized and can be observed as whitish lines in the

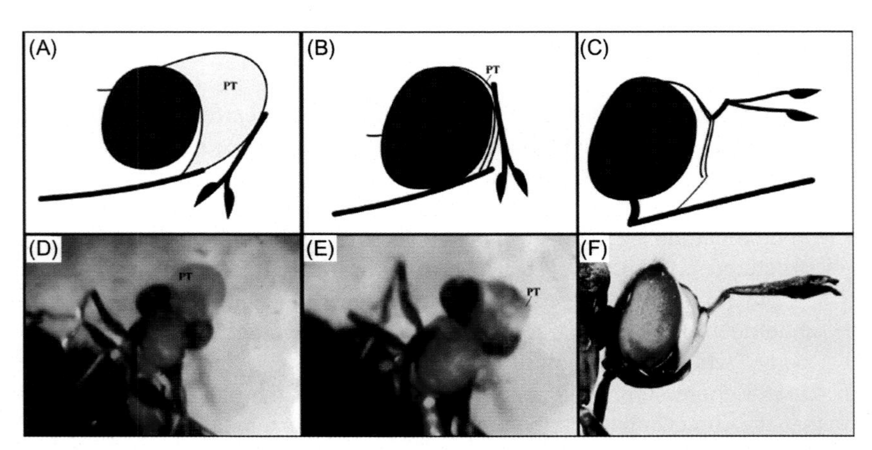

FIGURE 9.4 Ptilinum changes during the imaginal ecdysis of the fly *Physocephala tibialis*. (A) Lateral view with the ptilinum (PT) inflated. (B) Lateral view with the ptilinum deflated. (C) Lateral view after sclerotization. (D) Dorsal view with the ptilinum inflated. (E) Dorsal view with ptilinum deflated. (F) Lateral view after sclerotization. Images are extracted from a VIDEO recording from Gibson et al. (2014), with permission.

active insect. The rupture lines are not detrimental to the exoskeleton resistance because of their limited extent and because they are flanked by regions with complete cuticle. However, when the exuvia dries, they become lines of low resistance, so that the pressure generated is sufficient to tear them (Reynolds, 1980).

SCLEROTIZATION AND HARDENING

Immediately after the exuvia has been shed off, the insect's integument is pale and soft. It is time to expand the new cuticle as much as possible, for which the insect takes in air and drives the abdominal muscles to increase the internal pressure. In blowflies of the genus *Sarcophaga*, for example, the internal pressure can reach 95 mm of mercury, which allows the cuticle to expand. In the case of the imaginal molt, the expansion and stretching of the wings are especially delicate, involving great vulnerability, and it is energetically costly, since the insect must impel the hemolymph so that it arrives to the most remote and narrow body places, like the distal end of the wing veins. Once the body and its appendages have been distended, and the wings have been stretched in the case of the adult molt, the process culminates with the sclerotization and hardening of the new cuticle. This takes place simultaneously in all parts of the body although each local region of the cuticle is sclerotized in such a way and to such an extent that optimal functional properties are obtained. During cuticular sclerotization, two acyldopamines, *N*-acetyldopamine and *N*-β-alanyldopamine, are incorporated into the cuticular matrix (Andersen, 2010). The incorporation can occur before the ecdysis (preecdysial sclerotization) or soon after the ecdysis (postecdysial sclerotization).

The biochemical pathway leading to cuticle tanning has been described by Andersen (2010). Briefly, it comprises the hydroxylation of tyrosine to 3,4-dihydroxyphenylalanine (dopa) and decarboxylation of dopa to dopamine, which is the precursor for the formation of melanin, *N*-acetyldopamine, and *N*-β-alanyldopamine. These two catechols serve as precursors of quinones that can react with amino acid residues in the cuticular proteins and with chitin, thereby forming crosslinks and stabilizing and hardening the cuticle. In many species, sclerotization results in the formation of brownish colors, although nearly colorless and completely black cuticles based on melanin can also be produced.

The newly molted insect becomes sclerotized and acquires color in over the course of an hour after the ecdysis, but the hardening process, by which layers of chitin and other proteins are added, can last for several days and even weeks, until the next apolysis. An insect of remarkable cuticular hardness like the stag beetle *Lucanus cervus* can take about 3 weeks to complete the process of hardening after the imaginal molt, during which the insect triples the thickness of the chitin layer. Some insect species deposit the cuticle layer following one orientation of the chitin fibers at night and another orientation during the day. Thus if a cut of the cuticle is observed under polarized light, the different deposited layers can be counted, which may allow to determine the age in days of an adult insect (Nijhout, 1994).

ENDOCRINE REGULATION OF THE MOLTING PROCESS

Each molting process begins with the secretion of the prothoracicotropic hormone (PTTH), which triggers a pulse of ecdysone production in the PG, which is transformed into 20E in peripheral tissues. Remarkably, the successive phases of the ecdysteroid pulse are correlated with distinct activities in epidermal cells and the formation of the new cuticle (Charles, 2010; Nijhout, 1994). DNA synthesis or mitoses occur while ecdysteroid concentration is rising. Subsequently, the new epicuticle is deposited while ecdysteroids are at about peak levels. Finally, the bulk of the cuticle is deposited while the ecdysteroid titers are declining (Fig. 9.5).

Successful performance of the ecdysis stereotyped behavior is under the control of neuropeptides from the central nervous system (CNS) and peptide hormones produced by Inka cells (Mena et al., 2016; Roller et al.,

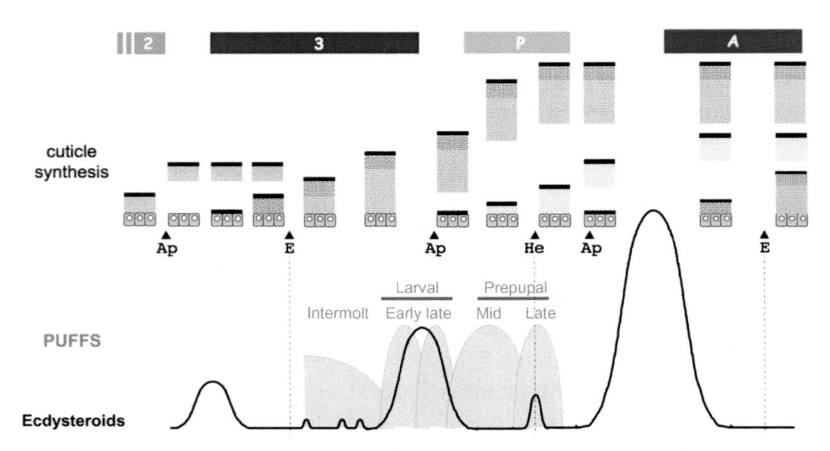

FIGURE 9.5 Cuticle synthesis and ecdysteroid titers during late postembryonic development of *Drosophila melanogaster*. Top: The bars represent the periods of cuticle synthesis in the developmental period considered (2 and 3: second and third larval instars; P: pupa; A: adult). Black arrowheads indicate key events of the molting cycles. Ap: apolysis; E: ecdysis; He: head eversion (typical of flies, but that coincides with larval–pupal ecdysis in other insects). The epicuticle, preecdysial cuticle, and postecdysial cuticle are represented with black boxes or with dotted and wavy patterns, respectively. Epidermal cells are shown as squares with round nuclei. The salivary gland puffs triggered by ecdysteroids are depicted in gray over the ecdysteroid titers. Cuticle synthesis (in the middle of the scheme, color bars) is interrupted when ecdysteroid titers are rising, with the exception of the third larval instar cuticle. From Charles (2010), with permission.

2010; Zitnan et al., 1996) (see Chapter 6: Hormones involved in the regulation of metamorphosis and Chapter 7: Molecular mechanisms regulating hormone production and action). The biosynthesis and release of these peptides are basically under the control of the fluctuating ecdysteroid concentrations, as well as transcription factors from the ecdysteroid signaling pathway. Then, Inka cells produce preecdysis and ecdysis-triggering hormones (ETH), which promote the ecdysis behavior via receptor-mediated actions on specific CNS neurons. The most intensive studies to determine the mechanisms of ETH expression and release from Inka cells and its action on the CNS have been carried out in the fly *Drosophila melanogaster* and the moth *M. sexta*. Thus most of the data available is based on these models. In general, although there are species-specific peculiarities, during the pulse of ecdysteroids, high levels of the hormone induce expression of ETH receptors in the CNS and promote increased ETH production in Inka cells, which coincides with the expression of the ecdysone receptor (EcR). Significantly, high ecdysteroid levels prevent ETH release from Inka cells. Acquisition of competence to release ETH by Inka cells requires the decline of ecdysteroid levels and the expression of the transcription factor β-Fushi tarazu-factor 1 (β-FTZ-F1), which occurs a few hours prior to ecdysis. Ecdysis behavior is initiated by ETH secretion into the hemolymph, which is controlled by two brain neuropeptides, corazonin and eclosion hormone (EH). Corazonin acts on its receptor in Inka cells to elicit low-level ETH secretion and the onset of the preecdysis sequence, while EH induces cyclic guanosine monophosphate (cGMP)-mediated ETH depletion and consequent activation of the ecdysis sequence. The activation of both behaviors is accomplished by ETH action on neurons of the CNS expressing ETH receptors. These neurons, especially crustacean cardioactive peptide (CCAP) neurons, produce excitatory or inhibitory neuropeptides that initiate or terminate different phases of the ecdysis sequence (Mena et al., 2016; Roller et al., 2010; Zitnan et al., 1996). Bursicon and partner of bursicon are among the peptides expressed in CCAP neurons, and at least the latter has been shown to play a role in the control of ecdysial behaviors. The data as a whole indicate that insect ecdysis is a very complex process divided into two main steps: (1) ecdysteroid-induced expression of receptors and transcription factors in the CNS and Inka cells and (2) release and interaction of Inka cell peptide hormones and multiple central neuropeptides to control consecutive phases of the ecdysis sequence (Mena et al., 2016; Roller et al., 2010; Zitnan et al., 1996; Zitnan and Adams, 2012).

The soft and vulnerable cuticle resulting after the ecdysis requires sclerotization and hardening, which is the last step of the molting process. Bursicon, the heterodimeric protein formed by the subunits bursicon and partner of bursicon, promotes both processes (Fig. 9.6). In addition to the essential role of bursicon in postecdysis tanning, it promotes wing

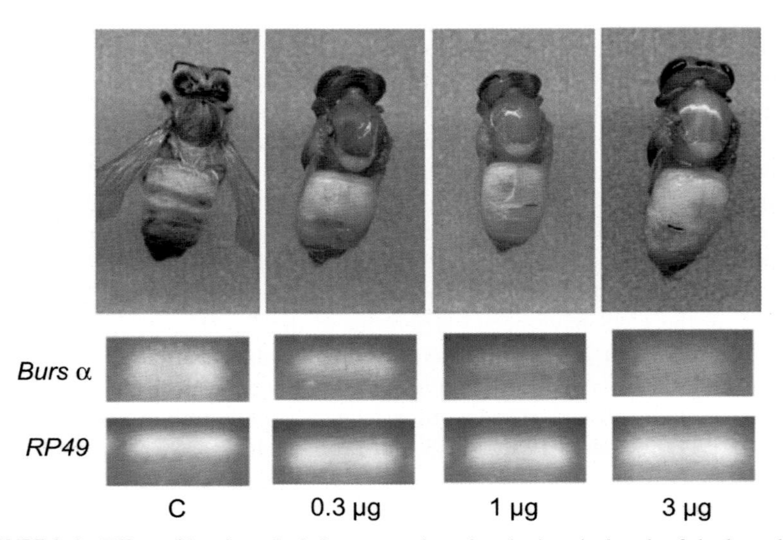

FIGURE 9.6 Effect of bursicon depletion on tanning after the imaginal molt of the honeybee *Apis mellifera*. The image shows the results of an experiment of RNAi depletion using three doses of dsRNA targeting the *Bursicon subunit α* (*Burs α*) that were applied in the pupal stage. Top: Aspect of the treated insects (0.3, 1, and 3 μg) compared to controls (C) after the imaginal molt. Bottom: Corresponding brain levels of *Burs α* mRNA compared with the expression of the housekeeping gene *RP49* used as a reference. From Pereira-Costa et al. (2016), with permission.

extension and maturation at the same stage (Honegger et al., 2008; Pereira-Costa et al., 2016) (see Chapter 6: Hormones involved in the regulation of metamorphosis and Chapter 7: Molecular mechanisms regulating hormone production and action).

THE METAMORPHIC MOLT

The molt that leads to the adult stage is qualitatively different from the juvenile molts, because the functional sense is not only to grow but implies an important morphological remodeling, especially in the holometabolan species. Furthermore, the metamorphic molt must occur once the insect has reached an appropriate size and a proper proportion between body parts to become an adult.

Control of developmental timing and adult body size

Using holometabolan models, two size-related premetamorphosis checkpoints have been identified: critical weight (defined as the weight at which nutrient restriction no longer delays pupation) and minimal viable weight (minimal mass required for metamorphosis to occur). In lepidopterans, these two checkpoints are sequential, whereas in *D. melanogaster*, these occur at the same weight (Mirth and Riddiford, 2007; Nijhout and Callier, 2015). In the end, the adult size is set by the rate of growth during juvenile stages, and the duration of this growth period (see (Boulan et al., 2015; Rewitz et al., 2013)). In this context, the insulin/Insulin growth factor signaling (IIS) and target of rapamycin (TOR) pathways essentially control growth rate, whereas the main regulator of growth duration is 20E (see Yamanaka et al., 2013). Obviously, nutrition is a necessary condition that influences IIS/TOR pathways and 20E signaling.

Regarding the role of the IIS/TOR pathways, and among recent studies in *D. melanogaster*, Agrawal et al. (2016) have reported a humoral link between the fat body and the brain insulin-producing cells (IPCs) relying on tumor necrosis factor α (TNF-α), which mediates an adaptive response to nutrient deprivation. Under nutrient shortage, a metalloprotease of the TNF-α converting enzyme (TACE) family becomes active in the fat body, which allows the cleavage and release of the *Drosophila* TNF-α Eiger into the hemolymph. Eiger activates the IPC receptor Grindelwald in the brain, which leads to Jun N-terminal kinase-dependent inhibition of insulin production (Agrawal et al., 2016).

Another factor that appears to act in a hormonal way in *D. melanogaster* is decapentaplegic (dpp), which can be transported from peripheral tissues, mainly the imaginal discs, to the PG where it inhibits ecdysone synthesis. As discs grow, dpp is thought to be trapped in the disc matrix lowering the level in circulation and thereby reducing the amount that reaches the PG. The reduced reception of dpp signal in the PG allows ecdysone production to ramp up and signal that critical weight has been achieved. If dpp signal is removed during the early third instar, then animals reach critical weight preciously and, if starved, will attempt pupariation without sufficient nutrient stores resulting in the formation of undersized pupa and death (Setiawan et al., 2018).

Also in *D. melanogaster*, knockdown of allatostatin A (AstA) and its receptor AstAR1 in PTTH-producing neurons delays the onset of metamorphosis by impairing PTTH production, which extends the growth period (Deveci et al., 2019). Moreover, AstAR1 is required in the brain IPCs to trigger insulin production. Interestingly, AstA/AstAR1 is homologous to KISS/GPR54, a ligand-receptor signal required for puberty in mammals, which suggests that an evolutionary conserved neural circuitry controls the onset of sexual maturation/metamorphosis (Deveci et al., 2019). A factor that plays important roles in regulating puberty timing in mammals is makorin RING finger protein 3 (MKRN3). In *D. melanogaster*, loss-of-function mutations of the orthologous gene *MKRN1*, which is expressed in the PG, delay the onset of pupariation, resulting in bigger pupae. Mutant larvae show lower expression levels of the steroidogenic gene *phantom*, and the 20E-dependent gene *E74*, which suggests that MKRN1 contributes to regulate the developmental timing and metamorphosis by affecting ecdysteroid production in *D. melanogaster* (Tran et al., 2018).

Factors regulating the cell cycle also contribute to control developmental timing. In *D. melanogaster*, Xie et al. (2015) have shown that the cyclin-dependent kinase 8 (CDK8) is affected by nutrient availability, and EcR-dependent transcription is regulated by CDK8 and its regulatory partner cyclin C (CycC). In the larva-pupa transition, CDK8 levels positively correlate with EcR and USP levels but inversely correlate with the activity of sterol regulatory element-binding protein (SREBP), which is the main regulator of intracellular lipid homeostasis. Starvation of early third larval instar prematurely increases CDK8, EcR, and USP levels, while SREBP activity is downregulated. In contrast, refeeding the starved larvae reduces CDK8 levels but increases SREBP activity, changes that correlate with the timing of larva-pupa transition. The whole data suggest that CDK8-CycC links nutrient intake with EcR and SREBP activities during the larva-pupa transition (Xie et al., 2015). In a similar line, Ohhara et al. (2017) have reported that endocycle progression in PG cells is coupled with nutrient checkpoint. If endocycle is blocked, then PG cells reduce ecdysone synthesis and larvae arrest development. Interestingly, inhibition of the nutrient sensor TOR in the PG during the checkpoint period inhibits the endocycle and triggers a developmental arrest, which is rescued by inducing additional rounds of endocycles by Cyclin E. The data suggest that a TOR-mediated cell cycle checkpoint in the PG provides a systemic growth checkpoint for metamorphosis (Ohhara et al., 2017).

Finally, Pan et al. (2019) have recently reported that nutrient restriction at the early (but not the late) third larval instar of *D. melanogaster* triggers autophagy in the PG. The timing of autophagy induction correlates with the nutritional checkpoints, which inhibit precocious metamorphosis during nutrient restriction in undersized larvae. Inhibition of autophagy impairs pupariation of the starved larvae, while forced autophagy induction triggers a developmental delay or arrest in normally fed insects. Importantly, induction of autophagy impairs the production of ecdysone at the time of pupariation by limiting the availability of cholesterol in the PG cells through a lipophagy mechanism. The function of this temporally regulated, PG-specific autophagy would be to prevent premature pupariation of *D. melanogaster* larvae experiencing nutritional stress. This enables the larvae to search for additional food sources so that they may acquire enough nutrient stores to achieve critical and minimal viable weight (Pan et al., 2019).

Regulation of right allometric growth

For undertaking a proper metamorphosis, it is not only important to reach an appropriate size but also a proper proportion between body parts. The experimental manipulation of IIS/TOR pathways results in larger or smaller insects but generally having proportionated organ and appendage size (Shingleton et al., 2007). This indicates the existence of mechanisms that establish communication between organs and factors that regulate such allometric growth.

Early work in *D. melanogaster* highlighted the role of *Drosophila* insulin-like peptide 8 (Dilp8), not only as a factor that promotes general growth but also a factor ensuring a proportionated growth (Colombani et al., 2012; Garelli et al., 2012). These studies indicate that Dilp8, in response to tissue damage, impairs ecdysone production, which slows development, thus ensuring extra time for tissue repair before metamorphosis. Subsequent studies showed that the leucine-rich repeat-containing G protein-coupled receptor 3 (Lgr3), acts as Dilp8 receptor, and that the Dilp8 though Lgr3 activates a neural circuit that interferes with PTTH neurons, thus attenuating growth and delaying metamorphosis (Colombani et al., 2015; Vallejo et al., 2015). Further studies in *D. melanogaster* have revealed that growth-impaired organs produce the stress response transcription factor Xrp1, which in turn triggers the production of Dilp8. Moreover, the small ribosomal subunit protein RpS12 is required to trigger the Xrp1 response. Briefly, the work unveils that RpS12-Xrp1-Dilp8 constitute an axis that contributes to regulate organ growth coordination during development and metamorphosis (Boulan et al., 2019).

Another recently discovered factor that contributes to the correct scaling of body and organ size is activin β (Actβ), a member of the TGF-β superfamily (Moss-Taylor et al., 2019). Loss-of-function mutations of *Actβ* lead to the formation of small larvae/pupae as well as undersized adults. These rare adult escapers show a disproportional reduction in abdominal size in comparison with other tissues. These data, and additional experiments involving tissue- and cell-specific knockdown and overexpression, indicated that Actβ, operating through the canonic activin branch of the TGF-β pathway and the mediator Smox, is involved in regulating proper allometric growth in *D. melanogaster*. Intriguingly, observations in larval tissues showed that Actβ, directly delivered to muscles by motoneurons, specifically enhances muscle mass. This contrasts with what is known in vertebrates, where activin-like factors, like myostatin, inhibit muscle mass increase (Moss-Taylor et al., 2019).

Histolysis of the prothoracic gland

An important characteristic of the metamorphic molt is that it represents the final molt of the life cycle. This is so in all insect orders, except in the Ephemeroptera, in which the metamorphic molt from the last nymphal instar to the subimago, is followed by a subsequent molt to the adult stage (see Chapter 4: The hemimetabolan development). The final molt is an innovation of the Pterygota associated to the emergence of hemimetabolan metamorphosis, which stops postembryonic development when the insect reaches adulthood (see Chapter 11: The origin of hemimetaboly). The final molt is determined by the disintegration of the PG at the beginning of the adult stage, which is mediated by apoptotic and autophagic processes leading to a programmed cell death (PCD) (Dai and Gilbert, 1999). Ecdysone has been proven to be the factor that triggers PCD in the PG, both in hemimetabolan and holometabolan species (Dai and Gilbert, 1999; Smith and Nijhout, 1983), whereas juvenile hormone (JH) protects PG from cell death (Dai and Gilbert, 1998). One of the species in which the PG disintegration has been best studied at molecular level is the German cockroach *Blattella germanica*, where the interplay of ecdysone signaling, JH, and the inhibitor of apoptosis1 (IAP1) regulates the cell death process. In *B. germanica*, the protein IAP1 is required to ensure PG viability during nymphal development, whereas PG disintegration during the nymphal–adult transition is triggered by ecdysone signaling that leads to a dramatic activation of the transcription factor FTZ-F1, which is key for the cell death process. The study in *B. germanica* has also showed that PG disintegration is prevented by JH, which inhibits *FTZ-F1* expression and tends to activate that of *IAP1* (Mané-Padrós et al., 2010). Further studies in *B. germanica* have shown that the expression of both *FTZ-F1* and *E93* in the PG peaks the last day of the last nymphal instar. Depletion of E93 prevents PG disintegration and depletion of the FTZ-F1 reduces de-expression of *E93* in the PG. The results suggest that the action of FTZ-F1 on PG cell death is mediated by E93 (Orathai Kamsoi and Xavier Belles, unpublished results).

SECRETION OF EMBRYONIC CUTICLES

A special case related with postembryonic molts is the secretion of successive cuticles during embryogenesis. Once the PG is formed, the embryo produces successive and discrete bursts of ecdysteroids that are associated with the secretion of successive embryonic cuticles. In Apterygota (ametabolan) species, exemplified by the silverfish *Thermobia domestica*, there are two successive cuticle depositions, whereas in Pterygota (hemimetabolan and holometabolan), there are generally three. Only the Diptera Brachycera, like *D. melanogaster*, have two. Therefore, the third cuticle deposition in the embryo of most Pterygota species gives the prolarva (hemimetabolan species) or pronymph (holometabolans). A number of differences between the formation of the cuticle in *T. domestica* and Pterygota embryos suggest that the Apterygota first nymphal instar was "embryonized" in Pterygota and became the embryonic pronymph. The secretion of two cuticles instead of three in Diptera Brachycera embryos is due to secondary loss of the second one (Konopová and Zrzavý, 2005).

REFERENCES

Agrawal, N., Delanoue, R., Mauri, A., Basco, D., Pasco, M., Thorens, B., et al., 2016. The *Drosophila* TNF Eiger is an adipokine that acts on insulin-producing cells to mediate nutrient response. Cell Metab. 23, 675–684. Available from: https://doi.org/10.1016/j.cmet.2016.03.003.

Andersen, S.O., 2010. Insect cuticular sclerotization: a review. Insect Biochem. Mol. Biol. 40, 166–178. Available from: https://doi.org/10.1016/j.ibmb.2009.10.007.

Boulan, L., Andersen, D., Colombani, J., Boone, E., Léopold, P., 2019. Inter-organ growth coordination is mediated by the Xrp1-Dilp8 axis in *Drosophila*. Dev. Cell 49, 811–818.e4. Available from: https://doi.org/10.1016/j.devcel.2019.03.016.

Boulan, L., Milán, M., Léopold, P., 2015. The Systemic Control of Growth. Cold Spring Harb. Perspect. Biol. 7, a019117. Available from: https://doi.org/10.1101/cshperspect.a019117.

Callier, V., Nijhout, H.F., 2011. Control of body size by oxygen supply reveals size-dependent and size-independent mechanisms of molting and metamorphosis. Proc. Natl. Acad. Sci. U.S.A. 108, 14664–14669. Available from: https://doi.org/10.1073/pnas.1106556108.

Charles, J.P., 2010. The regulation of expression of insect cuticle protein genes. Insect Biochem. Mol. Biol. 40, 205–213. Available from: https://doi.org/10.1016/j.ibmb.2009.12.005.

Colombani, J., Andersen, D.S., Boulan, L., Boone, E., Romero, N., Virolle, V., et al., 2015. *Drosophila* Lgr3 couples organ growth with maturation and ensures developmental stability. Curr. Biol. 25, 2723–2729. Available from: https://doi.org/10.1016/j.cub.2015.09.020.

Colombani, J., Andersen, D.S., Léopold, P., 2012. Secreted peptide Dilp8 coordinates *Drosophila* tissue growth with developmental timing. Science 336, 582–585. Available from: https://doi.org/10.1126/science.1216689.

Dai, J.D., Gilbert, L.I., 1998. Juvenile hormone prevents the onset of programmed cell death in the prothoracic glands of *Manduca sexta*. Gen. Comp. Endocrinol. 109, 155–165. Available from: https://doi.org/10.1006/gcen.1997.7022.

Dai, J.D., Gilbert, L.I., 1999. An in vitro analysis of ecdysteroid-elicited cell death in the prothoracic gland of *Manduca sexta*. Cell Tissue Res. 297, 319–327.

Davis, M.M., O'Keefe, S.L., Primrose, D.A., Hodgetts, R.B., 2007. A neuropeptide hormone cascade controls the precise onset of post-eclosion cuticular tanning in *Drosophila melanogaster*. Development 134, 4395–4404. Available from: https://doi.org/10.1242/dev.009902.

Ewer, J., Reynolds, S., 2002. Neuropeptide control of molting in insects. Hormones, Brain and Behavior. Elsevier, Boston, pp. 1–92. Available from: https://doi.org/10.1016/B978-012532104-4/50037-8.

Garelli, A., Gontijo, A.M., Miguela, V., Caparros, E., Dominguez, M., 2012. Imaginal discs secrete insulin-like peptide 8 to mediate plasticity of growth and maturation. Science 336, 579–582. Available from: https://doi.org/10.1126/science.1216735.

Gibson, J.F., Slatosky, A.D., Malfi, R.L., Roulston, T., Davis, S.E., 2014. Eclosion of *Physocephala tibialis* (Say) (Diptera: Conopidae) from a *Bombus* (Apidae: Hymenoptera) host: a video record. J. Ent. Soc. Ont. 145, 51–60.

Greenlee, K.J., Harrison, J.F., 2004. Development of respiratory function in the American locust *Schistocerca americana*. II. Within-instar effects. J. Exp. Biol. 207, 509–517.

Honegger, H.-W., Dewey, E.M., Ewer, J., 2008. Bursicon, the tanning hormone of insects: recent advances following the discovery of its molecular identity. J. Comp. Physiol. A: Neuroethol. Sens. Neural. Behav. Physiol. 194, 989–1005. Available from: https://doi.org/10.1007/s00359-008-0386-3.

Konopová, B., Zrzavý, J., 2005. Ultrastructure, development, and homology of insect embryonic cuticles. J. Morphol. 264, 339–362. Available from: https://doi.org/10.1002/jmor.10338.

Mané-Padrós, D., Cruz, J., Vilaplana, L., Nieva, C., Ureña, E., Belles, X., et al., 2010. The hormonal pathway controlling cell death during metamorphosis in a hemimetabolous insect. Dev. Biol. 346, 150–160. Available from: https://doi.org/10.1016/j.ydbio.2010.07.012.

Mena, W., Diegelmann, S., Wegener, C., Ewer, J., 2016. Stereotyped responses of *Drosophila* peptidergic neuronal ensemble depend on downstream neuromodulators. Elife 5, e19686. Available from: https://doi.org/10.7554/eLife.19686.

Mirth, C.K., Riddiford, L.M., 2007. Size assessment and growth control: how adult size is determined in insects. Bioessays 29, 344–355. Available from: https://doi.org/10.1002/bies.20552.

Moss-Taylor, L., Upadhyay, A., Pan, X., Kim, M.-J., O'Connor, M.B., 2019. Body size and tissue-scaling is regulated by motoneuron-derived activin ß in *Drosophila melanogaster*. Genetics 213, 1447–1464. Available from: https://doi.org/10.1534/genetics.119.302394.

Moussian, B., 2010. Recent advances in understanding mechanisms of insect cuticle differentiation. Insect Biochem. Mol. Biol. 40, 363–375. Available from: https://doi.org/10.1016/j.ibmb.2010.03.003.

Neville, A.C., 1975. Biology of the Arthropod Cuticle. Springer Berlin Heidelberg, Berlin, Heidelberg. Available from: https://doi.org/10.1007/978-3-642-80910-1.

Nijhout, H.F., 1994. Insect Hormones. Princeton University Press, Princeton, NJ.

Nijhout, H.F., Callier, V., 2015. Developmental mechanisms of body size and wing-body scaling in insects. Annu. Rev. Entomol. 60, 141–156.

Ohhara, Y., Kobayashi, S., Yamanaka, N., 2017. Nutrient-dependent endocycling in steroidogenic tissue dictates timing of metamorphosis in *Drosophila melanogaster*. PLoS Genet. 13, e1006583. Available from: https://doi.org/10.1371/journal.pgen.1006583.

Pan, X., Neufeld, T.P., O'Connor, M.B., 2019. A tissue- and temporal-specific autophagic switch controls *Drosophila* pre-metamorphic nutritional checkpoints. Curr. Biol 29, 2840–2851.e4. Available from: https://doi.org/10.1016/j.cub.2019.07.027.

Pereira-Costa, C., Elias-Neto, M., Falcon, T., Dallacqua, R.P., Martins, J.R., Bitondi, M.M.G., 2016. RNAi-mediated functional analysis of bursicon genes related to adult cuticle formation and tanning in the honeybee, *Apis mellifera*. PLoS One 11, e0167421. Available from: https://doi.org/10.1371/journal.pone.0167421.

Rewitz, K.F., Yamanaka, N., O'Connor, M.B., 2013. Developmental checkpoints and feedback circuits time insect maturation. Curr. Top. Dev. Biol. 103, 1–33. Available from: https://doi.org/10.1016/B978-0-12-385979-2.00001-0.

Reynolds, S.E., 1980. Integration of behaviour and physiology in ecdysis. Adv. Insect Physiol. 15, 475–595. Available from: https://doi.org/10.1016/S0065-2806(08)60144-7.

Roller, L., Zitnanova, I., Dai, L., Simo, L., Park, Y., Satake, H., et al., 2010. Ecdysis triggering hormone signaling in arthropods. Peptides 31, 429–441.

Setiawan, L., Pan, X., Woods, A.L., O'Connor, M.B., Hariharan, I.K., 2018. The BMP2/4 ortholog Dpp can function as an inter-organ signal that regulates developmental timing. Life Sci. Alliance 1, e201800216. Available from: https://doi.org/10.26508/lsa.201800216.

Shingleton, A.W., Frankino, W.A., Flatt, T., Nijhout, H.F., Emlen, D.J., 2007. Size and shape: the developmental regulation of static allometry in insects. Bioessays 29, 536–548. Available from: https://doi.org/10.1002/bies.20584.

Smith, W.A., Nijhout, H.F., 1983. In vitro stimulation of cell death in the moulting glands of *Oncopeltus fasciatus* by 20-hydroxyecdysone. J. Insect Physiol. 29, 169–176. Available from: https://doi.org/10.1016/0022-1910(83)90141-5.

Tran, H.T., Cho, E., Jeong, S., Jeong, E.B., Lee, H.S., Jeong, S.Y., et al., 2018. Makorin 1 regulates developmental timing in *Drosophila*. Mol. Cells 41, 1024–1032. Available from: https://doi.org/10.14348/molcells.2018.0367.

Truman, J.W., 2005. Hormonal control of insect ecdysis: endocrine cascades for coordinating behavior with physiology. Vitam. Horm. 73, 1–30. Available from: https://doi.org/10.1016/S0083-6729(05)73001-6.

Wigglesworth, V.B., 1983. The physiology of insect tracheoles. Adv. Insect Physiol. 17, 85–148. Available from: https://doi.org/10.1016/S0065-2806(08)60217-9.

Vallejo, D.M., Juarez-Carreño, S., Bolivar, J., Morante, J., Dominguez, M., 2015. A brain circuit that synchronizes growth and maturation revealed through Dilp8 binding to Lgr3. Science 350, aac6767. Available from: https://doi.org/10.1126/science.aac6767.

Xie, X.J., Hsu, F.N., Gao, X., Xu, W., Ni, J.Q., Xing, Y., et al., 2015. CDK8-cyclin C mediates nutritional regulation of developmental transitions through the ecdysone receptor in *Drosophila*. PLoS Biol. 13, e1002207. Available from: https://doi.org/10.1371/journal.pbio.1002207.

Yamanaka, N., Rewitz, K.F., O'Connor, M.B., 2013. Ecdysone control of developmental transitions: lessons from *Drosophila* research. Annu. Rev. Entomol. 58, 497–516. Available from: https://doi.org/10.1146/annurev-ento-120811-153608.

Zhang, J., Lu, A., Kong, L., Zhang, Q., Ling, E., 2014. Functional analysis of insect molting fluid proteins on the protection and regulation of ecdysis. J. Biol. Chem. 289, 35891–35906. Available from: https://doi.org/10.1074/jbc.M114.599597.

Zhu, K.Y., Merzendorfer, H., Zhang, W., Zhang, J., Muthukrishnan, S., 2016. Biosynthesis, turnover, and functions of chitin in insects. Annu. Rev. Entomol. 61, 177–196. Available from: https://doi.org/10.1146/annurev-ento-010715-023933.

Zitnan, D., Adams, M.E., 2012. Neuroendocrine regulation of ecdysis. Insect Endocrinology. Elsevier Science, Amsterdam, pp. 253–309.

Zitnan, D., Kingan, T.G., Hermesman, J.L., Adams, M.E., 1996. Identification of ecdysis-triggering hormone from an epitracheal endocrine system. Science 271, 88–91.

Zitnan, D., Kim, Y.-J., Zitnanová, I., Roller, L., Adams, M.E., 2007. Complex steroid-peptide-receptor cascade controls insect ecdysis. Gen. Comp. Endocrinol. 153, 88–96. Available from: https://doi.org/10.1016/j.ygcen.2007.04.002.

Chapter 10

Regulation of ametabolan, hemimetabolan, and holometabolan development

The most important hormones regulating metamorphosis are ecdysone and its most bioactive derivative, 20-hydroxyecdysone (20E), and juvenile hormone (JH), as discussed in Chapter 6 (Hormones involved in the regulation of metamorphosis). 20E promotes molts, including those occurring in the embryo and the metamorphic molts of the postembryonic development, whereas JH inhibits metamorphosis, maintaining the nymphal or larval morphology and general juvenile growth. The mechanisms underlying the hormonal action, that is, how the hormonal signal is transduced, have been reviewed in Chapter 7 (Molecular mechanisms regulating hormone production and action). In more direct relation with the regulation of metamorphosis, E93 and Broad-complex (BR-C), which are 20E-dependent, and Krüppel homolog 1 (Kr-h1), which is induced by JH, stand out for their importance as hormonal signal effectors.

E93 is the factor that triggers metamorphosis, while Kr-h1 inhibits it by repressing *E93* expression. The fall of JH that occurs in the premetamorphic stage results in a fall of Kr-h1, with the consequent disinhibition of *E93*, with which metamorphosis proceeds. BR-C has a crucial role in holometabolan metamorphosis, since it determines the formation of the pupal stage. In the context of regulating metamorphosis, Kr-h1, E93, and BR-C interact with each other in a similar way in all species studied, essentially in accordance with the MEKRE93 pathway (Belles and Santos, 2014) (see Chapter 7: Molecular mechanisms regulating hormone production and action). However, although the MEKRE93 pathway applies to all metamorphosing insects, different species show specificities, variations on the same theme, particularly referring to the timing of the interactions among E93, Kr-h1, and BR-C. Therefore the best way to have an appropriate vision of these variations is to describe the patterns and interactions in each of the best studied species. After all, each species has its own story to tell.

Insect Metamorphosis. DOI: https://doi.org/10.1016/B978-0-12-813020-9.00010-7

HORMONES AND EMBRYOGENESIS

The question of what are the functions of ecdysteroids and JH during embryo development has been the subject of several studies. The role of ecdysteroids appears to be related to the secretion of the successive cuticles that are deposited during the embryogenesis, as well as with specific developmental functions. On the contrary, the role of JH is not well understood, although the available information seems to indicate that it may be of less importance than previously believed, especially in holometabolan species.

Ecdysteroids

In hemimetabolan insects, the levels of ecdysteroids in the embryo have been studied in detail in the locust *Locusta migratoria* (Lagueux et al., 1979) and in the cockroach *Blattella germanica* (Maestro et al., 2005). In both cases, the hormonal levels fluctuate, and successive peaks can be distinguished (Fig. 10.1). As for holometabolan models, the levels of ecdysteroids in the embryo are fluctuating in the silkworm *Bombyx mori* (Mizuno et al., 1981), whereas a single large peak is observed in the fly *Drosophila melanogaster* (Maróy et al., 1988) (Fig. 10.1). The peaks observed in *L. migratoria*, *B. germanica*, and *B. mori* can be correlated with the secretion of the embryonic cuticles. In *D. melanogaster*, the single large peak of ecdysteroids covers most of the embryonic development, while the deposition of the last embryonic cuticle occurs between 50% and 67% development time (Hillman and Lesnik, 1970). However, the functions of embryonic ecdysteroids are related to the cuticle deposition also in this species. For example, it has been shown that the ecdysteroid-dependent gene *Fushi tarazu-factor 1* (*FTZ-F1*) plays a role in cuticle formation during late embryogenesis, not only in *D. melanogaster*, but also in the beetle *Tribolium castaneum* (Heffer et al., 2013). In addition to be involved in the secretion of embryonic cuticles, ecdysteroid signaling plays also relevant roles in embryo development, from early patterning to morphogenesis and organogenesis, as well as in the development of the immune system (Chavoshi et al., 2010; Cheatle Jarvela and Pick, 2017; Tan et al., 2014).

Juvenile hormone

The study of embryonic JH has raised much interest because of its possible relation to the origin of holometaboly (see Chapter 12: The evolution of metamorphosis). On the basis of data from the locusts *L. migratoria* and *Schistocerca gregaria* (Temin et al., 1986) and the tobacco hornworm *Manduca sexta* (Bergot et al., 1981), it has been considered that the production of JH begins earlier in the holometabolan embryo than in the hemimetabolan, and this advance in the JH secretion was thought to be instrumental to

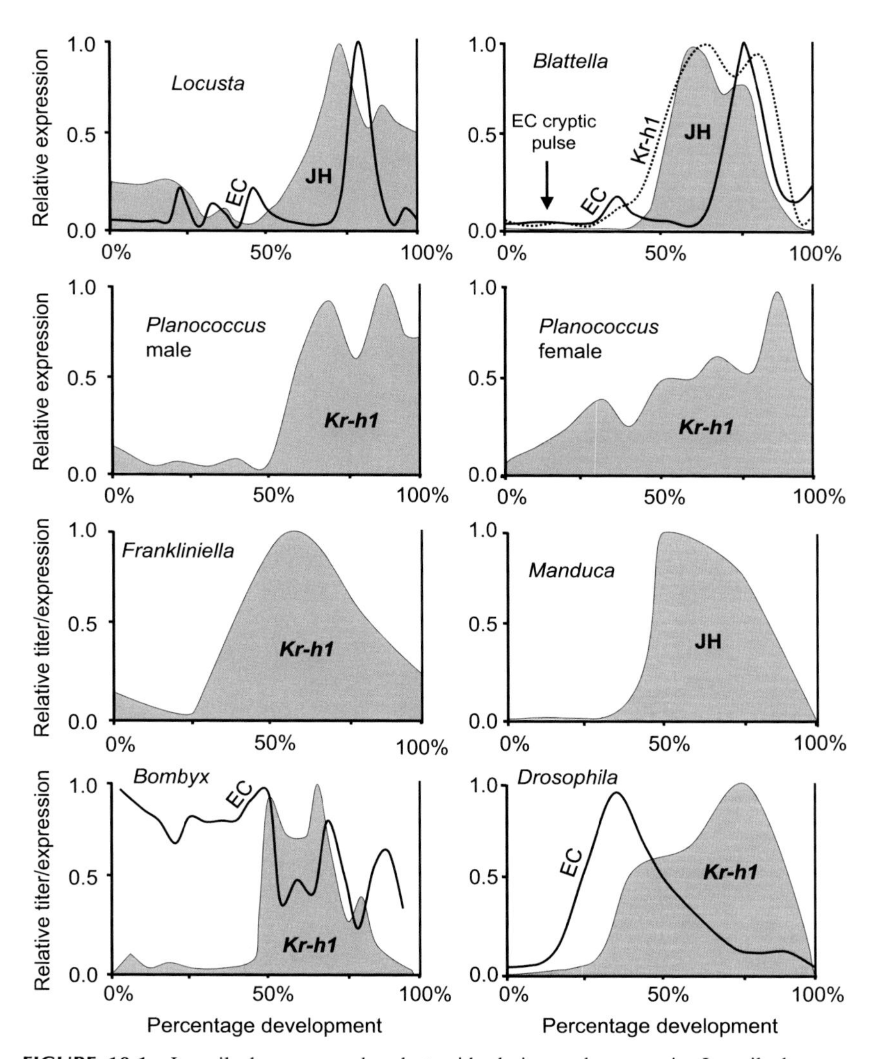

FIGURE 10.1 Juvenile hormone and ecdysteroids during embryogenesis. Juvenile hormone (JH) titers are indicated with a gray area, those of ecdysteroids (EC) with a continuous line, and the expression of *Krüppel homolog 1* (*Kr-h1*), which should mirror the JH levels, with a gray area or a dashed line. The titers or the expression levels are indicated in relation to the maximum value (=1). The origin of the data is as follows: *Locusta migratoria*: Lagueux et al. (1979) (EC) and Temin et al. (1986) (JH); *Blattella germanica*: Maestro et al. (2005) (EC) and Maestro et al. (2010) (JH); *Planococcus kraunhiae*: Vea et al. (2016); *Frankliniella occidentalis*: Minakuchi et al. (2011); *Manduca sexta*: Bergot et al. (1981); *Bombyx mori*: Mizuno et al. (1981) (EC) and Daimon et al. (2015) (*Kr-h1*); *Drosophila melanogaster*: Maróy et al. (1988) (EC) and Ylla et al. (2017) (*Kr-h1*); the EC cryptic pulse in *B. germanica* was identified by an expression pulse of *E75A* (Mané-Padrós et al., 2008), an early EC-dependent gene.

changing the nymphal program of hemimetabolan embryogenesis into the holometabolan larval program (Truman and Riddiford, 1999).

Nevertheless, the study of new insect models suggests that patterns of JH levels are quite diverse and depend on the species, which makes it difficult to establish generalizations. In those few species with consistent data available, JH usually begins to be detected toward 50% of development, as in the hemimetabolans *L. migratoria* (Temin et al., 1986) and *B. germanica* (Maestro et al., 2010), and in the holometabolan *M. sexta* (Bergot et al., 1981) (Fig. 10.1). However, subsequent levels may increase rapidly, as in *B. germanica*, *M. sexta*, and *B. mori* (in *B. mori* by considering the expression levels of the JH-dependent gene *Kr-h1* as representative of the circulating JH), or rather gradually, as in *L. migratoria* (Fig. 10.1). In *D. melanogaster* (Pecasse et al., 2000; Ylla et al., 2018) and in the hemimetabolan thysanopteran *Frankliniella occidentalis* (Minakuchi et al., 2011), *Kr-h1* expression levels begin to increase before 50% of development (Fig. 10.1), while in another hemimetabolan, the hemipteran scale insect *Planococcus kraunhiae*, *Kr-h1* mRNA levels rapidly increase from 50% of development in the male, while in the female, they begin to increase progressively, albeit fluctuating, practically from the beginning of embryogenesis (Vea et al., 2016) (Fig. 10.1).

To assess the importance of JH in embryogenesis, a number of experiments increasing the hormonal concentration have been carried out. Thus treatment with JH mimics early in embryo development resulted in growth arrest and premature tanning of mandibles and formation of the nymphal cuticle features in embryos of hemimetabolan species, like the locust *S. gregaria* and the cricket *Acheta domestica* (Erezyilmaz et al., 2004; Novák, 1969; Truman and Riddiford, 2002). Application of JH mimics early in embryogenesis in holometabolan species, like the lepidopterans *Hyalophora cecropia*, *M. sexta*, and *B. mori*, impairs katatrepsis, so that the embryos assume abnormal positions in the egg (Truman and Riddiford, 2002). However, experiments applying a JH mimic are difficult to interpret because it is administered outside the normal physiological context and at doses that are pharmacological rather than physiological. In addition, some of the JH mimics used, such as pyriproxyfen (a propylene glycol with a 2-pyridyl group at the O-1 position and a 4-phenoxyphenyl group at the O-3 position), are structurally very different from JH and, although it does interact with the JH receptor (Jindra et al., 2015), might produce additional toxic effects.

Experiments decreasing JH levels have also been reported in locusts and cockroaches, where the period of JH production extends from shortly after the dorsal closure, which occurs at about 45% embryogenesis (Li, 2007), to almost hatching. In embryos of *L. migratoria*, JH was eliminated with a precocene compound (which destroys the corpora allata, CA, the JH-producing glands) applied in late (65%) embryogenesis. No apparent anomalies were observed until after the second postembryonic molt, when nymphs arrested, showing signs of precocious metamorphosis (Aboulafia-Baginsky et al.,

1984). In the cockroach *Nauphoeta cinerea*, treatments of eggs with precocene just after the dorsal closure depleted the JH concentration in the embryo and resulted in reduced vitality, a delay in development, and deficiencies in pigmentation and formation of hairs and bristles in the cuticle. Moreover, the fat body disintegrated and the midgut did not differentiate properly. Importantly, application of JH to these eggs attenuated the deficiencies provoked by the precocene treatment (Brüning and Lanzrein, 1987). These results suggest that JH promotes embryo vitality and plays roles in the formation of the nymphal cuticle. JH appears to influence also fat body and midgut development, at least in *N. cinerea*.

In the cockroach *B. germanica*, maternal RNAi has been used to deplete JH acid *O*-methyl transferase (JHAMT, a key enzyme catalyzing the last step of JH biosynthesis), the JH receptor Methoprene tolerant (Met), and the JH signaling transducer Kr-h1 (Fernandez-Nicolas and Belles, 2017). Intriguingly, significant amounts of transcripts of the three factors were observed early in embryogenesis, when the CA is not yet formed and JH not produced (Maestro et al., 2010). This suggests functions of these factors in early embryogenesis, likely unrelated to JH. Consistent with this, c.40% of the JH signaling-depleted embryos were unable to form the germ-band anlage, while c.20% showed the development interrupted before mid-embryogenesis, involving defects related to dorsal closure. The remaining 40% developed in the period of JH production and resulted in two types of phenotypes (with approximately 20% incidence each): embryos with intensely tanned cuticle (showing a premature upregulation of laccase 2, a promoter of cuticle tanning), and embryos featuring the first nymphal instar but with reduced vitality, unable to hatch (Fig. 10.2) (Fernandez-Nicolas and

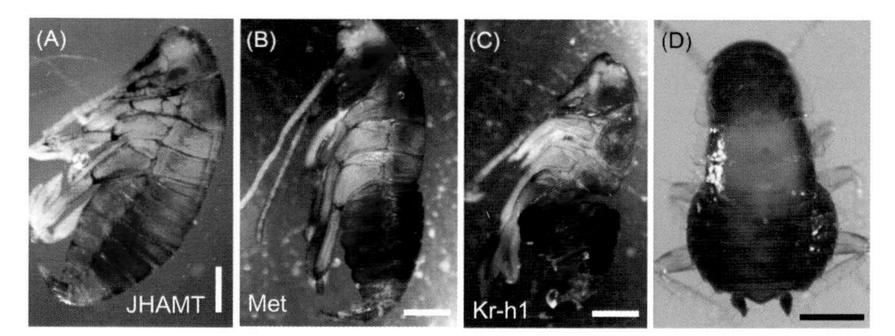

FIGURE 10.2 Effects of depleting JH signaling in the cockroach *Blattella germanica*. (A–C) Embryo after the deposition of the third cuticle, which is intensely tanned, resulting from depletion of JH acid *O*-methyl transferase (JHAMT), Methoprene tolerant (Met), and Krüppel homolog 1 (Kr-h1), respectively. (D) First nymphal instar that was unable to hatch, but whose hatching was facilitated by artificially breaking the eggshell. This phenotype is common to JHAMT, Met, or Kr-h1 depletions. Scale bar: 0.4 mm. Phenotypes are described by Fernandez-Nicolas and Belles (2017). Photos: Ana Fernandez-Nicolas and Xavier Belles.

Belles, 2017). In both species, *N. cinerea* and *B. germanica*, the experiments impaired the formation of a normal cuticle, however, no premature tanning was observed in *N. cinerea*.

In holometabolan embryos, null mutants for *JHAMT* and for the JH receptors *Met1* and *Met2*, have been generated through genome editing in the silkworm *B. mori* (Daimon et al., 2015). Results showed that $JHAMT^{-/-}$ embryos lost *Kr-h1* expression, but they still develop normal larvae. The only anomaly observed was that embryos without JH had hatching problems, which were corrected by artificially breaking the egg-shell or by applying JH. Embryos without Met1 and Met2 also developed into normal larvae (Daimon et al., 2015). Equivalent observations have been recently made in the mosquito *Aedes aegypti*, where knockout of Met did not affect embryogenesis (Zhu et al., 2019). Other experiments have reported that double-null mutants of *D. melanogaster* with a complete zygotic loss of *Met* and its paralog *germ cell-expressed* (*gce*) die no earlier than at the onset of metamorphosis (Abdou et al., 2011; Jindra et al., 2015). *D. melanogaster* embryos homozygous for certain alleles of *Kr-h1* have hatching problems, but the larva develops normally and even null *Kr-h1* mutants predominantly die in the prepupal stage (Beck et al., 2004).

REGULATION OF POSTEMBRYONIC DEVELOPMENT IN AMETABOLANS

The best characterized functions of JH in ametabolan insects are related to oocyte maturation. In females of the Zygentoma *Thermobia domestica*, JH induces the expression of vitellogenin, the major egg yolk protein precursor in insects. As described in Chapter 3 (The ametabolan development), ametabolans continue molting during adult life, and molts are interspersed between cycles of oocyte maturation. A cycle from one ecdysis to the next of *T. domestica* takes 10 days, during which a burst of JH production (Bitsch et al., 1985) triggers the expression of a wave of vitellogenin (Rousset et al., 1987), which is incorporated into the maturing oocytes. Then, a next molt is triggered by a pulse of ecdysteroids (de la Paz et al., 1983), and the eggs corresponding to the previous gonadotrophic cycle are laid on the fifth day after the ecdysis (Fig. 10.3).

Although the Zygentoma do not metamorphose, they do undergo changes throughout their life cycle, the most apparent of which is the formation of scales in early instars. In *T. domestica* scales appear toward the middle of the third nymphal instar (Delany, 1957), and integument transplantation experiments, as well as CA volume and activity measurements (Watson, 1967), suggest that scales appear shortly after a transient decrease in JH. According to Watson (1967), when a fragment of integument from a first nymphal instar of *T. domestica* is implanted into an adult that was going to molt, scales form in the implanted fragment, suggesting that the scale formation is determined by humoral factors. CA measurements showed that the minimal gland volume is found in the early–mid third nymphal instar, in other words prior to scale formation. It is worth noting that, in general, CA volume correlates approximately with JH production, as demonstrated in a number of models by measuring JH biosynthesis (see, e.g., Belles et al., 1987). Finally, the JH activity of the CA, when tested using the Gilbert and Schneiderman bioassay (1960), was shown to be minimal in the third nymphal instar (Watson, 1967). The data as a whole suggest that the formation of scales in early nymphs of *T. domestica* might be repressed by JH.

REGULATION OF POSTEMBRYONIC DEVELOPMENT IN HEMIMETABOLANS

The behaviors of Kr-h1, E93, and BR-C, in relation to the regulation of hemimetabolan metamorphosis, have been studied in detail in selected model species of palaeopterans, polyneopterans, and paraneopterans.

Palaeoptera

There is not much information in palaeopterans, but the available data suggest that the regulation of metamorphosis is based on the MEKRE93 pathway. Regarding Odonata, the damselfly species *Ischnura senegalensis* is being established as an experimental model for physiological studies (Okude et al., 2017b), in which an electroporation-mediated RNA interference technique has been implemented (Okude et al., 2017a). Using this technique, Okude et al. (2019) have shown that E93 is essential for adult morphogenesis and that Kr-h1 represses *E93* expression, as in other more classical model insects.

Regarding Ephemeroptera, a transcriptomic study conducted on *Cloeon viridulum* revealed that the expression of *Kr-h1* decreases progressively from young to mature nymphs and to the subimago and the adult (Si et al., 2017) (Fig. 10.3), suggesting that *Kr-h1* represses metamorphosis in mayflies. The expression pattern of all *BR-C* isoforms and *BR-C Z1* is similar to that of *Kr-h1*, thus decreasing in preadult stages, as is usual in other hemimetabolan insects. The authors also report a *BR-C Z6* isoform, whose expression increases from young to mature nymphs, the subimago, and the adult. The data suggest that this intriguing isoform might be associated with wing development. Unfortunately, expression of *E93* was not reported by Si et al. (2017). A species that is becoming a mayfly model for developmental studies is *Cloeon dipterum* (Almudi et al., 2019). In females of *C. dipterum*, *Kr-h1* expression is high until the penultimate nymphal instar, then it decreases in this instar, keeps low values during most of the last nymphal instar, increases again the last day of it, and keeps high values in the subimago and adult. At the same time, the expression of *E93* is very low or absent until the penultimate nymphal instar, when it begins to increase, coinciding with the decrease of *Kr-h1* expression. Then *E93* expression remains high in the last nymphal instar, increasing dramatically in

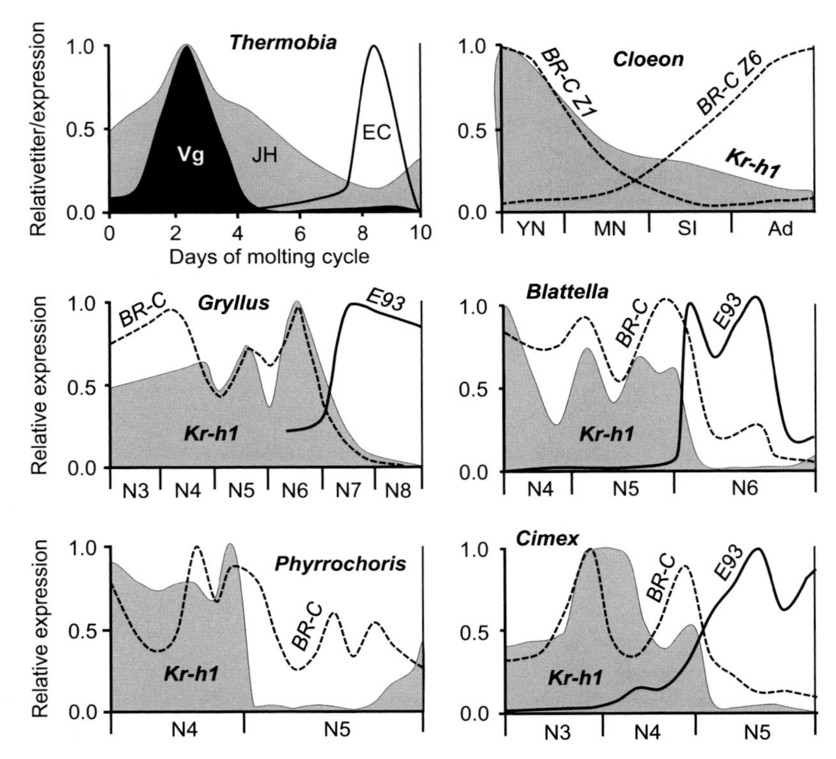

FIGURE 10.3 Juvenile hormone and ecdysteroids during postembryonic development and metamorphosis in ametabolan and hemimetabolan species. In the ametabolan *Thermobia domestica*, molts regulated by ecdysteroids (EC) are interspersed between cycles of oocyte maturation regulated by juvenile hormone (JH), which promotes the production of vitellogenin (Vg). Data are from Bitsch et al. (1985) (JH), Rousset et al. (1987) (Vg), and de la Paz et al. (1983) (EC). In hemimetabolan species, the expression of *Kruppel homolog 1* (*Kr-h1*), *Broad-complex* (*BR-C*), and *E93* are shown with a gray area, a dashed line, and a continuous line, respectively. The origin of the data is as follows. *Cloeon viridulum*: Si et al. (2017); *Gryllus bimaculatus*: Ishimaru et al. (2019); *Blattella germanica*: Huang et al. (2013) (*BR-C*), Lozano and Belles (2011) (*Kr-h1*), Ureña et al. (2014) and Belles and Santos (2014) (*E93*); *Pyrrhocoris apterus*: Konopová et al. (2011); *Cimex lectularius*: Gujar and Palli (2016). The titers or the expression levels are indicated in relation to the maximum value (=1). Different nymphal instars (N3–N8) are shown for every species; the last nymphal instar shown is the final nymphal instar in all cases; in the mayfly *C. viridulum*, the following stages are shown: young nymph (YN), mature nymph (MN), subimago (SI), and adult (Ad), according to Si et al. (2017).

the subimago and the adult. Regarding *BR-C*, its expression levels increase progressively until the penultimate nymphal instar, and then decrease until practically vanishing at the end of the last nymphal instar, thus preceding the most important development of the wing primordia (Orathai Kamsoi, Fernando Casares, Isabel Almudi, and Xavier Belles, unpublished). These patterns suggest that E93 triggers the transition to the subimaginal stage and that Kr-h1 represses

E93, whereas BR-C appears to be involved in wing development. Since *C. dipterum* is viviparous, and the eggs mature from the last nymphal instar until the adult stage (Almudi et al., 2019), the dramatic upregulation of *Kr-h1* expression from the last day of the last nymphal instar could be related to gonadotrophic functions. These data suggest that the subimago is indeed a first adult phase that is needed to allow necessary growth and to develop reproductive capacity, as conjectured by Sehnal et al. (1996). This notion is also supported by the transcriptomic data of *C. viridulum* discussed above, which show that the closest similarity of the subimago is with the adult (Si et al., 2017).

Polyneoptera

Most of the studies on the metamorphosis of polyneopterans have been conducted in cockroaches, locusts, and crickets. The studies performed in the cockroach *B. germanica* and the cricket *Gryllus bimaculatus* are especially detailed, including data on the behavior of the components of the MEKRE93 pathway.

In *B. germanica*, the main components of the MEKRE93 pathway have been studied, determining the expression patterns and carrying out RNAi functional studies. Thus data on Kr-h1, E93, the two components of the JH receptor, Met and Taiman, and BR-C are available. *Kr-h1* expression maintains high, although oscillating, levels until the first day of the last nymphal instar (N6), when the levels suddenly decrease, whereas the *E93* expression dramatically increases in parallel (Fig. 10.3). *BR-C* expression follows a pattern similar to that of *Kr-h1*, which is not surprising as in *B. germanica*, *BR-C* expression is enhanced by JH (Huang et al., 2013). Depletion of Kr-h1 in N4 or N5 triggers a precocious metamorphosis in N6 (Lozano and Belles, 2011), and similar results are obtained when depleting Met (Lozano and Belles, 2014) and Taiman (Lozano et al., 2014) (see Chapter 7: Molecular mechanisms regulating hormone production and action). Depletion of E93 in the last nymphal instar prevents adult morphogenesis and instead causes the formation of supernumerary nymphs (Belles and Santos, 2014; Ureña et al., 2014). Depletion of BR-C in nymphal instars impairs wing development, resulting in smaller wings with vein defects in the adult (Huang et al., 2013) (see Chapter 7: Molecular mechanisms regulating hormone production and action). Regarding the epistatic relationships between these factors, RNAi studies have shown that Kr-h1 represses *E93* expression (Belles and Santos, 2014), while E93 represses that of *Kr-h1* (Belles and Santos, 2014; Ureña et al., 2014). These data fit the MEKRE93 pathway as an essential regulator of metamorphosis, which, in fact, was proposed based on the information obtained in *B. germanica* (Belles and Santos, 2014) (see Chapter 7: Molecular mechanisms regulating hormone production and action).

In *G. bimaculatus*, expression and functional data on Kr-h1 and E93, as well as BR-C, have been reported by Ishimaru et al. (2019). The studies show that *Kr-h1* expression keeps high, oscillating levels until the penultimate nymphal instar (N7), when the levels decrease quite rapidly, being practically undetectable in the last nymphal instar (N8) (Fig. 10.3). Depletion of Kr-h1 and BR-C in N6 upregulates *E93* expression and triggers the formation of precocious adults after the next molt. Conversely, E93 depletion in N6 prevents the N8–adult transition and triggers the formation of supernumerary nymphs (Ishimaru et al., 2019). The data of *G. bimaculatus* fit the MEKRE93 pathway and reveal that depletion of BR-C results in a downregulation of *Kr-h1* and an upregulation of *E93*, which may explain the formation of precocious adults when depleting BR-C in juvenile stages.

Regarding the factors that determine the fall of JH and Kr-h1 toward the end of the nymphal period, Ishimaru et al. (2016) showed that myoglianin (myo), a ligand in the TGF-β signaling pathway, is crucial in this regard. In the CA, high *myo* expression during the last nymphal instars represses the expression of *JHAMT*, which is crucial in JH biosynthesis, thus resulting in the observed decrease in JH production that occurs at the end of the nymphal period. The inhibitory effect of myoglianin upon JHAMT in the CA also operates in the penultimate nymphal instar (N5) of *B. germanica*. At the same time, a sharp pulse of *myo* expression in the prothoracic gland of this species in N5 indirectly stimulates the expression of steroidogenic genes, enhancing the production of the metamorphic ecdysone pulse in the last nymphal instar (N6) (Kamsoi and Belles, 2019).

Paraneoptera

This is the most diverse superorder among the hemimetabolan insects, not only numerically (c.69,000 species, which represents approximately 50% of all living and described hemimetabolans), but also in terms of diversity of habits and life cycles. Heteropterans show the usual hemimetabolan life cycle, equivalent to that of polyneopterans. The best known heteropteran models studied from the point of view of the regulation of metamorphosis on a molecular scale are the kissing bug, *Rhodnius prolixus*, and the linden bug, *Pyrrhocoris apterus*. More recently, data from the bed bug, *Cimex lectularius*, have also been reported. In *R. prolixus*, RNAi depletion of Kr-h1 (Konopová et al., 2011) and Met (Villalobos-Sambucaro et al., 2015) in juveniles triggers the formation of precocious adult traits. In *P. apterus*, the expression of *Kr-h1* falls dramatically on the first day of the last nymphal instar (N5), whereas *BR-C* continues to be expressed in an oscillating way during N5 (Konopová et al., 2011) (Fig. 10.3). RNAi studies on *P. apterus* revealed that Met transduces the JH antimetamorphic signal maintaining a high expression of *Kr-h1*, thus preventing premature metamorphosis. RNAi experiments targeting *BR-C* transcripts showed that these transcription factors are involved in wing development, as in other hemimetabolans (Konopová et al., 2011) (see Chapter 7: Molecular mechanisms regulating hormone production and action). More recent work in *P. apterus* has shown that E93 triggers adult morphogenesis, as its RNAi depletion in late nymphal instars prevents adult morphogenesis and triggers the formation of supernumerary nymphs (Vlastimil Smykal and Marek Jindra, in Jindra, 2019). Equivalent expression studies gave similar results in another heteropteran, the bed bug *C. lectularius* (Gujar and Palli, 2016). In this species, *Kr-h1* expression also falls on the first day of the last nymphal instar, which coincides with a dramatic increase of *E93* transcript levels (Fig. 10.3), while RNAi experiments showed the inhibitory role of Kr-h1 and the inducing role of E93 on metamorphosis.

Other paraneopteran groups, such as Thysanoptera (thrips) and some Hemiptera Sternorrhyncha (such as whiteflies and scale insects), offer very unique cases of postembryonic development with quiescent preadult stages reminiscent of the holometabolan pupa. The expression of *Kr-h1* and *BR-C* has been monitored in two species of thrips: *F. occidentalis* and *Haplothrips brevitubus*. They are representative of the two suborders of Thysanoptera: Terebrantia (*F. occidentalis*) and Tubulifera (*H. brevitubus*). In the Terebrantia, the third and fourth nymphal instars are nonfeeding and quiescent, being called "propupa" and "pupa," respectively. The life cycle of the Tubulifera contains three quiescent instars after the two mobile first nymphal instars (N1 and N2), which are called the "propupa," the first "pupa" and the second "pupa" ("pupa 2") (Moritz, 2006). The external wing pads emerge in these quiescent instars, the genitalia is being formed, and internal structures are remodeled (see Chapter 4: The hemimetabolan development). In *F. occidentalis* and *H. brevitubus*, the expression of *Kr-h1* is high and oscillating until the "pupa," instar when it drops. In the case of *BR-C*, a broad peak of expression covers the second nymphal instar (N2) and the "propupa" and then declines to minimal levels in the "pupa" instar (Minakuchi et al., 2011) (Fig. 10.4). Treatment with a JH mimic in different developmental stages causes death in the "pupa" of *F. occidentalis* or in the "pupa 2" of *H. brevitubus*, and an equivalent treatment to freshly ecdysed "propupa" causes prolonged expression of *Kr-h1* and *BR-C* in both species (Minakuchi et al., 2011). The results as a whole suggest that JH represses metamorphosis and that it stimulates *Kr-h1* (and *BR-C*) expression. Although there is no data on *E93*, it might be presumed that its expression would increase in parallel to the decline of *Kr-h1* expression, thus triggering adult morphogenesis in thrips species. As for *BR-C*, its expression appears to be enhanced by JH, and its function may consist, at least in part, of promoting wing development, as in heteropterans. Other functions of BR-C related to general remodeling in the transition from nymph to "propupa" and "pupa" cannot be ruled out. If they existed, then it would represent an exaptation for the pupal determinant functions that BR-C has in holometabolan species.

The life cycles of whiteflies and scale insects, within the Hemiptera Sternorrhyncha, also include quiescent stages. Data on the molecular

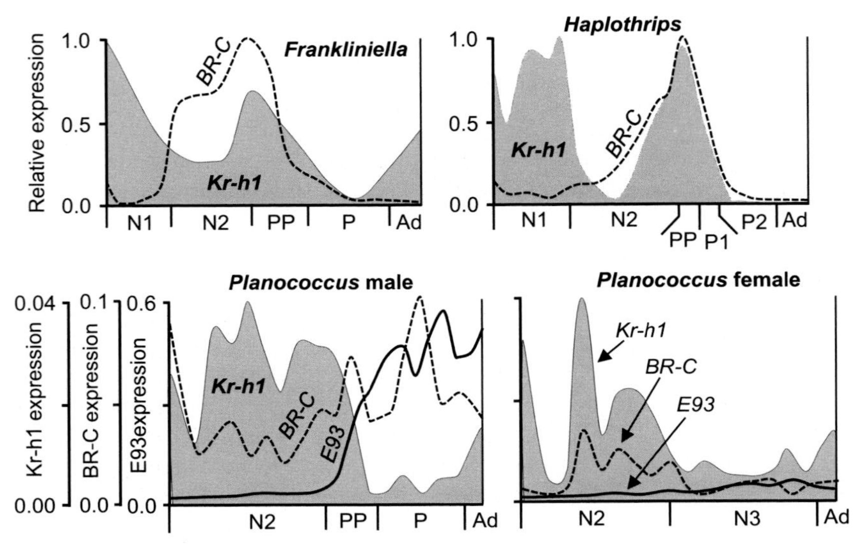

FIGURE 10.4 Expression of *Kruppel homolog 1* (*Kr-h1*), *Broad-complex* (*BR-C*), and *E93* during metamorphosis in thrips and scale insects. The thrips examined are *Frankliniella occidentalis* and *Haplothrips brevitubus*, with data from Minakuchi et al. (2011). The scale insect represented is *Planococcus kraunhiae*, with data from Vea et al. (2016) (*Kr-h1*, *BR-C* isoform 3, which is the most representative) and Vea et al. (2019) (*E93*, isoform C, which is the most representative). *Kr-h1* is indicated with a gray area, *BR-C* with a dashed line, and *E93* with a continuous line. In *F. occidentalis* and *H. brevitubus*, the qRT-PCR levels of expression are represented in relation to the maximum value (=1), whereas in *P. kraunhiae*, the absolute qRT-PCR values are indicated, in order to facilitate the comparison of males and females (a specific ordinate for each gene is shown to the left of the diagram). The stages and instars represented are first, second, third nymphal instar (N1, N2, N3), "Propupa" (PP), "Pupa" (P), "Pupa 1" (P1), "Pupa 2" (P2), and adult (Ad).

regulation of metamorphosis are available for a scale insect, the Pseudococcidae *P. kraunhiae* (Vea et al., 2016). As in other scale insects, *P. kraunhiae* exhibits an extreme sexual dimorphism. The life cycle of the male includes two crawling nymphal instars (N1 and N2) followed by two quiescent instars, the "prepupa" and the "pupa," after which it molts to the adult stage. In contrast, the female goes through three wingless nymphal instars (N1, N2, and N3), and N3 molts into a larviform neotenic adult, sexually mature but retaining juvenile features. In males, the expression pattern of *Kr-h1* shows high and oscillating levels until the stage of "prepupa," when it drops to minimal levels. In parallel, *E93* expression is dramatically upregulated in this stage. *BR-C* expression shows relatively high levels in N2, but it subsequently peaks in the "propupa" and in the "pupa" instars (Fig. 10.4). Treatment of male juveniles with a JH mimic prolonged the expression of *Kr-h1* and *BR-C* and triggered the formation of a supernumerary "pupa" (Vea et al., 2016). Further studies showed that, in parallel to *Kr-h1* upregulation, JH mimic treatment of male "prepupae" resulted in reduced levels of *E93* transcripts (Vea et al., 2019). These results fit with a regulation based on the MEKRE93 pathway, that is, adult morphogenesis would be triggered by E93 in the "pupa," while the *Kr-h1* expression sustained by the JH would inhibit it by suppressing *E93* activity in previous instars.

In the female of *P. kraunhiae*, the situation is very different. The expression of *Kr-h1* is relatively high and fluctuating in N2, and it subsequently declines, a pattern similar to that of the male. In contrast, *BR-C* and *E93* mRNA levels do not increase beyond N2. JH mimic treatment in female early N3 triggers a modest increase of *Kr-h1* expression, but it is significantly lower than that induced in males (Vea et al., 2019). This is possibly because the two components of the JH receptor, Met and Taiman, are expressed at low levels compared with their expression in males (Vea et al., 2016). Consequently, the treatment with the JH mimic practically do not affect the expression of *E93* in females. The fact that the expression of *BR-C* and *E93* remains low beyond N2 explains why wing development and adult morphogenesis do not proceed in the female of *P. kraunhiae*, but the question of why the expression of these two genes is persistently low in females remains unsolved.

REGULATION OF POSTEMBRYONIC DEVELOPMENT IN HOLOMETABOLANS

The most important difference in the regulation of the postembryonic development of the holometabolan species with respect to the hemimetabolans is the expression dynamics of *BR-C* and its functions as pupal specifier. The best genetically studied species are the coleopteran *T. castaneum*, the lepidopteran *B. mori*, and the dipteran *D. melanogaster*.

Tribolium castaneum

Expression studies in the red flour beetle *T. castaneum* show that *Kr-h1* mRNA levels fluctuate, experiencing a dramatic drop in the last larval instar (usually L7), are upregulated toward the end of that stage, coinciding with the stage of prepupa, and fall to undetectable levels at the beginning of the pupa (Minakuchi et al., 2009). In parallel, *E93* shows very low expression levels at the beginning of L7, experiences a very slight increase toward the end of this instar, coinciding with the fall of *Kr-h1* expression, and a sustained increase from the prepupal stage in L7, peaking in the pupa (Ureña et al., 2014) (Fig. 10.5). The expression pattern of *BR-C* is simpler, consisting of a strong peak centered on the prepupal stage (Minakuchi et al., 2009) (Fig. 10.5).

Early RNAi studies in *T. castaneum* showed that JH represses metamorphosis via Met (Konopová and Jindra, 2008). Using the same

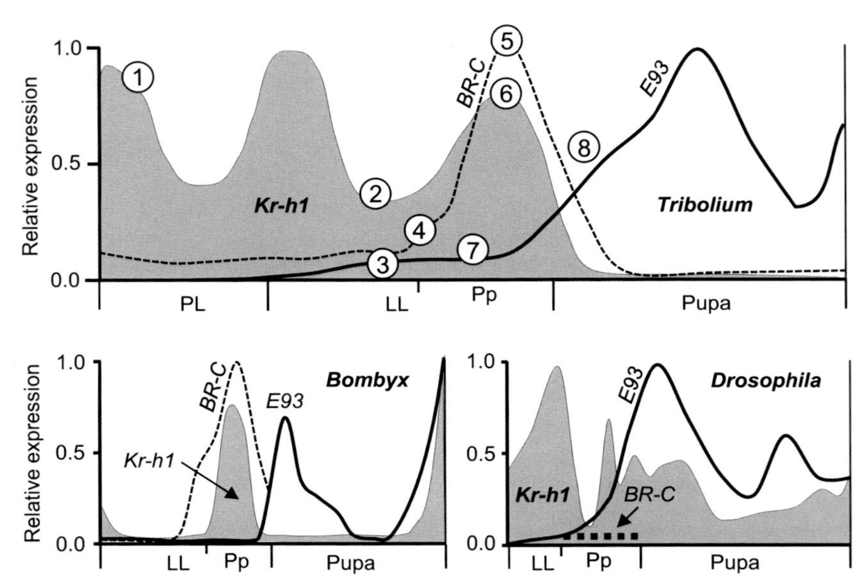

FIGURE 10.5 Expression of *Kruppel homolog 1* (*Kr-h1*), *Broad-complex* (*BR-C*), and *E93* during metamorphosis of holometabolan insects. The species examined are the following: *Tribolium castaneum*, with data from Minakuchi et al. (2009 and 2011) (*Kr-h1*, *BR-C*), and Ureña et al. (2014) (*E93*). *Bombyx mori*, with data from Daimon et al. (2015) (*BR-C*), Kayukawa et al. (2014) (*Kr-h1*), and Kayukawa et al. (2017) (*E93B* in male epidermis). *Drosophila melanogaster*, with data from Karim and Thummel (1992) (*BR-C*), Minakuchi et al. (2008) (*Kr-h1*), and Ureña et al. (2014) (*E93B*, the most abundant isoform). *Kr-h1* is indicated with a gray area, *BR-C* with a dashed line, and *E93* with a continuous line. Expression levels are represented in relation to the maximum value (=1), except in the case of *BR-C* of *D. melanogaster*, where only the period of expression is indicated. The penultimate (PL) and last (LL) larval instars, the prepupal stage (Pp), and the pupa are shown for every species. In *T. castaneum*, the succession of regulatory events is indicated as follows: 1: *Kr-h1* expression maintains the larval morphology; 2: transient decrease of *Kr-h1* expression; 3: slight increase of *E93* expression; 4: upregulation of *BR-C* expression; 5: *BR-C* expression peak, formation of the pupa; 6: *Kr-h1* expression increases again; 7: no further increase of *E93* expression in the prepupal stage, and thus adult morphogenesis does not proceed; 8: expression of *BR-C* and *Kr-h1* decline and that of *E93* increases, and thus adult morphogenesis proceeds.

technique combined with treatments with a JH mimic, it was shown that BR-C is inequivocally required for pupal morphogenesis (see Chapter 7: Molecular mechanisms regulating hormone production and action). Interestingly, similar results were obtained when using the lacewing *Chrysopa perla* (Konopová and Jindra, 2007), which belongs to the early-branching endopterygote order Neuroptera. This suggests that the role of BR-C as a pupal specifier is conserved throughout all holometabolan species. The relationships between JH and BR-C in *T. castaneum* are intriguing, as in late larval instars JH through Met represses the expression of *BR-C* (and the formation of the pupa), whereas JH treatment in the early pupa promotes *BR-C* expression and prevents the formation of the adult (Konopová and Jindra, 2008; Suzuki et al., 2008). The individuals obtained by depleting BR-C in last larval instar showed mixed features of the larva, pupa, and adult, which indicates that BR-C not only promotes pupal morphogenesis but also represses the formation of adult characters. Additionally, results of specific depletion of BR-C isoforms suggest that there is some degree of redundancy, although certain isoforms appear to have a more prominent influence in given functions. For example, BR-C Z2 appears to be consistently involved in wing development (Konopová and Jindra, 2008; Suzuki et al., 2008), which is reminiscent of the role of BR-C in hemimetabolans.

Related with these results, Minakuchi et al. (2009) demonstrated that in late larval instars, Kr-h1 acts downstream of the JH receptor Met, whereas in the pupal stage, it acts downstream of Met and upstream of BR-C, in both cases repressing metamorphosis. The role of E93 as an inducer of adult morphogenesis has also been demonstrated in *T. castaneum*, as RNAi transcript depletion in pupae inhibits the formation of the adult and triggers the formation of supernumerary pupae (Ureña et al., 2014). Further studies (Chafino et al., 2019; Ureña et al., 2016) showed that the peak of *Kr-h1* expression occurring at the end of the last larval instar (Fig. 10.5) is key for pupal morphogenesis, as depleting it results in a precocious upregulation of *E93* expression and a direct transition from larva to adult, thus bypassing the pupal stage. These studies also showed that the Kr-h1-dependent repression of *E93* is crucial to allow the dramatic upregulation of *BR-C* expression that triggers the formation of the pupa. Finally, depletion of E93 in the last larval instar prevented this upregulation of *BR-C* expression, which indicates that the subtle increase of *E93* expression that occurs at the beginning of the last larval instar triggers that of *BR-C* (Chafino et al., 2019; Ureña et al., 2016). The data available would be compatible with the following sequence of events underlying the regulation of the holometabolan metamorphosis in *T. castaneum*: (1) JH through Met and Kr-h1 maintains the larval status until the last larval instar; (2) at the beginning of the last larval instar, a transient decrease of *Kr-h1* expression results in a slight increase of *E93* expression, and an upregulation of *BR-C* expression, which peaks in the prepupal stage thus triggering the formation of the pupa; (3) the BR-C peak and a rebound of *Kr-h1* expression prevent a further increase of *E93* expression in the prepupal stage, thus preventing adult morphogenesis; and (4) the decline of *BR-C* and *Kr-h1* expression at the beginning of the pupal stage disinhibits the expression of *E93*, which dramatically increases, thus triggering adult morphogenesis.

The functional data based on RNAi experiments, the epistatic relationships among Kr-h1, BR-C, and E93, and the expression patterns (Fig. 10.5) fit into the general frame of the MEKRE93 pathway, as adapted to holometabolan insects (Belles, 2019).

Bombyx mori

The expression patterns of *Kr-h1*, *BR-C*, and *E93* in the silkworm *B. mori* (Fig. 10.5) are relatively similar to those of *T. castaneum*. In the epidermis, *Kr-h1* expression is high in larvae, decreases in the last larval instar (L5), transiently increases in the prepupal stage, and vanishes in the pupa (Kayukawa et al., 2014). *BR-C* expression shows an acute peak centered in the prepupal stage (Daimon et al., 2015; Kayukawa et al., 2014), whereas *E93* expression shows a peak at the beginning of the pupal stage, then decreases and increases again in the transition to the adult stage (Kayukawa et al., 2017).

Causal evidence that a JH receptor Met1 in *B. mori* (there are two Met paralogs in lepidopterans) transduces the antimetamorphic JH signal has been obtained in knockout silkworms with null mutations in *Met1* and *Met2* (Daimon et al., 2015). The *Met2* mutants develop without apparent anomalies until the adult stage, but those of *Met1* develop normally until L3 or L4, after which they molt to individuals with precocious adult features. This indicates that JH represses metamorphosis through Met1, although this mechanism operates when larvae are competent to metamorphose (see Chapter 7: Molecular mechanisms regulating hormone production and action). Intriguingly, the expression levels of *BR-C*, which is repressed by JH (see below), remain at low basal levels during the early larval instars of JH-deficient or JH signaling-deficient knockouts (Daimon et al., 2015), which suggests that *BR-C* expression cannot be highly upregulated until the larva becomes competent to metamorphose (see Smykal et al., 2014).

Experiments of CA removal and JH mimic treatment have shown that JH induces the expression of *Kr-h1* through Met in *B. mori*. Indeed a JH response element containing a canonical E-box sequence to which Met is likely to bind has been located in the promoter region of *B. mori Kr-h1* (see Chapter 7: Molecular mechanisms regulating hormone production and action) (Kayukawa et al., 2012). Moreover, genetic experiments have shown that transgenic silkworms overexpressing *Kr-h1* grow normally until the spinning stage, but they arrest development at the prepupal stage (Kayukawa et al., 2014), which indicates that the decline of *Kr-h1* expression is required to form the pupa. In the silkworm, JH inhibits the 20E-dependent induction of *BR-C* in larval epidermis (Muramatsu et al., 2008), an effect presumably mediated by Kr-h1, as it has been shown that Kr-h1 directly binds to a specific binding site (KBS) within the *BR-C* promoter, preventing *BR-C* activation (Kayukawa et al., 2016).

More recently, Kayukawa et al. (2017) have reported that JH suppresses E93 expression in the epidermis of the silkworm and in a *B. mori* cell line, whereas reporter assays in the cell line revealed that the JH-dependent suppression is mediated by Kr-h1. Genome-wide analysis of chromatin immunoprecipitation sequencing (ChIP-seq) identified a consensus KBS in the *E93* promoter region, to which Kr-h1 directly binds (Kayukawa et al., 2017) (see Chapter 7: Molecular mechanisms regulating hormone production and action). Although the functional data are limited, the expression patterns (Fig. 10.5) suggest that the interactions of Kr-h1, BR-C, and E93 regulating *B. mori* metamorphosis are equivalent to those observed in *T. castaneum*.

Drosophila melanogaster

In the fruit fly *D.* melanogaster, *Kr-h1* expresses three isoforms: α, ß, and γ. The most abundant is the isoform γ, whose transcripts are observed during most of the last larval instar (L3). Then its expression decreases at the beginning of the prepupal stage, subsequently forms an acute and transient peak centered in the prepupal stage, maintains intermediate levels at the beginning of the pupa, and decreases again toward the middle of this stage (Minakuchi et al., 2008). *E93* expresses two isoforms, A and B, both showing a similar expression pattern: low levels in the last larval instar, L3, increasing in the prepupal stage and peaking at the beginning of the pupa, and subsequently keeping quite high and oscillating values (Ureña et al., 2014). The expression of *BR-C* centers in the prepupal period, as shown, for example, by Karim and Thummel (1992) (Fig. 10.5).

The antimetamorphic effects of Kr-h1 were discovered in *D. melanogaster* using the abdominal integument at pupariation as an experimental system. JH applied at pupariation inhibited bristle formation and caused pupal cuticle formation in the abdomen, while the expression of *BR-C* was prolonged and *Kr-h1* was one of the genes upregulated in the JH-treated abdominal integument. Importantly, ectopic expression of *Kr-h1* in the abdominal epidermis resulted in missing or short bristles in the dorsal midline, a phenotype similar to that obtained after a treatment of JH at low dose, or after misexpression of *BR-C* some 24 hours after puparium formation (Minakuchi et al., 2008). These results indicate that Kr-h1 conveys the antimetamorphic signal of JH in the abdominal epidermis of *D. melanogaster*, and that it is upstream of *BR-C* in the pathway. The *BR-C* gene and its function as a trigger of the pupal stage were also discovered in *D. melanogaster* (Kiss et al., 1988; von Kalm et al., 1994) (see Chapter 7: Molecular mechanisms regulating hormone production and action). *E93* was found while studying the degeneration of the salivary glands during *D. melanogaster* metamorphosis (Baehrecke and Thummel, 1995). More recently, it has been shown that the action of E93 in metamorphosis is not restricted to regulation of cell death processes, but it is also a morphogenetic factor that regulates adult patterning during metamorphosis (Mou et al., 2012) (Ureña et al., 2014) (see Chapter 7: Molecular mechanisms regulating hormone production and action). The functional data available and the expression patterns (Fig. 10.5) suggest that the interactions and expression patterns of *Kr-h1*, *E93*, and *BR-C* in *D. melanogaster* during metamorphosis are similar to those of *T. castaneum*.

Mechanisms regulating holometabolan neoteny: *Xenos vesparum*

Within the holometabolan insects, strepsipterans offer a classical example of endoparasitism with dramatic differences between the sexes. Female larvae and adults live within the host, but mature females are neotenic, retaining a larva-like morphology while being reproductively competent. In contrast, adult males metamorphose from larva to pupa and then to adult, which leaves the host as a typical free-flying insect (Kathirithamby, 2009) (see Chapter 5: The holometabolan development). The first molecular studies were carried out on the species *Xenos vesparum* by monitoring the expression of *BR-C* during the transition from larvae to pupa in males and from larva to reproductively competent larviform females. Results revealed elevated levels of expression of total *BR-C* and the isoforms Z1, Z3, and Z4 in the last larval instar of males but not in females (Erezyilmaz et al., 2014). The fact that *BR-C* expression does not increase beyond the last larval instar explains why the females do not transform into the pupal stage.

Further studies in *X. vesparum* revealed that while a clear increase of *E93* expression is measured in the pupal and adult stages of males, persistent low levels of *E93* expression are observed in females (Chafino et al., 2018). This explains why the female of *X. vesparum*, unlike what happens in the male, cannot undertake a complete metamorphosis. The persistent low expression of *E93* in the female could also explain why the expression of *BR-C* remains also low, if the mechanism of induction of *BR-C* expression by moderate levels of E93, which has been observed in the red flour beetle *T. castaneum* (Chafino et al., 2019), also operates in *X. vesparum*. As discussed in Chapter 5 (The holometabolan development), mating occurs with the female inside the host, from where it extrudes the cephalothorax, through which the male inserts the sperm. Intriguingly, the female of *X. vesparum*, despite maintaining a larviform shape, undergoes some morphological transformation and tanning in the cephalothorax. In relation to this, a specific increase of *E93* expression was observed in the cephalothorax of females in late fourth larval instar. *BR-C* expression also tended to be upregulated in the cephalothorax, compared with the abdomen (Chafino et al., 2018). The persistent low expression of *BR-C* and *E93* explains why metamorphosis does not proceed in the female of *X. vesparum*, especially in the abdomen. However, further research will be needed to answer the question of why the expression of *E93* and *BR-C* remains low specifically in females.

REFERENCES

Abdou, M.A., He, Q., Wen, D., Zyaan, O., Wang, J., Xu, J., et al., 2011. *Drosophila* Met and Gce are partially redundant in transducing juvenile hormone action. Insect Biochem. Mol. Biol. 41, 938–945. Available from: https://doi.org/10.1016/j.ibmb.2011.09.003.

Aboulafia-Baginsky, N., Pener, M.P., Staal, G.B., 1984. Chemical allatectomy of late *Locusta* embryos by a synthetic precocene and its effect on hopper morphogenesis. J. Insect Physiol. 30, 839–852. Available from: https://doi.org/10.1016/0022-1910(84)90057-X.

Almudi, I., Martín-Blanco, C.A., García-Fernandez, I.M., López-Catalina, A., Davie, K., Aerts, S., et al., 2019. Establishment of the mayfly *Cloeon dipterum* as a new model system to investigate insect evolution. Evodevo 10, 6. Available from: https://doi.org/10.1186/s13227-019-0120-y.

Baehrecke, E.H., Thummel, C.S., 1995. The *Drosophila* E93 gene from the 93F early puff displays stage- and tissue-specific regulation by 20-hydroxyecdysone. Dev. Biol. 171, 85–97. Available from: https://doi.org/10.1006/dbio.1995.1262.

Beck, Y., Pecasse, F., Richards, G., 2004. Krüppel-homolog is essential for the coordination of regulatory gene hierarchies in early *Drosophila* development. Dev. Biol. 268, 64–75. Available from: https://doi.org/10.1016/j.ydbio.2003.12.017.

Belles, X., 2019. Krüppel homolog 1 and E93: the doorkeeper and the key to insect metamorphosis. Arch. Insect Biochem. Physiol. e21609. Available from: https://doi.org/10.1002/arch.21609.

Belles, X., Santos, C.G., 2014. The MEKRE93 (Methoprene tolerant-Krüppel homolog 1-E93) pathway in the regulation of insect metamorphosis, and the homology of the pupal stage. Insect Biochem. Mol. Biol. 52, 60–68. Available from: https://doi.org/10.1016/j.ibmb.2014.06.009.

Belles, X., Casas, J., Messeguer, A., Piulachs, M.D., 1987. In vitro biosynthesis of JH III by the corpora allata of adult females of *Blattella germanica* (L). Insect Biochem. 17, 1007–1010. Available from: https://doi.org/10.1016/0020-1790(87)90111-9.

Bergot, B.J., Baker, F.C., Cerf, D.C., Jamieson, G., Schooley, D.A., 1981. Qualitative and quantitative aspects of JH titers in developing embryos of several insect species: discovery of a new JH-like substance extracted from eggs of *Manduca sexta*. In: Pratt, G.E., Brooks, G.T. (Eds.), Juvenile Hormone Biochemistry. Elsevier, Amsterdam, pp. 33–45.

Bitsch, C., Baehr, J.C., Bitsch, J., 1985. Juvenile hormones in *Thermobia domestica* females: identification and quantification during biological cycles and after precocene application. Experientia 41, 409–410. Available from: https://doi.org/10.1007/BF02004532.

Brüning, E., Lanzrein, B., 1987. Function of juvenile hormone III in embryonic development of the cockroach, *Nauphoeta cinerea*. Int. J. Invertebr. Reprod. Dev. 12, 29–44.

Chafino, S., López-Escardó, D., Benelli, G., Kovac, H., Casacuberta, E., Franch-Marro, X., et al., 2018. Differential expression of the adult specifier E93 in the strepsipteran *Xenos vesparum* Rossi suggests a role in female neoteny. Sci. Rep. 8, 14176. Available from: https://doi.org/10.1038/s41598-018-32611-y.

Chafino, S., Ureña, E., Casanova, J., Casacuberta, E., Franch-Marro, X., Martín, D., 2019. Upregulation of E93 gene expression acts as the trigger for metamorphosis independently of the threshold size in the beetle *Tribolium castaneum*. Cell Rep. 27, 1039–1049.e2. Available from: https://doi.org/10.1016/j.celrep.2019.03.094.

Chavoshi, T.M., Moussian, B., Uv, A., 2010. Tissue-autonomous EcR functions are required for concurrent organ morphogenesis in the *Drosophila* embryo. Mech. Dev. 127, 308–319. Available from: https://doi.org/10.1016/j.mod.2010.01.003.

Cheatle Jarvela, A.M., Pick, L., 2017. The function and evolution of nuclear receptors in insect embryonic development. Curr. Top. Dev. Biol. 125, 39–70. Available from: https://doi.org/10.1016/BS.CTDB.2017.01.003.

Daimon, T., Uchibori, M., Nakao, H., Sezutsu, H., Shinoda, T., 2015. Knockout silkworms reveal a dispensable role for juvenile hormones in holometabolous life cycle. Proc. Natl.

Acad. Sci. U.S.A. 112, E4226–E4235. Available from: https://doi.org/10.1073/pnas.1506645112.

Delany, M.J., 1957. Life histories in the Thysanura. Acta Zool. Cracovien. 2, 61–90.

de la Paz, A.R., Delbecque, J.P., Bitsch, J., Delachambre, J., 1983. Ecdysteroids in the haemolymph and the ovaries of the firebrat *Thermobia domestica* (Packard) (Insecta, Thysanura): correlations with integumental and ovarian cycles. J. Insect Physiol. 29, 323–329. Available from: https://doi.org/10.1016/0022-1910(83)90033-1.

Erezyilmaz, D., Riddiford, L., Truman, J., 2004. Juvenile hormone acts at embryonic molts and induces the nymphal cuticle in the direct-developing cricket. Dev. Genes Evol. 214, 313–323. Available from: https://doi.org/10.1007/s00427-004-0408-2.

Erezyilmaz, D.F., Hayward, A., Huang, Y., Paps, J., Acs, Z., Delgado, J.A., et al., 2014. Expression of the pupal determinant broad during metamorphic and neotenic development of the strepsipteran *Xenos vesparum* Rossi. PLoS One 9, e93614. Available from: https://doi.org/10.1371/journal.pone.0093614.

Fernandez-Nicolas, A., Belles, X., 2017. Juvenile hormone signaling in short germ-band hemimetabolan embryos. Development 144, 4637–4644. Available from: https://doi.org/10.1242/dev.152827.

Gilbert, L.I., Schneiderman, H.A., 1960. The development of a bioassay for the juvenile hormone of insects. Trans. Am. Micr. Soc. 79, 38–67.

Gujar, H., Palli, S.R., 2016. Krüppel homolog 1 and E93 mediate Juvenile hormone regulation of metamorphosis in the common bed bug, *Cimex lectularius*. Sci. Rep. 6, 26092. Available from: https://doi.org/10.1038/srep26092.

Heffer, A., Grubbs, N., Mahaffey, J., Pick, L., 2013. The evolving role of the orphan nuclear receptor ftz-f1, a pair-rule segmentation gene. Evol. Dev. 15, 406–417. Available from: https://doi.org/10.1111/ede.12050.

Hillman, R., Lesnik, L.H., 1970. Cuticle formation in the embryo of *Drosophila melanogaster*. J. Morphol. 131, 383–395. Available from: https://doi.org/10.1002/jmor.1051310403.

Huang, J.-H., Lozano, J., Belles, X., 2013. Broad-complex functions in postembryonic development of the cockroach *Blattella germanica* shed new light on the evolution of insect metamorphosis. Biochim. Biophys. Acta Gen. Subj. 1830, 2178–2187. Available from: https://doi.org/10.1016/j.bbagen.2012.09.025.

Ishimaru, Y., Tomonari, S., Matsuoka, Y., Watanabe, T., Miyawaki, K., Bando, T., et al., 2016. TGF-β signaling in insects regulates metamorphosis via juvenile hormone biosynthesis. Proc. Natl. Acad. Sci. U.S.A. 113, 5634–5639. Available from: https://doi.org/10.1073/pnas.1600612113.

Ishimaru, Y., Tomonari, S., Watanabe, T., Noji, S., Mito, T., 2019. Regulatory mechanisms underlying the specification of the pupal-homologous stage in a hemimetabolous insect. Philos. Trans. R. Soc. Lond. B. Biol. Sci. 374, 20190225. Available from: https://doi.org/10.1098/rstb.2019.0225.

Jindra, M., 2019. Where did the pupa come from? The timing of juvenile hormone signalling supports homology between stages of hemimetabolous and holometabolous insects. Philos. Trans. R. Soc. Lond. B. Biol. Sci. 374,. Available from: https://doi.org/10.1098/rstb.2019.006420190064374.

Jindra, M., Uhlirova, M., Charles, J.-P., Smykal, V., Hill, R.J., 2015. Genetic evidence for function of the bHLH-PAS protein Gce/Met as a juvenile hormone receptor. PLoS Genet. 11, e1005394. Available from: https://doi.org/10.1371/journal.pgen.1005394.

Kamsoi, O., Belles, X., 2019. Myoglianin triggers the premetamorphosis stage in hemimetabolan insects. FASEB J. 33, 3659–3669. Available from: https://doi.org/10.1096/fj.201801511R.

Karim, F.D., Thummel, C.S., 1992. Temporal coordination of regulatory gene expression by the steroid hormone ecdysone. EMBO J. 11, 4083–4093.

Kathirithamby, J., 2009. Host-parasitoid associations in Strepsiptera. Annu. Rev. Entomol. 54, 227–249. Available from: https://doi.org/10.1146/annurev.ento.54.110807.090525.

Kayukawa, T., Minakuchi, C., Namiki, T., Togawa, T., Yoshiyama, M., Kamimura, M., et al., 2012. Transcriptional regulation of juvenile hormone-mediated induction of Krüppel homolog 1, a repressor of insect metamorphosis. Proc. Natl. Acad. Sci. U.S.A. 109, 11729–11734. Available from: https://doi.org/10.1073/pnas.1204951109.

Kayukawa, T., Murata, M., Kobayashi, I., Muramatsu, D., Okada, C., Uchino, K., et al., 2014. Hormonal regulation and developmental role of Krüppel homolog 1, a repressor of metamorphosis, in the silkworm *Bombyx mori*. Dev. Biol. 388, 48–56. Available from: https://doi.org/10.1016/j.ydbio.2014.01.022.

Kayukawa, T., Nagamine, K., Ito, Y., Nishita, Y., Ishikawa, Y., Shinoda, T., 2016. Krüppel homolog 1 inhibits insect metamorphosis via direct transcriptional repression of Broad-Complex, a pupal specifier gene. J. Biol. Chem. 291, 1751–1762. Available from: https://doi.org/10.1074/jbc.M115.686121.

Kayukawa, T., Jouraku, A., Ito, Y., Shinoda, T., 2017. Molecular mechanism underlying juvenile hormone-mediated repression of precocious larval-adult metamorphosis. Proc. Natl. Acad. Sci. U.S.A. 114, 1057–1062. Available from: https://doi.org/10.1073/pnas.1615423114.

Kiss, I., Beaton, A.H., Tardiff, J., Fristrom, D., Fristrom, J.W., 1988. Interactions and developmental effects of mutations in the Broad-Complex of *Drosophila melanogaster*. Genetics 118, 247–259.

Konopová, B., Jindra, M., 2007. Juvenile hormone resistance gene Methoprene-tolerant controls entry into metamorphosis in the beetle *Tribolium castaneum*. Proc. Natl. Acad. Sci. U.S.A. 104, 10488–10493. Available from: https://doi.org/10.1073/pnas.0703719104.

Konopová, B., Jindra, M., 2008. Broad-Complex acts downstream of Met in juvenile hormone signaling to coordinate primitive holometabolan metamorphosis. Development 135, 559–568. Available from: https://doi.org/10.1242/dev.016097.

Konopová, B., Smykal, V., Jindra, M., 2011. Common and distinct roles of juvenile hormone signaling genes in metamorphosis of holometabolous and hemimetabolous insects. PLoS One 6, e28728. Available from: https://doi.org/10.1371/journal.pone.0028728.

Lagueux, M., Hetru, C., Goltzene, F., Kappler, C., Hoffmann, J.A., 1979. Ecdysone titre and metabolism in relation to cuticulogenesis in embryos of *Locusta migratoria*. J. Insect Physiol. 25, 709–723. Available from: https://doi.org/10.1016/0022-1910(79)90123-9.

Li, X., 2007. Juvenile hormone and methyl farnesoate production in cockroach embryos in relation to dorsal closure and the reproductive modes of different species of cockroaches. Arch. Insect Biochem. Physiol. 66, 159–168. Available from: https://doi.org/10.1002/arch.20207.

Lozano, J., Belles, X., 2011. Conserved repressive function of Krüppel homolog 1 on insect metamorphosis in hemimetabolous and holometabolous species. Sci. Rep. 1, 163. Available from: https://doi.org/10.1038/srep00163.

Lozano, J., Belles, X., 2014. Role of methoprene-tolerant (Met) in adult morphogenesis and in adult ecdysis of *Blattella germanica*. PLoS One 9, e103614. Available from: https://doi.org/10.1371/journal.pone.0103614.

Lozano, J., Kayukawa, T., Shinoda, T., Belles, X., 2014. A role for Taiman in insect metamorphosis. PLoS Genet. 10, e1004769. Available from: https://doi.org/10.1371/journal.pgen.1004769.

Maestro, O., Cruz, J., Pascual, N., Martín, D., Belles, X., 2005. Differential expression of two RXR/ultraspiracle isoforms during the life cycle of the hemimetabolous insect *Blattella germanica* (Dictyoptera, Blattellidae). Mol. Cell. Endocrinol. 238, 27–37. Available from: https://doi.org/10.1016/j.mce.2005.04.004.

Maestro, J.L., Pascual, N., Treiblmayr, K., Lozano, J., Belles, X., 2010. Juvenile hormone and allatostatins in the German cockroach embryo. Insect Biochem. Mol. Biol. 40, 660–665. Available from: https://doi.org/10.1016/j.ibmb.2010.06.006.

Mané-Padrós, D., Cruz, J., Vilaplana, L., Pascual, N., Belles, X., Martín, D., 2008. The nuclear hormone receptor BgE75 links molting and developmental progression in the direct-developing insect *Blattella germanica*. Dev. Biol. 315, 147–160. Available from: https://doi.org/10.1016/j.ydbio.2007.12.015.

Maróy, P., Kaufmann, G., Dübendorfer, A., 1988. Embryonic ecdysteroids of *Drosophila melanogaster*. J. Insect Physiol. 34, 633–637. Available from: https://doi.org/10.1016/0022-1910(88)90071-6.

Minakuchi, C., Zhou, X., Riddiford, L.M., 2008. Krüppel homolog 1 (Kr-h1) mediates juvenile hormone action during metamorphosis of *Drosophila melanogaster*. Mech. Dev. 125, 91–105. Available from: https://doi.org/10.1016/j.mod.2007.10.002.

Minakuchi, C., Namiki, T., Shinoda, T., 2009. Krüppel homolog 1, an early juvenile hormone-response gene downstream of Methoprene-tolerant, mediates its anti-metamorphic action in the red flour beetle *Tribolium castaneum*. Dev. Biol. 325, 341–350. Available from: https://doi.org/10.1016/j.ydbio.2008.10.016.

Minakuchi, C., Tanaka, M., Miura, K., Tanaka, T., 2011. Developmental profile and hormonal regulation of the transcription factors broad and Krüppel homolog 1 in hemimetabolous thrips. Insect Biochem. Mol. Biol. 41, 125–134. Available from: https://doi.org/10.1016/j.ibmb.2010.11.004.

Mizuno, T., Watanabe, K., Ohnishi, E., 1981. Developmental changes of ecdysteroids in the eggs of the silkworm, *Bombyx mori*. Dev. Growth Differ. 23, 543–552. Available from: https://doi.org/10.1111/j.1440-169X.1981.00543.x.

Moritz, G., 2006. Thripse: Fransenflügler, Thysanoptera. Die Neue Brehm-Bücherei. Westarp-Verlag, Hohenwarsleben.

Mou, X., Duncan, D.M., Baehrecke, E.H., Duncan, I., 2012. Control of target gene specificity during metamorphosis by the steroid response gene E93. Proc. Natl. Acad. Sci. U.S.A. 109, 2949–2954. Available from: https://doi.org/10.1073/pnas.1117559109.

Muramatsu, D., Kinjoh, T., Shinoda, T., Hiruma, K., 2008. The role of 20-hydroxyecdysone and juvenile hormone in pupal commitment of the epidermis of the silkworm, *Bombyx mori*. Mech. Dev. 125, 411–420. Available from: https://doi.org/10.1016/j.mod.2008.02.001.

Novák, V.J., 1969. Morphogenetic analysis of the effects of juvenile hormone analogues and other morphogenetically active substances on embryos of *Schistocerca gregaria* (Forskål). J. Embryol. Exp. Morphol. 21, 1–21.

Okude, G., Futahashi, R., Kawahara-Miki, R., Yoshitake, K., Yajima, S., Fukatsu, T., 2017a. Electroporation-mediated RNA interference reveals a role of the multicopper oxidase 2 gene in dragonfly cuticular pigmentation. Appl. Entomol. Zool. 52, 379–387. Available from: https://doi.org/10.1007/s13355-017-0489-9.

Okude, G., Futahashi, R., Tanahashi, M., Fukatsu, T., 2017b. Laboratory rearing system for *Ischnura senegalensis* (Insecta: Odonata) enables detailed description of larval development and morphogenesis in dragonfly. Zool. Sci. 34, 386–397. Available from: https://doi.org/10.2108/zs170051.

Okude, G., Futahashi, R., Fukatsu, T., 2019. Molecular mechanisms of metamorphosis in dragonflies. In: Commun. 4th Int. Insect Horm. Work., Kolymbari, Crete, Greece, p. 43.

Pecasse, F., Beck, Y., Ruiz, C., Richards, G., 2000. Krüppel-homolog, a stage-specific modulator of the prepupal ecdysone response, is essential for *Drosophila* metamorphosis. Dev. Biol. 221, 53–67. Available from: https://doi.org/10.1006/dbio.2000.9687.

Rousset, A., Bitsch, C., Bitsch, J., 1987. Vitellogenins in the firebrat, *Thermobia domestica* (Packard): immunological quantification during reproductive cycles and in relation to insemination (Thysanura: Lepismatidae). J. Insect Physiol. 33, 593–601. Available from: https://doi.org/10.1016/0022-1910(87)90075-8.

Sehnal, F., Švácha, P., Zrzavý, J., 1996. Evolution of insect metamorphosis. In: Gilbert, L.I., Tata, J.R., Atkinson, B.G. (Eds.), Metamorphosis. Postembryonic Reprogramming of Gene Expression in Amphibian and Insect Cells. Academic Press, San Diego, CA, pp. 3–58.

Si, Q., Luo, J.-Y., Hu, Z., Zhang, W., Zhou, C.-F., 2017. De novo transcriptome of the mayfly *Cloeon viridulum* and transcriptional signatures of Prometabola. PLoS One 12, e0179083. Available from: https://doi.org/10.1371/journal.pone.0179083.

Smykal, V., Daimon, T., Kayukawa, T., Takaki, K., Shinoda, T., Jindra, M., 2014. Importance of juvenile hormone signaling arises with competence of insect larvae to metamorphose. Dev. Biol. 390, 221–230. Available from: https://doi.org/10.1016/j.ydbio.2014.03.006.

Suzuki, Y., Truman, J.W., Riddiford, L.M., 2008. The role of Broad in the development of *Tribolium castaneum*: implications for the evolution of the holometabolous insect pupa. Development 135, 569–577. Available from: https://doi.org/10.1242/dev.015263.

Tan, K.L., Vlisidou, I., Wood, W., 2014. Ecdysone mediates the development of immunity in the *Drosophila* embryo. Curr. Biol. 24, 1145–1152. Available from: https://doi.org/10.1016/j.cub.2014.03.062.

Temin, G., Zander, M., Roussel, J.-P., 1986. Physico-chemical (GC-MS) measurements of juvenile hormone III titres during embryogenesis of *Locusta migratoria*. Int. J. Invertebr. Reprod. Dev. 9, 105–112. Available from: https://doi.org/10.1080/01688170.1986.10510184.

Truman, J.W., Riddiford, L.M., 1999. The origins of insect metamorphosis. Nature 401, 447–452. Available from: https://doi.org/10.1038/46737.

Truman, J.W., Riddiford, L.M., 2002. Endocrine insights into the evolution of metamorphosis in insects. Annu. Rev. Entomol. 47, 467–500. Available from: https://doi.org/10.1146/annurev.ento.47.091201.145230.

Ureña, E., Manjón, C., Franch-Marro, X., Martín, D., 2014. Transcription factor E93 specifies adult metamorphosis in hemimetabolous and holometabolous insects. Proc. Natl. Acad. Sci. U.S.A. 111, 7024–7029. Available from: https://doi.org/10.1073/pnas.1401478111.

Ureña, E., Chafino, S., Manjón, C., Franch-Marro, X., Martín, D., 2016. The occurrence of the holometabolous pupal stage requires the interaction between E93, Krüppel-homolog 1 and Broad-Complex. PLoS Genet. 12, e1006020. Available from: https://doi.org/10.1371/journal.pgen.1006020.

Vea, I.M., Tanaka, S., Shiotsuki, T., Jouraku, A., Tanaka, T., Minakuchi, C., 2016. Differential juvenile hormone variations in scale insect extreme sexual dimorphism. PLoS One 11, e0149459. Available from: https://doi.org/10.1371/journal.pone.0149459.

Vea, I.M., Tanaka, S., Tsuji, T., Shiotsuki, T., Jouraku, A., Minakuchi, C., 2019. E93 expression and links to the juvenile hormone in hemipteran mealybugs with insights on female neoteny. Insect Biochem. Mol. Biol. 104, 65–72. Available from: https://doi.org/10.1016/j.ibmb.2018.11.008.

Villalobos-Sambucaro, M.J., Riccillo, F.L., Calderón-Fernández, G.M., Sterkel, M., Diambra, L. A., Ronderos, J.R., 2015. Genomic and functional characterization of a methoprene-tolerant gene in the kissing-bug *Rhodnius prolixus*. Gen. Comp. Endocrinol. 216, 1–8. Available from: https://doi.org/10.1016/j.ygcen.2015.04.018.

von Kalm, L., Crossgrove, K., Von Seggern, D., Guild, G.M., Beckendorf, S.K., 1994. The Broad-Complex directly controls a tissue-specific response to the steroid hormone ecdysone at the onset of *Drosophila* metamorphosis. EMBO J. 13, 3505–3516.

Watson, J.A.L., 1967. The growth and activity of the corpora allata in the larval firebrat, *Thermobia domestica* (Packard) (Thysanura, Lepismatidae). Biol. Bull. 132, 277–291. Available from: https://doi.org/10.2307/1539895.

Ylla, G., Piulachs, M.-D., Belles, X., 2017. Comparative analysis of miRNA expression during the development of insects of different metamorphosis modes and germ-band types. BMC Genom. 18, 774. Available from: https://doi.org/10.1186/s12864-017-4177-5.

Ylla, G., Piulachs, M.D., Belles, X., 2018. Comparative transcriptomics in two extreme neopterans reveals general trends in the evolution of modern insects. iScience 4, 164–179.

Zhu, G.-H., Jiao, Y., Chereddy, S.C.R.R., Noh, M.Y., Palli, S.R., 2019. Knockout of juvenile hormone receptor, Methoprene-tolerant, induces black larval phenotype in the yellow fever mosquito. *Aedes aegypti, Proc. Natl. Acad. Sci. U. S. A* 116, 21501–21507. Available from: https://doi.org/10.1073/pnas.1905729116.

Chapter 11

The origin of hemimetaboly

Fossil evidence and phylogenetic reconstructions indicate that ametaboly is the ancestral mode of postembryonic development in insects. Ametaboly gave rise to hemimetaboly, which in turn gave rise to holometaboly (Belles, 2011). Often holometaboly is considered the only significant type of metamorphosis (Clapham et al., 2016), and some authors even consider metamorphosis and holometaboly to be synonymous (Danley et al., 2007). Probably for this reason, studies have focused on the evolution of holometaboly (Sehnal et al., 1996; Truman and Riddiford, 1999). However, the innovation of hemimetaboly represents the origin of insect metamorphosis and a necessary step in the evolution of holometaboly.

Hemimetaboly emerged with the clade Pterygota (Misof et al., 2014; Wang et al., 2016), and the innovation of hemimetabolan metamorphosis after the innovation of wings is probably not coincidental, but rather hemimetaboly could have emerged as a consequence of wing acquisition. Ametabolan species, like the firebrat *Thermobia domestica* (Zygentoma), undergo morphological modifications throughout their life cycle, including the formation of scales between the third and the fourth nymphal instars (Delany, 1957) (see Chapter 3: The ametabolan development). Therefore the most genuine innovation after the emergence of wings is the final molt, which marks the transition from the last nymphal instar to the adult. The final molt is mainly achieved through the disintegration of the prothoracic gland (PG), which produces the molting hormone, and perhaps also through the adult commitment of epidermal cells, which become incapable of producing a new cuticle. Possibly, the most important selective advantage of the final molt is associated with the wing innovation, as molting of a membranous wing can be complicated by mechanical problems during the ecdysis and exuviation.

THE ORIGIN OF WINGS AND THE EMERGENCE OF HEMIMETABOLY

According to phylogenetic and paleontological data, wings and hemimetaboly originated during the early–middle Devonian, some 400 Mya (Misof et al., 2014; Wang et al., 2016). The fact that extant pterygotes are

Insect Metamorphosis. DOI: https://doi.org/10.1016/B978-0-12-813020-9.00011-9

monophyletic, and that the wing types of the different insect orders are formed in the same way, both in extant and fossil species (Prokop et al., 2017), strongly suggest that wings evolved only once. The monophyly of Pterygota, in turn, suggests that there was a single origin of hemimetaboly, and that this could be associated with the emergence of wings.

Fossils are unique materials that can provide direct information on the origin and morphological evolution of metamorphosis, but studies examining fossils of successive developmental stages of the same species are rare (Haug et al., 2016). The most complete series of Paleozoic fossils of successive nymphal instars are of Megasecoptera and Ephemeroptera, as reported, for example, by Kukalová-Peck (1978, 1991, 1997). These fossils show how the wings of early nymphs were small and curved backward, whereas, with successive molts, the wings increased in length, became less curved, and tended to adopt a position perpendicular to the body axis. In the adult, the wings were more elongated, with the basal region narrower than the distal, and practically perpendicular to the body axis (Haug et al., 2016; Kukalová-Peck, 1978, 1991, 1997) (Fig. 11.1).

Regarding the articulation of the wings in the juveniles, the data have been largely controversial. Kukalová-Peck (1978, 1983, 1985, 1991) argues that the wings of Paleozoic juveniles are fully articulated, although this view has been challenged by some authors, like Wootton (1981), who contended that evidence for articulation of wings in Paleozoic insect nymphs is not widespread. Recent observations on fossil samples in which the preservation of the articulation is sufficiently three-dimensional have allowed stereo images and surface reconstructions that provide more detailed information (Haug et al., 2016). Some of these samples correspond to the holotype of *Herdina mirificus* (Fig. 11.2A), a Pterygote species of uncertain systematic

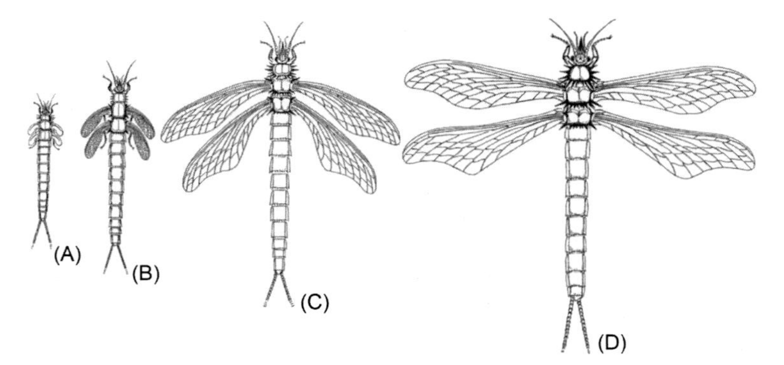

FIGURE 11.1 Selected developmental stages of Paleozoic *Mischoptera* (Megasecoptera) species. (A) Young nymph with apparently articulated wings arched backward, with wing venation still absent. (B) Older nymph with larger wings already showing wing venation. (C) A mature nymph or subadult with straighter wings. (D) Adult with the wings practically perpendicular to the body axis. Modified from Kukalová-Peck (1991), with permission.

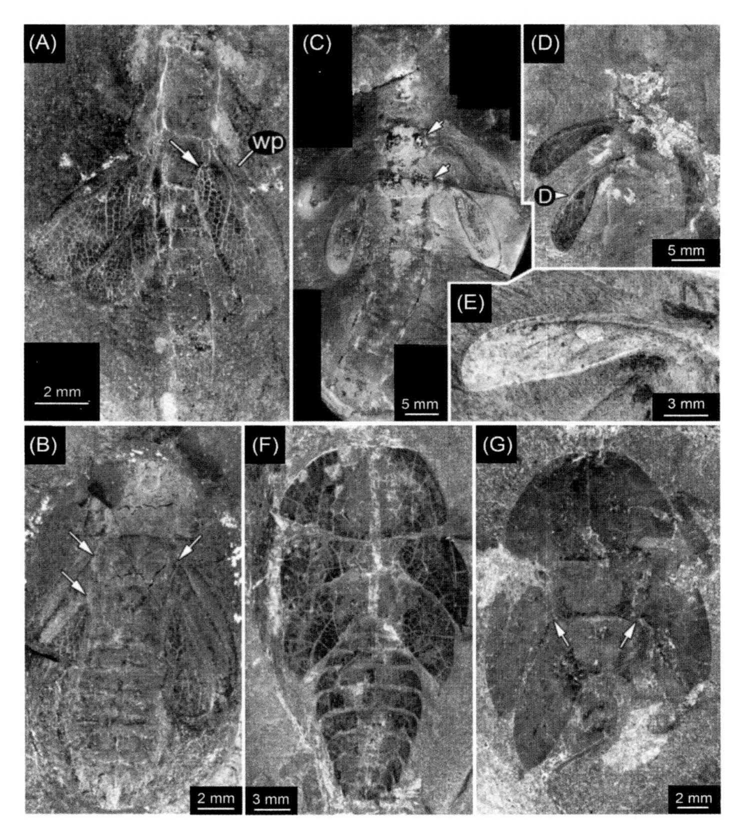

FIGURE 11.2 Paleozoic insect nymphs. (A) Holotype of *Herdina mirificus*; the arrow indicates uneven edge of wing pad (wp), suggesting that the structure is ripped off. (B) *Herdina*-like nymph showing the folds setting off the winglets from the tergum (arrows). (C) *Mischoptera*-like fossil showing the lines separating the winglets from the tergum (arrows). (D) Holotype of *Mischoptera? (= Lamereeites) curvipennes*; note the exactly matching shape and arrangement of the winglets compared to the specimen shown in panel (C). (E) Detail of the metathoracic winglet of the specimen shown in panel (D). (F) Late nymphal instar of a roachoid. (G) Mature roachoid nymph showing very long wing pads that appear separated from the tergum (arrows). From Haug et al. (2016), with permission.

position. The specimen shows mesothoracic and metathoracic winglets partly folded over the abdomen. These winglet features and the general shape of the insect led Kukalová-Peck (1998) to consider it as a nymph with articulated winglets. The study of this specimen and other *Herdina*-like fossil nymphs (Fig. 11.2B) suggests that the winglets were articulated along the entire lateral edge of the mesonotum and metanotum, which would limit the axes of movement (Haug et al., 2016). This differs from the wing of modern adult insects, which articulates through a narrow series of sclerites (the pteralia), thus allowing a more versatile movement. Similar observations and conclusions were obtained with the study of the Megasecopteran *Mischoptera douglassi* and allied species, like *Mischoptera?* (= *Lamereeites*) *curvipennes*, whose nymphal winglets protrude laterally from the mesothorax and metathorax, from which they are separated by an apparent line, and curve backward (Fig. 11.2C–E) (Haug et al., 2016). As observed by Kukalová-Peck (1998), the younger nymphs show more posteriorly directed wing pads than the later stages. Regarding the wing articulation, the point of attachment spans the entire lateral margin of the tergum (as in *Herdina* nymphs), which clearly differs from the narrow wing joint area observed in the adult (Haug et al., 2016).

Another type of wing development is found in Paleozoic roachoid nymphs, early representatives of the Blattodea lineage, which broadly resemble extant cockroach nymphs (Fig. 11.2F and G). In extant cockroaches, the wing pads develop posteriorly, and the attachment is lateroposterior, whereas in Carboniferous forms, the wing pads protruded to the sides and the attachment area was restricted to the lateral tergum (Haug et al., 2016). Some Paleozoic mature nymphs show long wing pads that differentiate from the tergum by a conspicuous line or furrow (Fig. 11.2F and G) (Haug et al., 2016).

In those fossil groups with external nymphal winglets, such as the Ephemeroptera and Megasecoptera, selective pressures might have operated facilitating the concealment of these winglets in juveniles. As proposed by Kukalová-Peck (1978), winglets concealment was achieved by retracting them backward and attaching them to the body. Possibly, winglets concealment facilitated that the insect could hide from predators, as well as molting, by reducing the mechanical problems involved in the ecdysis and shedding off the exuvia. If the nymphal winglets were originally mobile, then concealment evolution implies that the winglets would end up being immobile, attached to the body (Fig. 11.3). In Paleozoic roachoids, the evolution could

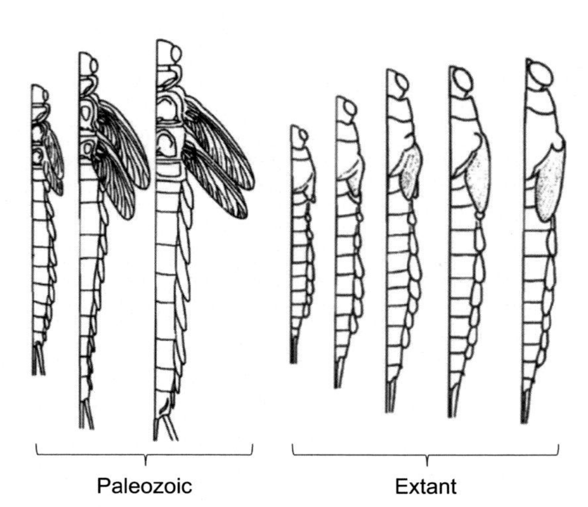

FIGURE 11.3 Wing development in Paleozoic and extant Ephemeroptera. Note the articulated and detached winglets in Paleozoic mayflies and the winglets attached to the body in extant species. *Modified from Kukalová-Peck (1991), with permission.*

have followed the same tendency of winglets concealment, via retracting them backward and attaching them to the body, which would be accompanied by loss of mobility. Although it is difficult to discern, the wing pads of Paleozoic roaches suggest that the wing primordium is not external and membranous. It reminds the wing primordia of modern cockroaches, which are enclosed in a cuticular pocket or pterotheca. If this were the case for Paleozoic roaches, then the formation of pterothecae might have been an early cockroach innovation.

THE FINAL MOLT: A CRUCIAL INNOVATION

Nymph fossils suggest that the life cycle of Carboniferous insects comprised a high number of molts, but a general evolutionary trend led to reducing that number. In this context, Kukalová-Peck (1991) proposed that some late nymphal instars could have become merged into the adult stage, and also younger nymphs would have undergone various degrees of merging. Moreover, the differences of juvenile/imaginal specialization between juveniles and adults in the different stages would have produced the life cycle diversity observed in the fossils. Remarkably, these ideas on the evolution of the life cycle based on the paleontological data are similar to those of Hinton (1963) based on the information of extant species.

On the basis of the disparate nymphal morphologies (such as, e.g., between mayfly and roachoid nymphs), and the different life cycles shown by the Paleozoic fossils, Kukalova-Peck (1978, 1991) suggested that metamorphosis had multiple origins that evolved independently and at different rates in different hemimetabolan lineages. However, all hemimetabolan life cycles, although they can be very different in terms of number of molts and nymphal morphologies, have in common that postembryonic development ends with a final molt, which is consubstantial with the origin of the hemimetabolan metamorphosis. Moreover, given that the origin of the Pterygota is monophyletic, and that all the lineages and main clades of Pterygota have a final molt, the most parsimonious explanation is to consider that the final molt has a monophyletic origin (Belles, 2019). This monophyly is also supported by the data indicating that the mechanisms that regulate hemimetabolan metamorphosis and the final molt are common to all extant hemimetabolan species (see below and Chapter 10: Regulation of ametabolan, hemimetabolan and holometabolan development).

The hypothesis of a monophyletic origin of the final molt and hemimetabolan metamorphosis is compatible with the diverse life cycles observed in the different hemimetabolan groups, which would have evolved independently in each lineage. It is also compatible with the scenarios of life cycle evolution by the merging of nymphal and subimaginal stages proposed by Kukalová-Peck (1991). The final molt may have been fixed immediately after the formation of the mature wings, although there could be intermediate subimaginal stages, with imperfectly flying membranous wings. The extant Ephemeroptera offer us an example of this, since their life cycle includes a winged subimaginal stage prior to the adult stage. The functional sense of the subimago of the Ephemeroptera is controversial (Edmunds and McCafferty, 1988), as discussed in Chapter 4 (The hemimetabolan development). However, whatever the function of the subimago is, this stage probably represents a relic of one or more preadult winged instars of ancestral mayflies.

POSSIBLE MECHANISMS UNDERLYING THE TRANSITION FROM AMETABOLY TO HEMIMETABOLY

In extant insects, the PG disintegrates shortly after the imaginal molt. The histolysis of the PG is associated with the hormonal and molecular mechanisms that regulate metamorphosis, essentially operating through the MEKRE93 pathway. In the hemimetabolan metamorphosis, this pathway is simple: the juvenile hormone (JH) bound to its receptor Methoprene-tolerant (Met) induces the expression of the transcription factor Krüppel homolog 1 (Kr-h1), which represses the expression of the transcription factor E93, the trigger of metamorphosis. Early in the last nymphal stage, the production of JH vanishes, the expression of *Kr-h1* dramatically falls, and that of *E93* increases, triggering the metamorphic molt (Belles and Santos, 2014). The expression of *E93* is induced by the cascade of transcription factors initiated by the ecdysone/20-hydroxyecdysone (20E) pulse that occurs in the penultimate nymphal stage. Another 20E-related transducer that is important for metamorphosis is the transcription factor Fushi tarazu-factor 1 (FTZ-F1), expressed when the pulse of 20E declines. In the cockroach *Blattella germanica*, the acute expression of *FTZ-F1* in the PG at the end of the last nymphal instar is crucial for the gland histolysis (Mané-Padrós et al., 2010). Among other roles, FTZ-F1 enhances the expression of *E93* (Orathai Kamsoi and Xavier Belles, unpublished), which finally triggers the disintegration of the PG (see Chapter 9: Molting: the basis for growing and for changing the form).

In the hemimetabolan metamorphosis, the MEKRE93 pathway operates in Polyneoptera (Belles and Santos, 2014; Ureña et al., 2014) and Paraneoptera (Gujar and Palli, 2016). Regarding the clade Palaeoptera, an RNAi method combined with electroporation has been used in the damselfly *Ischnura senegalensis* (Okude et al., 2017) to functionally study *Kr-h1* and *E93*. Results have shown that E93 is essential for adult morphogenesis and that Kr-h1 represses *E93* expression (Okude et al., 2019), which fits with the MEKRE93 pathway. Regarding Ephemeroptera, a transcriptomic analysis conducted in *Cloeon viridulum* (Si et al., 2017) indicates that the expression of *Kr-h1* decreases progressively from young to mature nymphs and to the subimago, which suggests that Kr-h1 represses metamorphosis. The same has been observed in the species *Cloeon dipterum*, in which, moreover, *E93* starts to increase its expression in the subimago (Orathai Kamsoi, Isabel Almudi, Fernando Casares and Xavier Belles, unpublished). The recent establishment of a continuous rearing system of *C. dipterum* (Almudi et al., 2019) paves the way for the use of this mayfly species for metamorphosis research. Taken together, the available data indicate that E93, through the MEKRE93 pathway, triggers metamorphosis in the three hemimetabolan divisions, Palaeoptera, Polyneoptera, and Paraneoptera. This suggests that the functions of E93 in relation to metamorphosis already existed in the last common ancestor of these groups, at the origin of Pterygota, and that it was instrumental in the origin of hemimetaboly.

Considering these antecedents, one of the first innovations in the last common ancestor of the Pterygota might have been the upregulation of *E93* expression in mature nymphs, mediated by the ecdysone signaling cascade (Belles, 2019). An increase in the *E93* expression might have triggered the maturation of the wings (see Uyehara et al., 2017) and, in turn, the disintegration of the PG (Orathai Kamsoi and Xavier Belles, unpublished). It has also been reported that E93 acts as a chromatin modifier, enabling or preventing expression in certain genetic regions (Uyehara et al., 2017). If this mode of action is ancestral, then it could have been a versatile way to turn genes on at appropriate times for new regulatory contexts.

Another candidate to play the double function of promoting at the same time wing development and PG disintegration in the last common ancestor of pterygote insects would be Broad-complex (BR-C). The function of BR-C as a promoter of wing development in juveniles of early-branching neopterans, like the cockroach *B. germanica*, has been clearly shown (Huang et al., 2013). Moreover, BR-C controls the ecdysone-induced expression of the death regulator Nedd2-like caspase during hormone-dependent programmed cell death in *D. melanogaster* (Cakouros et al., 2002). Finally, it has been demonstrated in this same fly that overexpression of each of the BR-C isoforms causes degeneration of the PG cells, which indicates that BR-C is a signal for the degeneration of the PG cells that normally occur during the pupal–adult transition (Zhou et al., 2004).

THE COMPONENTS OF THE MEKRE93 PATHWAY AND BROAD-COMPLEX PREDATE THE ORIGIN OF PTERYGOTA

All genes involved in the MEKRE93 pathway are present in Zygentoma, the sister group of pterygote insects (Misof et al., 2014; Wang et al., 2016). *E93* has been recently reported in the firebrat *T. domestica* (Belles, 2019), which also has the *Kr-h1* and *Met* genes (Konopová et al., 2011), the other components of the MEKRE93 pathway. The degree of conservation of the three corresponding proteins, Met, Kr-h1, and E93 in *T. domestica*, as compared with the corresponding orthologs of pterygote insects is very high, which suggests conservation of function. The Zygentoma do not metamorphose, but they undergo more or less subtle changes throughout the life cycle (see Chapter 3: The ametabolan development). The most apparent change is the formation of scales, a process in which JH might be involved. It has been reported that the formation of scales in *T. domestica* occurs in the third nymphal instar, just after a transient decrease in JH that occurs at the beginning of the instar (Watson, 1967). This suggests that the formation of scales might be repressed by JH, and we can speculate that this effect could be mediated by Kr-h1 and E93. An additional possibility is that JH, Kr-h1, and E93 might be involved in the formation of the mature genitalia that occurs in the eighth or ninth nymphal instar. Regarding *BR-C*, it has been studied in the embryo of *T. domestica*, where it is constitutively expressed (Erezyilmaz et al., 2009). However, no functional studies have been conducted, and therefore its role in apterous insects is unknown. The recently developed CRISPR/Cas9-based heritable targeted mutagenesis in *T. domestica* (Ohde et al., 2018) can allow functional studies and testing these hypotheses. If the MEKRE93 pathway, including BR-C, regulated morphogenetic processes in Zygentoma, then it could represent an exaptation for the acquisition of wings and the emergence of the final molt, and the hemimetaboly.

REFERENCES

Almudi, I., Martín-Blanco, C.A., García-Fernandez, I.M., López-Catalina, A., Davie, K., Aerts, S., et al., 2019. Establishment of the mayfly *Cloeon dipterum* as a new model system to investigate insect evolution. Evodevo 10, 6. Available from: https://doi.org/10.1186/s13227-019-0120-y.

Belles, X., 2011. Origin and evolution of insect metamorphosis. Encyclopedia of Life Sciences (ELS). John Wiley & Sons, Ltd, Chichester, UK. Available from: https://doi.org/10.1002/9780470015902.a0022854.

Belles, X., 2019. The innovation of the final molt and the origin of insect metamorphosis. Philos. Trans. R. Soc. Biol. Sci. 374, 20180415.

Belles, X., Santos, C.G., 2014. The MEKRE93 (Methoprene tolerant-Krüppel homolog 1-E93) pathway in the regulation of insect metamorphosis, and the homology of the pupal stage. Insect Biochem. Mol. Biol. 52, 60–68. Available from: https://doi.org/10.1016/j.ibmb.2014.06.009.

Cakouros, D., Daish, T., Martin, D., Baehrecke, E.H., Kumar, S., 2002. Ecdysone-induced expression of the caspase DRONC during hormone-dependent programmed cell death in *Drosophila* is regulated by Broad-complex. J. Cell Biol. 157, 985–995. Available from: https://doi.org/10.1083/jcb.200201034.

Clapham, M.E., Karr, J.A., Nicholson, D.B., Ross, A.J., Mayhew, P.J., 2016. Ancient origin of high taxonomic richness among insects. Proc. R. Soc. B: Biol. Sci. 283, 20152476. Available from: https://doi.org/10.1098/rspb.2015.2476.

Danley, P.D., Mullen, S.P., Liu, F., Nene, V., Quackenbush, J., Shaw, K.L., 2007. A cricket Gene Index: a genomic resource for studying neurobiology, speciation, and molecular evolution. BMC Genom. 8, 109. Available from: https://doi.org/10.1186/1471-2164-8-109.

Delany, M.J., 1957. Life histories in the Thysanura. Acta Zool. Cracoviensia 2, 61–90.

Edmunds, G.F., McCafferty, W.P., 1988. The mayfly subimago. Annu. Rev. Entomol. 33, 509–527. Available from: https://doi.org/10.1146/annurev.en.33.010188.002453.

Erezyilmaz, D.F., Rynerson, M.R., Truman, J.W., Riddiford, L.M., 2009. The role of the pupal determinant broad during embryonic development of a direct-developing insect. Dev. Genes Evol. 219, 535–544. Available from: https://doi.org/10.1007/s00427-009-0315-7.

Gujar, H., Palli, S.R., 2016. Krüppel homolog 1 and E93 mediate juvenile hormone regulation of metamorphosis in the common bed bug, *Cimex lectularius*. Sci. Rep. 6, 26092. Available from: https://doi.org/10.1038/srep26092.

Haug, J.T., Haug, C., Garwood, R.J., 2016. Evolution of insect wings and development – new details from Palaeozoic nymphs. Biol. Rev. 91, 53–69. Available from: https://doi.org/10.1111/brv.12159.

Hinton, H.E., 1963. The origin and function of the pupal stage. Proc. R. Entomol. Soc. Lond. 38, 77–85.

Huang, J.-H., Lozano, J., Belles, X., 2013. Broad-complex functions in postembryonic development of the cockroach *Blattella germanica* shed new light on the evolution of insect metamorphosis. Biochim. Biophys. Acta Gen. Subj. 1830, 2178–2187. Available from: https://doi.org/10.1016/j.bbagen.2012.09.025.

Konopová, B., Smykal, V., Jindra, M., 2011. Common and distinct roles of juvenile hormone signaling genes in metamorphosis of holometabolous and hemimetabolous insects. PLoS One 6, e28728. Available from: https://doi.org/10.1371/journal.pone.0028728.

Kukalová-Peck, J., 1978. Origin and evolution of insect wings and their relation to metamorphosis, as documented by the fossil record. J. Morphol. 156, 53–125. Available from: https://doi.org/10.1002/jmor.1051560104.

Kukalová-Peck, J., 1983. Origin of the insect wing and wing articulation from the arthropodan leg. Can. J. Zool. 61, 1618–1669. Available from: https://doi.org/10.1139/z83-217.

Kukalová-Peck, J., 1985. Ephemeroid wing venation based upon new gigantic Carboniferous mayflies and basic morphology, phylogeny, and metamorphosis of pterygote insects (Insecta, Ephemerida). Can. J. Zool. 63, 933–955. Available from: https://doi.org/10.1139/z85-139.

Kukalová-Peck, J., 1991. Fossil history and the evolution of hexapod structures. The Insects of Australia. Melbourne University Press, Carlton, pp. 141–179.

Kukalová-Peck, J., 1997. Mazon Creek insect fossils. The origin of insect wings and clues about the origin of insect metamorphosis. In: Shabica, C.W., Hay, A.A. (Eds.), Richardson's Guide to the Fossil Fauna of Mazon Creek. Northeastern Illinois University, Chicago, IL, pp. 194–207.

Kukalová-Peck, J., 1998. Arthropod phylogeny and "basal" morphological structure. In: Fortey, R.A., Thomas, R.H. (Eds.), Arthropod Relationships. Chapman & Hall, London, pp. 249–268.

Mané-Padrós, D., Cruz, J., Vilaplana, L., Nieva, C., Ureña, E., Belles, X., et al., 2010. The hormonal pathway controlling cell death during metamorphosis in a hemimetabolous insect. Dev. Biol. 346, 150–160. Available from: https://doi.org/10.1016/j.ydbio.2010.07.012.

Misof, B., Liu, S., Meusemann, K., Peters, R.S., Donath, A., Mayer, C., et al., 2014. Phylogenomics resolves the timing and pattern of insect evolution. Science 346, 763–767.

Ohde, T., Takehana, Y., Shiotsuki, T., Niimi, T., 2018. CRISPR/Cas9-based heritable targeted mutagenesis in *Thermobia domestica*: a genetic tool in an apterygote development model of wing evolution. Arthropod. Struct. Dev. 47, 362–369.

Okude, G., Futahashi, R., Kawahara-Miki, R., Yoshitake, K., Yajima, S., Fukatsu, T., 2017. Electroporation-mediated RNA interference reveals a role of the multicopper oxidase 2 gene in dragonfly cuticular pigmentation. Appl. Entomol. Zool. 52, 379–387. Available from: https://doi.org/10.1007/s13355-017-0489-9.

Okude, G., Futahashi, R., Fukatsu, T., 2019. Molecular mechanisms of metamorphosis in dragonflies. In: Commun. 4th Int. Insect Horm. Work, Kolymbari, Crete, Greece, p. 43.

Prokop, J., Pecharová, M., Nel, A., Hörnschemeyer, T., Krzemińska, E., Krzemiński, W., et al., 2017. Paleozoic nymphal wing pads support dual model of insect wing origins. Curr. Biol. 27, 263–269. Available from: https://doi.org/10.1016/j.cub.2016.11.021.

Sehnal, F., Švácha, P., Zrzavý, J., 1996. Evolution of insect metamorphosis. In: Gilbert, L.I., Tata, J.R., Atkinson, B.G. (Eds.), Metamorphosis. Postembryonic Reprogramming of Gene Expression in Amphibian and Insect Cells. Academic Press, San Diego, CA, pp. 3–58.

Si, Q., Luo, J.-Y., Hu, Z., Zhang, W., Zhou, C.-F., 2017. De novo transcriptome of the mayfly *Cloeon viridulum* and transcriptional signatures of Prometabola. PLoS One 12, e0179083. Available from: https://doi.org/10.1371/journal.pone.0179083.

Truman, J.W., Riddiford, L.M., 1999. The origins of insect metamorphosis. Nature 401, 447–452. Available from: https://doi.org/10.1038/46737.

Ureña, E., Manjón, C., Franch-Marro, X., Martín, D., 2014. Transcription factor E93 specifies adult metamorphosis in hemimetabolous and holometabolous insects. Proc. Natl. Acad. Sci. U.S.A. 111, 7024–7029. Available from: https://doi.org/10.1073/pnas.1401478111.

Uyehara, C.M., Nystrom, S.L., Niederhuber, M.J., Leatham-Jensen, M., Ma, Y., Buttitta, L.A., et al., 2017. Hormone-dependent control of developmental timing through regulation of chromatin accessibility. Genes Dev. 31, 862–875. Available from: https://doi.org/10.1101/gad.298182.117.

Wang, Y.-H., Engel, M.S., Rafael, J.A., Wu, H.-Y., Rédei, D., Xie, Q., et al., 2016. Fossil record of stem groups employed in evaluating the chronogram of insects (Arthropoda: Hexapoda). Sci. Rep. 6, 38939. Available from: https://doi.org/10.1038/srep38939.

Watson, J.A.L., 1967. The growth and activity of the corpora allata in the larval firebrat, *Thermobia domestica* (Packard) (Thysanura, Lepismatidae). Biol. Bull. 132, 277–291. Available from: https://doi.org/10.2307/1539895.

Wootton, R.J., 1981. Palaeozoic insects. Annu. Rev. Entomol. 26, 319–344. Available from: https://doi.org/10.1146/annurev.en.26.010181.001535.

Zhou, X., Zhou, B., Truman, J.W., Riddiford, L.M., 2004. Overexpression of broad: a new insight into its role in the *Drosophila* prothoracic gland cells. J. Exp. Biol. 207, 1151–1161. Available from: https://doi.org/10.1242/jeb.00855.

Chapter 12

The evolution of metamorphosis

After the emergence of hemimetaboly, the most important innovation was the evolutionary transition to holometaboly. It occurred during the early Carboniferous (Mississippian, 360–325 Mya), according to molecular phylogenetic analyses (Misof et al., 2014; Wang et al., 2016), although the earliest known holometabolan fossils are a little later, from mid–late Carboniferous (Pennsylvanian, 325–230 Mya) (Nel et al., 2013). Holometaboly has been tremendously successful, as 85% of insect species follow this mode of metamorphosis. Despite this success, the selective advantage that may explain it still is a matter of debate (Rolff et al., 2019). A widespread point of view is that the greatest advantage of holometaboly is the absence of competition between the juveniles and the adults of the same species, as they exploit different resources (Carpenter, 1953; Mayhew, 2007). This makes sense, as even among the hemimetabolans, there are remarkably successful orders that have pronounced differences between nymphs, which are aquatic, and adults, which are terrestrial, like the Plecoptera (c.3700 species), Odonata (c.6000 species), and Ephemeroptera (c.3500 species) (see Chapter 4: The hemimetabolan development). Among the holometabolan orders, the most diverse are Coleoptera (c.390,000 species), Lepidoptera (c.165,000 species), Diptera (c.140,000 species), and Hymenoptera (c.125,000 species). The species of these four orders represent 97% of all extant holometabolans, while the remaining 3% is divided into seven orders, among which the Raphidioptera (c.170 species) and Megaloptera (c.300 species) stand out as the least diverse. These barely diverse early-branched holometabolan orders contrast with the success of some hemimetabolan orders, such as the Hemiptera, with some 70,000 species and with the same ecological niche in immatures and adults. These data suggest that the success, measured in terms of diversity, is not only due to the absence of competition between juveniles and adults, but also to factors specific of each group.

SELECTIVE ADVANTAGES OF HOLOMETABOLY

An important driving force in holometabolan evolution must have been the opportunity to exploit new resources. The argument is not new, as in the

Insect Metamorphosis. DOI: https://doi.org/10.1016/B978-0-12-813020-9.00012-0

19th century Auguste Lameere reasoned that the "raison d'être" of complete metamorphosis was that larvae could have access to new resources, such as the pulp of a fruit (Lameere, 1899). It is probably no coincidence that the greatest radiation of holometabolan insects concurred with important enrichments of terrestrial floras during the Devonian, culminating with the emergence of the seed plants around the Carboniferous boundary (see Chapter 2: A spectacular diversity of forms and developmental modes). The exploitation of these new resources, such as the pulp of a fruit, may have acted as selective pressure triggering the internal concealment of the wing primordia (which might hinder access to them) during the juvenile period (Sehnal et al., 1996). Other adaptive body modifications would have developed, such as the specialization of mouthparts that chronologically followed the increase of available vegetal diversity (Nel et al., 2018). Morphological modifications could be useful not only to gain access to new resources but also to hide from predators (Downes, 1987). Improved accessibility to food resources could in turn result in metabolic optimizations, increased growth rates (Cole, 1980; Ferral et al., 2020), and life cycle shortening (Sehnal, 1985). Holometaboly could have also facilitated more effective mechanisms for controlling development, even at the cost of losing flexibility. Extreme examples are provided by cyclorrhaphan flies such as *Drosophila melanogaster*, which exhibit a very rapid development (10 days from egg to adult at 25°C), but through a number of larval instars (three) that is inflexible.

THE LARVA AND THE PUPA

In the larva, typically adult structures are not formed. Instead the larva maintains the imaginal precursors (of wings, compound eyes, etc.) internalized, until these develop their adult potential in the pupa. In parallel with these secondary simplifications, structures adaptive for specialized lifestyles emerge. In general, larvae of diverse groups show different degrees of modification, from the larvae of the Neuropterida, which are very similar to the adults, to the worm-shaped larvae of Diptera (Costa et al., 2006) (see Chapter 5: The holometabolan development). Among the Neuropterida, the aquatic larva of Megaloptera presents the least amount of differentiation with respect to the pupa (which is fully motile, having large mandibles that can be used for defense against predators), and the adult (Fig. 12.1A and B). The larvae of other Neuropterida, like those of Neuroptera and Raphidioptera, resemble the adults, while they almost look like nymphs (Fig. 12.1C–F).

For larvae to achieve the adult morphology, a more or less profound remodeling is necessary, and this process takes place in the typically quiescent pupal stage. The degree of modification of the larva is mirrored by the degree of remodeling in the pupa, from the least modified pupae of the Neuropterida mentioned above, which are relatively active and similar to the larva and adult (Fig. 12.1), to those of more modified groups, like the

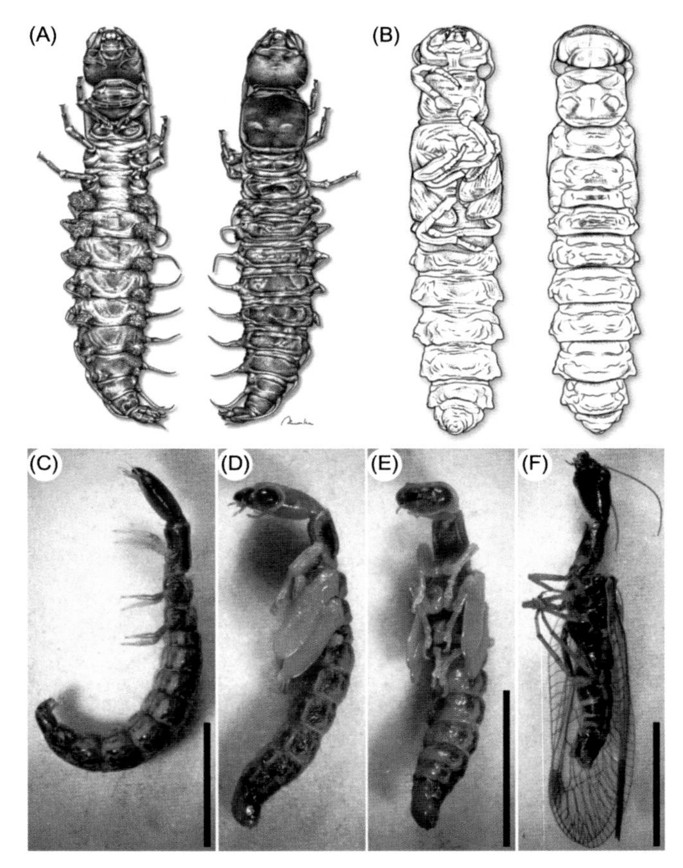

FIGURE 12.1 Life stages of Neuropterida. (A and B) Mature larva (A) and pupa (B) of a Corydalidae (Megaloptera) in ventral and dorsal view. (C–F) Mature larva (C), pupa in lateral and ventro-lateral view (D and E) and adult (F) of *Phaeostigma notata* (Raphidioptera). Scale bar: 5 mm. *(A and B) from Costa et al. (2006), with permission; (C–F) photos courtesy of Marek Jindra, from Jindra (2019), with permission.*

Lepidoptera and Diptera, where the pupa is very different from both the larva and the adult (see Chapter 5: The holometabolan development). An exception in this apparent correlation between degree of modification and phylogenetic position can be found in the Hymenoptera Apocrita (ants, bees, wasps), an early-branching group of Endopterygota with very modified, worm-like larvae. In the end, the degree of change occurring during post-embryonic development and the degree of remodeling at the pupal stage depend on how far the juveniles and adults have diverged from each other (Fig. 12.2).

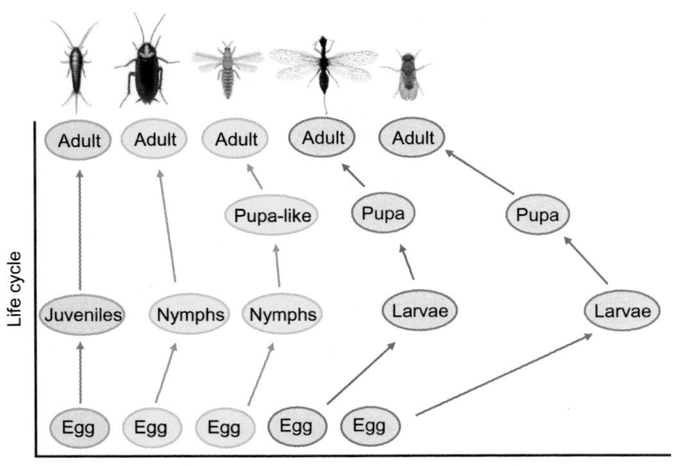

FIGURE 12.2 Diagram showing the relationship between the type of postembryonic development and the increased specialization and divergence from the ancestral morphology of the juveniles. The green color represents the ametabolan mode, the orange the hemimetabolan, including the paraneopteran cases where there are pupal-like quiescent instars (thrips, male scale insects, whiteflies), and the blue the holometabolan mode, represented by a basal case (like a megalopteran or raphidiopteran), where the larva is relatively similar to the adult, and by a modified case (like a lepidoptera or a dipteran), where the larva is very different from the adult.

The pupa is a defining feature, an autapomorphy, of the Endopterygota. The most obvious functional sense of the pupa is to serve as a bridge between the larva and the adult. Some authors have proposed that the pupa, being a protected stage, could represent a selective advantage of holometaboly as it allows greater buffering from environmental variability (Ross et al., 2000). However, the argument is the opposite: due to its immobility, the pupa is a vulnerable stage, which represents rather a weakness of the holometaboly, and which required the selection of specific protection mechanisms. These include a relatively hard cuticle, the production of a silk cocoon, and concealment in the substrate. Pupae of some species are capable of producing sounds that can scare away potential predators or produce toxic or repellent chemical secretions (Lindstedt et al., 2019). Endopterygotes also have a special type of aquaporins that may have been selected to protect the pupa from dehydration (Finn et al., 2015).

TWO THEORIES TO EXPLAIN THE EVOLUTION OF METAMORPHOSIS

The first scientific attempt to explain the evolution of metamorphosis came from John Lubbock, who proposed that the embryo of holometabolan species hatches at an earlier stage than the hemimetabolan (Lubbock, 1873).

Lubbock's theory of "deembryonization" was further formalized by Antonio Berlese (1913) and accepted widely until the 1960s, when growing evidence suggested that the juvenile stages of the hemimetabolan insects were equivalent to those of the holometabolan. This theory of direct homology between stages, championed by Howard Hinton (1963), was prevalent until 1999, when James Truman and Lynn Riddiford revived the "deembryonization" concepts, modernizing them as the pronymph theory on the basis of morphophysiological and endocrinological arguments (Truman and Riddiford, 1999).

The pronymph theory

The pronymph theory, announced by Truman and Riddiford in 1999, is rooted in the "deembryonization" concepts proposed by Lubbock in 1873 and further formalized by Berlese in 1913. Truman and Riddiford (1999, 2002) argued that three successive embryonic cuticles, EC1, EC2, and EC3, are sequentially deposited during the hemimetabolan embryogenesis, and that there is a cryptic stage (called the pronymph) formed after EC2 deposition, from which the first nymphal instar would derive. They contended that the EC3 is not deposited in holometabolans, so instead of forming the pronymph, the insect hatches prematurely as a larva. Later the insect would resume the arrested development in the pupa, in which adult-like structures would be formed. The above hypotheses led to the proposal that the hemimetabolan pronymph would be homologous to the set of holometabolan larval instars, while the set of hemimetabolan nymphal instars would be homologous to the holometabolan pupa (Truman and Riddiford, 1999, 2002) (Fig. 12.3A). The idea that holometabolan embryos deposit only two cuticles was refuted later (see below) (Konopová and Zrzavý, 2005). However, Truman and Riddiford adapted their pronymph theory to that evidence, emphasizing the idea that holometabolan embryos arrest development at a given time and resume it at the pupal stage and defending the homology between the pronymph and the holometabolan larvae and between the hemimetabolan nymphs and the pupa (Truman, 2019; Truman and Riddiford, 2019).

Of importance in the pronymph theory is the causal explanation, which hypothesizes that a shift in juvenile hormone (JH) secretion during embryogenesis was instrumental in the evolution of holometaboly (Truman and Riddiford, 1999). Accordingly, in hemimetabolan species (mainly exemplified by locusts), the embryonic deposition of EC1 and EC2 would occur in the absence of JH, while the deposition of EC3 would occur in its presence. In contrast, holometabolan embryos (exemplified by lepidopterans) would start producing JH earlier, during EC1. In line with this idea, the authors suggested that an advance in JH secretion toward early embryogenesis informed the change of the pronymph into a larva (Truman and Riddiford, 1999).

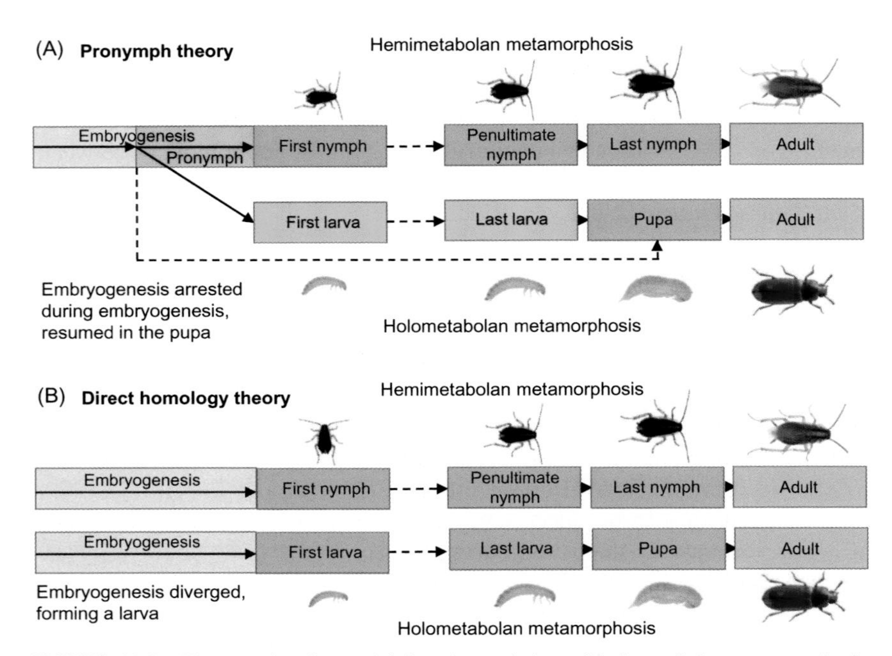

FIGURE 12.3 The two theories explaining the evolution of holometabolan metamorphosis. (A) The pronymph theory considers that holometabolan embryogenesis is "arrested" at the pronymph stage and resumed at the pupal stage. The pronymph would be homologous to all the holometabolan larval instars, while the set of hemimetabolan nymphs would be homologous to the pupa. (B) The alternative theory of direct homology between stages suggests that during holometabolan embryogenesis, the ancestral nymphal morphogenetic program is not "arrested" but continues with the larval program. Here, the nymphal instars would be homologous to the larval instars, so the pupa is only a modified nymph.

Moreover, the 20-hydroxyecdysone (20E)-dependent gene *Broad-complex* (*BR-C*) has been invoked by the proponents of the pronymph theory to support the pronymph-to-larva and nymph-to-pupa homologies (Fig. 12.3A) and to explain the mechanisms that triggered the transition from nymphs to larvae (Erezyilmaz et al., 2006, 2009). Accordingly, the earlier appearance of JH in the holometabolan ancestors would have suppressed the onset of *BR-C* expression during embryonic development. Then, the absence of BR-C at the pronymphal molt "would freeze the proportions of the pronymph, resulting in a larva with more embryonic proportions" (Erezyilmaz et al., 2009).

The theory of direct homology between stages

This theory is rooted in the concepts formalized by Éraste Poyarkoff in 1914, as opposed to Berlese's theories of 1913. For Poyarkoff, the pupa originated as a kind of mold required for the development of certain adult skeletal muscles (Poyarkoff, 1914). Poyarkoff proposed two important ideas. One considers the pupa as being homologous to the hemimetabolan adult, therefore contending that the pupa arose by subdividing the adult stage into two phases. The other is that the juveniles of both hemimetabolan and holometabolan species hatch from the egg at a comparable stage of development (Poyarkoff, 1914). Hinton initially favored Poyarkoff's ideas against those of Berlese (Hinton, 1948). However, after studying the metamorphosis of the flight muscles in several species, Hinton dismissed the idea of the pupa being the mold substage of the adult. Instead he came to the conclusion that all holometabolan juvenile stages were homologous to those of the hemimetabolans, including the pupa, which would be homologous to the last nymphal instar (Fig. 12.3B). At the same time, Hinton agreed with Poyarkoff's proposal that the juveniles hatch at a comparable stage of development in hemimetabolans and holometabolans alike (Hinton, 1963).

To Hinton's contributions, comprehensive phylogenetic, morphological, and physiological arguments were added later by Sehnal et al. (1996). The phylogenetic context is an important part of the modern version of the theory by these authors, including findings on the paleontology of metamorphosis (Kukalová-Peck, 1991). As for morphology, Sehnal et al. (1996) argued that holometaboly evolved via the specialization of larvae and their morphological divergence from the adults. In general, the holometabolan larva would be characterized by a delay in the appearance of external wing primordia until the pupa, which was considered a resting juvenile instar. In holometabolan larvae, the wing primordia usually form as epidermal evaginations under the cuticle, which in more derived species develop as imaginal discs (Švácha, 1992). The remodeling needed in the preadult stage explains the emergence of the pupa, which would be homologous to the last nymphal instar. Regarding embryo development, on the basis of the studies of Polivanova (1979), Sehnal et al. (1996) contend that embryogenesis is equivalent in hemimetabolan and holometabolan species. From an endocrine point of view, hemimetabolan metamorphosis occurs when the production of JH falls dramatically (and definitively within the juvenile period) in the last nymphal instar. In holometabolan species, JH decreases twice, the first time in the last larval instar, which allows the formation of the pupa, and the second time in the pupa, which leads to the formation of the adult (Fig. 12.4). As the definitive fall of JH that triggers the adult morphogenesis occurs in the last nymphal instar in hemimetabolan species and in the holometabolan pupa, Sehnal et al. (1996) propose that this supports the respective homology.

TESTING HYPOTHESES

Both theories make assumptions that are empirically testable. The observations and experiments reported in this regard can be summarized as follows.

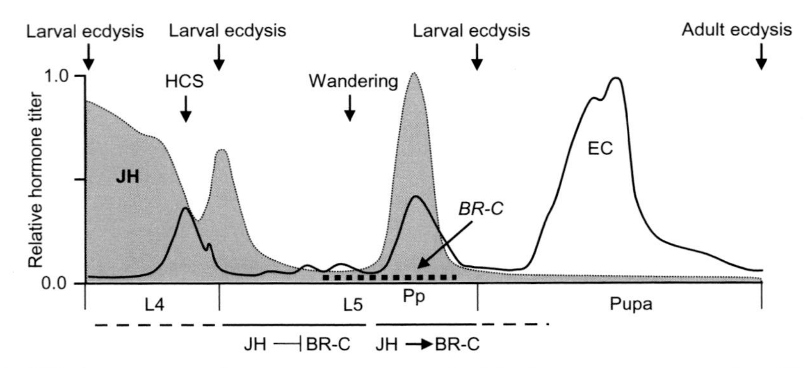

FIGURE 12.4 Juvenile hormone (JH), ecdysteroid (EC) titers, and *Broad-complex* (*BR-C*) expression during postembryonic development of *Manduca sexta.* The penultimate and last larval instars (L4 and L5, including the prepupal, Pp, stage) and the pupal stage are shown. The titers of JH and EC are represented by a gray area and a continuous line, respectively, whereas *BR-C* expression is indicated with a dashed line. JH represses *BR-C* expression until the end of the feeding stage (Zhou and Riddiford, 2001), whereas beyond the prepupal stage, JH stimulates *BR-C* (Zhou and Riddiford, 2002). HCS, head capsule slippage (c.29 hours before larval ecdysis). *Data from Zhou et al. (1998).*

Embryonic cuticles

A systematic examination of representatives of seven insect orders showed that both hemimetabolan and holometabolan embryos deposit three cuticles, EC1, EC2, and EC3, although EC2 may be simplified (with only a reduced epicuticle) in embryos of more derived holometabolans or even absent in the most evolutionarily modified flies (Konopová and Zrzavý, 2005). This allows to definitively discard the classic ideas of Lubbock and Berlese that explain the origin of holometaboly by a "deembryonization" process.

Embryo development

The pronymph theory proposes that holometabolan embryogenesis "arrests" development, which is resumed at the pupal stage. Evidence compatible with the "arrest" concept stems from the development of the central nervous system (CNS), visual system, and legs (Truman, 2019; Truman and Riddiford, 2019). The development of the CNS follows a conserved sequence starting with stem cells that give rise to neuroblasts, which in turn generate a characteristic set of neurons (Thomas et al., 1984). In hemimetabolans, exemplified by the locust *Schistocerca gregaria*, the neuroblasts generate all of their neurons during embryogenesis (Shepherd and Bate, 1990), whereas in the examined holometabolans (belonging to the considerably modified orders Lepidoptera and Diptera), the CNS development appears to pause at a particular moment of embryogenesis. Therefore the neurons of the newly hatched locust nymph show the basic form and connectivity seen in the adult. By contrast, the larval neurons of moths and flies have a modified anatomy and connectivity adapted to larval needs; they only acquire the adult form and connectivity when remodeled at metamorphosis (Truman, 1996).

As for the development of the visual system, the holometabolan larva has one to several pairs of stemmata, rather than the compound eyes of hemimetabolan nymphs and adults. The eye of the nymph arises from primordia situated on the lateral margins of the embryonic head. The ommatidia first form along the posterior border of the primordium with successive rows added anteriorly across it (Heming, 2003; Paulus, 1989; Truman and Riddiford, 2019). In holometabolan embryogenesis, eye development also starts from embryonic head primordia, but only two small stemmata are formed. In the larva, this typically results in the paired stemmata mentioned above or, exceptionally such as in larvae of the Mecoptera and in Diptera Chaoboridae, in a simplified compound eye, which becomes remodeled at metamorphosis (Friedrich, 2003, 2006) (see Chapter 5: The holometabolan development).

In crickets, which are hemimetabolans, the leg structure is established by the sequential action of proximal–distal patterning genes, starting with *extradenticle* and *Distal-less*, which promote the formation of the leg bud and define the proximal–distal axis. Then the expression of *dachshund* establishes the leg middle regions, and *bric-a-brac* expresses in the tarsal region and determines the tarsomeres. In contrast, in the extremely modified legless larvae of the holometabolan *D. melanogaster*, the sequence involves *extradenticle* and *Distal-less* only. The remainder of the patterning gene cascade is then expressed at metamorphosis. This interrupted development would fit well with the hypothesis of arrest in the embryo and resumption in the pupa proposed in the pronymph theory. However, development in lepidopteran caterpillars suggests that the process can be flexible, leading to the formation of modified thoracic legs. Thus in the tobacco hornworm, *Manduca sexta*, the embryonic patterning of the leg bud progresses through a basal ring of *dachshund* expression and a single *bric-a-brac* domain but then directs the development of the caterpillar leg. Then the remaining sequence of genes is expressed at metamorphosis, thus forming the adult leg (Cohen, 1993; Inoue et al., 2002; Tanaka and Truman, 2007).

Although the above evidence has been invoked in support of the pronymph theory (Truman, 2019; Truman and Riddiford, 2019), it is also compatible with the theory of direct homology between stages. Strictly speaking, the embryonic development in holometabolan species does not arrest but diverges from that of the hemimetabolan mode, producing unique adaptive forms. Comparative data of leg development in different holometabolan species represent an instructive example of how different species have achieved specific leg adaptations (including complete leg absence) from an interruption and then a divergence, in different points of the basic, adult-like genetic program. The study of representatives of early-branching holometabolan orders, like Megaloptera (Fig. 12.1), whose larva is similar to the adult, would possibly show that their embryo development is more similar to that of a typical hemimetabolan than that of a more evolutionarily modified holometabolan. It follows that embryogenesis would end with a first juvenile instar homologous in hemimetabolan and holometabolan species.

Juvenile hormone and embryogenesis

As discussed in Chapter 10 (Regulation of ametabolan, hemimetabolan and holometabolan development), JH production in the embryo generally starts around 50% development, but the temporal patterns are quite diverse depending on the species. It is true that in the hemimetabolan *L. migratoria* JH production increases around 60% development, which is later than in the holometabolans *Manduca sexta* and *Bombyx mori*, in which it increases just before 50% development. However, the temporal patterns of *M. sexta* and *B. mori* are quite similar to that of the hemimetabolan *Blattella germanica*, whereas in the thrips *Frankliniella occidentalis* (Minakuchi et al., 2011) and the hemipteran scale insect *Planococcus kraunhiae* (Vea et al., 2016), both hemimetabolan species, JH production starts much earlier that 50% of development, significanly earlier than in the lepidopterans *M. sexta* (Bergot et al., 1981) and *B. mori* (Daimon et al., 2015). This is at odds with the idea that JH production is advanced in holometabolan embryos, as argued in the pronymph theory (Truman and Riddiford, 1999).

With respect to the role of JH in embryogenesis, the experiments applying JH on eggs (Erezyilmaz et al., 2004; Novák, 1969; Truman and Riddiford, 1999, 2002) or treating the eggs with CA-destroying compounds (precocenes) (Aboulafia-Baginsky et al., 1984; Brüning and Lanzrein, 1987), as well as those depleting the expression of genes involved in JH synthesis and signaling, have been discussed in Chapter 10 (Regulation of ametabolan, hemimetabolan, and holometabolan development). Regarding cockroaches, results of treating *Nauphoeta cinerea* eggs with precocenes just after the embryo dorsal closure suggest that JH contributes to the correct formation of the nymphal cuticle (Brüning and Lanzrein, 1987). In *B. germanica*, maternal RNAi approaches were used to deplete the JH biosynthesis enzyme JH acid *O*-methyl transferase (JHAMT), the JH receptor Methoprene-tolerant (Met), and the JH signaling transducer Krüppel homolog 1 (Kr-h1) (Fernandez-Nicolas and Belles, 2017). JH is secreted along the latter half of embryogenesis, after the dorsal closure, and during this period, the maternal RNAi experiments suggest that JH controls proper cuticle tanning through the regulation of the *laccase 2* gene and also promotes embryo fitness (Fernandez-Nicolas and Belles, 2017). In the holometabolan *B. mori*, a genome editing approach was used to generate null mutants for JHAMT and for Met1 and Met2 (Daimon et al., 2015). The results showed that JH had little functional relevance, being merely involved in the hatching process but not required for embryogenesis itself. Recently reported observations in *Aedes aegypti* also indicate that knockout of Met does not affect embryogenesis in mosquitoes (Zhu et al., 2019). In conclusion, data available appear to relate JH with the formation of a proper nymphal cuticle in orthopteroid embryos, although further work is needed to fully characterize the embryonic functions of this important hormone in hemimetabolans. In lepidopterans and dipterans, the results obtained with the experiments involving genome editing allow to conclude that JH has a very limited role in holometabolan embryogenesis.

Broad-complex in the embryo

In hemimetabolan embryogenesis, like in the milkweed bug *Oncopeltus fasciatus*, *BR-C* expression is observed around the stage of germ band invagination and later during the last half of development, comprising the deposition of EC2 and EC3 (Erezyilmaz et al., 2009). This pattern is similar to that reported in the cockroach *B. germanica*, where *BR-C* is expressed in the early embryo, and then in the second half of embryogenesis (Piulachs et al., 2010), when JH is being produced (Maestro et al., 2010). As for holometabolan embryogenesis, immunocytochemical studies performed in *D. melanogaster* have shown that BR-C first appears in the CNS of late stage 12 embryos (45% development), just before the completion of the germ band retraction, and then the expression continues until hatching (Zhou et al., 2009). Mature embryos of the tobacco hornworm, *M. sexta*, also show high levels of BR-C in the CNS (Zhou et al., 2009). In the flour beetle *Tribolium castaneum*, moderate levels of *BR-C* expression can be observed throughout embryogenesis, although the abundance is higher in the last two-thirds of development (Konopová and Jindra, 2008). In the silkworm, *B. mori*, *BR-C* mRNAs are relatively abundant in early embryo (from 0% to 15% development), and then *BR-C* reexpresses after the dorsal closure (from 50% to 100% development) (Daimon et al., 2015). Whether the expression of *BR-C* in *T. castaneum* and *B. mori* embryos is restricted to the CNS was not determined.

Maternal RNAi carried out in *O. fasciatus* (Erezyilmaz et al., 2009) and *B. germanica* (Piulachs et al., 2010) has indicated that during mid–late embryogenesis, when JH is present, BR-C depletion mainly affects the development of the abdomen and eventually nymph hatching. Regarding holometabolan species, classical studies of *BR-C* mutants in *D. melanogaster* have shown that they die at metamorphosis, not during embryogenesis (Kiss et al., 1988; Zhimulev et al., 1995), which suggests that the zygotic function of *BR-C* is not required until metamorphosis. More recently, genome editing experiments targeting the core region of BR-C and applied in early embryos of *B. mori* only reduced embryo viability/hatchability (Daimon et al., 2015).

Therefore, in hemimetabolan embryogenesis and during the period of JH production, BR-C plays roles at least related to the development of the abdomen and eventually to enhanced vitality and nymph hatching. The functions on abdomen development appear related with the short or intermediate germ-band type of development of the studied species, rather than to the mode of metamorphosis. Indeed, more studies would be needed to assess the possible role of BR-C on other developmental processes, for example, in the formation of wing primordia. In holometabolan species, embryonic BR-C does not seem involved in morphogenesis, and in the species where the protein or the mRNA has been spatially localized, it appears largely confined to the CNS.

Broad-complex and postembryonic development

In *O. fasciatus, BR-C* is expressed throughout the nymphal instars until the last one, when it declines (Erezyilmaz et al., 2006). The same is true in *B. germanica*, in which *BR-C* is continuously expressed from the mid-embryogenesis until the last nymphal instar (Huang et al., 2013). During the holometabolan postembryonic development, high levels of *BR-C* start being expressed in the last days of the last larval instar, in the prepupal stage, which determine the formation of the pupa, as shown in *D. melanogaster*, *B. mori*, *T. castaneum*, and the lacewing *Chrysopa perla* (Bayer et al., 1996; Konopová and Jindra, 2008; Parthasarathy et al., 2008; Suzuki et al., 2008; Uhlirova et al., 2003). Detailed transcript measurements and immunocytochemical examination indicate that *BR-C* is expressed in most tissues in the mid-to-late final larval instar of both *M. sexta* (Zhou et al., 1998; Zhou and Riddiford, 2001) and *D. melanogaster* (Emery et al., 1994; Zhou et al., 2009). In the latter species, immunocytochemical observations show that BR-C proteins are present in most tissues about 18 hours before pupariation.

In hemimetabolans, *BR-C* sustains wing development during the nymphal period, as shown in *O. fasciatus* (Erezyilmaz et al., 2006), *Pyrrhocoris apterus* (Konopová and Jindra, 2008), and *B. germanica* (Huang et al., 2013). In *O. fasciatus*, which has colored wings, *BR-C* was shown to be necessary for nymphal heteromorphosis in pigmentation (Erezyilmaz et al., 2006). Importantly, *BR-C* expression is enhanced by JH (Huang et al., 2013), which is consistent with the observation that *BR-C* expression vanishes in the last nymphal instar (see Chapter 10, Regulation of ametabolan, hemimetabolan, and holometabolan development), suggesting that BR-C could impact metamorphosis. However, the RNAi depletion of *BR-C* in nymphs of *O. fasciatus*, *P. apterus*, and *B. germanica* never led to a premature metamorphosis. Intriguingly, the equivalent experiment carried out on nymphs of the cricket *Gryllus bimaculatus* triggered the precocious formation of adult features (Ishimaru et al., 2019). This study also revealed that depletion of BR-C results in a downregulation of *Kr-h1* and an upregulation of *E93* expression. Therefore the inhibitory action of BR-C upon *E93* would explain the premature adult morphogenesis observed when depleting BR-C. It is not known whether BR-C depletion in *O. fasciatus*, *P. apterus*, and *B. germanica* would downregulate *Kr-h1* and upregulate *E93* expression. If so, then the ectopic upregulation of *E93* resulting from BR-C depletion perhaps did not reach the minimum levels necessary to trigger adult morphogenesis.

In holometabolans, *BR-C* expressed toward the end of the last larval instar, in the prepupal stage, determines the formation of the pupa (see Chapter 7: Molecular mechanisms regulating hormone production and action). Interestingly, depletion of BR-C isoforms Z2 and Z3 in the basal holometabolan *T. castaneum* results in pupae with shortened wings (Konopová and Jindra, 2008; Suzuki et al., 2008), while in the very modified *D. melanogaster,* mutants for one of the complementation groups of *BR-C* develop abnormally "broad" (hence the gene name), oval wings (Morgan et al., 1925). Therefore, the functions of *BR-C* related to controlling wing size and shape appear to be ancestral and conserved from cockroaches to flies. In contrast with the hemimetabolan species, where JH enhances *BR-C* expression in nymphs, JH represses *BR-C* expression during the larval period of holometabolans, as shown, for example, in *M. sexta* (Zhou et al., 1998) and *B. mori* (Reza et al., 2004). Thus high levels of JH correlate with low levels of *BR-C* expression (Kayukawa et al., 2014; Konopová et al., 2011; Smykal et al., 2014; Zhou et al., 1998; Zhou and Riddiford, 2002). In *M. sexta,* JH represses *BR-C* expression in larvae until the end of the feeding stage in the last larval instar (Zhou and Riddiford, 2001). However, when *BR-C* mRNA levels decrease at the beginning of the pupal stage where JH is naturally absent (which allows the formation of the adult), the administration of ectopic JH can induce reexpression of *BR-C* (Zhou and Riddiford, 2002) and therewith block the adult program (Fig. 12.4). Although this *BR-C* reexpression never happens in natural conditions, the underlying mechanism is intriguing.

The expression of *BR-C* in all nymphal instars of hemimetabolan species, contrasts with its expression specifically at the end of the last larval instar in holometabolans, which would be at odds with the idea that the entire nymphal period is homologous to the pupa, as proposed in the pronymph theory (Truman and Riddiford, 1999, 2019; Truman, 2019). On the other hand, the functional observations are compatible with the proposal that a shift in JH action on *BR-C* expression, from stimulatory (hemimetabolan nymphs) to inhibitory (holometabolan larvae), might have been instrumental in the evolution toward holometaboly (Huang et al., 2013), as detailed later in the chapter.

Molecular mechanisms regulating metamorphosis

The molecular mechanisms underlying the hormonal action have been argued in support of the homology of the hemimetabolan last nymphal instar and the holometabolan pupa (Belles and Santos, 2014; Huang et al., 2013; Konopová et al., 2011) and the theory of direct homology between the stages in general (Jindra, 2019). The molecular mechanisms that regulate metamorphosis are condensed in the MEKRE93 pathway (Belles and Santos, 2014), which encompasses three transcription factors, E93, which triggers adult morphogenesis, Kr-h1, which inhibits precocious adult morphogenesis by repressing *E93*, and Met, which receives the JH signal, directly inducing the expression of *Kr-h1*. Before the formation of the adult, JH production ceases, expression of *Kr-h1* falls, and that of *E93* increases, making metamorphosis proceed (see Chapter 7: Molecular mechanisms regulating hormone production and action). Thus the last nymphal instar and the pupa are equally characterized by a dramatic decline of *Kr-h1* and a concomitant upregulation of *E93* expression, which gives support for a homology of these stages, as pointed out by Belles and Santos (2014) and Jindra (2019).

From a nomenclatural point of view, the homology of nymphs and larvae would suggest that the two names are not needed to distinguish the juvenile stages of hemimetabolans and holometabolans. Thus some defenders of the theory of direct homology between stages use the term "larva" for both cases (Hinton, 1963; Jindra, 2019; Rédei and Štys, 2016; Sehnal et al., 1996), while the term "pupa" for the holometabolan preadult stage is generally agreed. However, the term "nymph" for hemimetabolan and "larva" for holometabolan species have been used throughout this book. This nomenclature appears more practical because when the term "nymph" or "larva" is mentioned, it is implicitly understood that we are talking about a hemimetabolan or a holometabolan species, respectively. More importantly, the nymph/larva nomenclature is justified in morphological and evolutionary terms, since the larva is a modified nymph, as is the pupa, which already has its specific term.

THE BROAD-COMPLEX HYPOTHESIS. A MECHANISM TO EXPLAIN THE ORIGIN OF HOLOMETABOLY

If the evolutionary transition to holometaboly was driven by the selective pressure favoring internalization of adult structures in juveniles, arguably starting with the wings, then this process requires the corresponding underlying mechanism. Ideally, such a mechanism should explain at the same time (1) the qualitative transformation of the nymph into a larva and (2) the emergence of the pupal stage. In the frame of the theory of direct homology between stages, a mechanism that fits these conditions, based on a change in sensitivity of *BR-C* to JH, (Huang et al., 2013) can be described as follows.

In the hemimetabolan nymphs, JH enhances *BR-C* expression, which in turn promotes wing development. In contrast, JH represses *BR-C* in holometabolan larvae, so that *BR-C* only begins to be prominently expressed toward the end of the last larval instar, concomitantly with a decrease in JH production. BR-C then triggers pupal morphogenesis, including the formation of the wings. A mechanism that can explain the origin of holometaboly and is compatible with these observations would be a switch in the action of JH on *BR-C* expression from stimulatory to inhibitory. This switch, operating in the last common ancestor of the endopterygotes, would trigger at the same time (1) the transformation of a nymph bearing external wing pads into an externally apterous larva and (2) the determination of the pupal stage, with the formation of the external wings, triggered by the phasic expression of *BR-C* at the end of the last larval instar, when the production of JH transiently falls. Subsequently, BR-C would expand its functions, from one specialized in wing development to a larger array of morphogenetic roles, which would culminate with the pupal specifier function that operates in extant holometabolan species (Fig. 12.5) (Huang et al., 2013).

In the Thysanoptera (thrips) and male Coccomorpha (scale insects), which are hemimetabolans but whose life cycle includes quiescent pupal-like stages, BR-C is upregulated just before the molt to these stages (Minakuchi et al., 2011; Vea et al., 2016) (see Chapter 10: Regulation of ametabolan, hemimetabolan, and holometabolan development). In these species, *BR-C* might simply trigger the formation of the wing pads, which just appear in the respective pupal-like stages. But BR-C might even determine the formation of these pupal-like stages, which would represent an exaptation of the *BR-C* pupal specifier function. Studies depleting or suppressing BR-C will

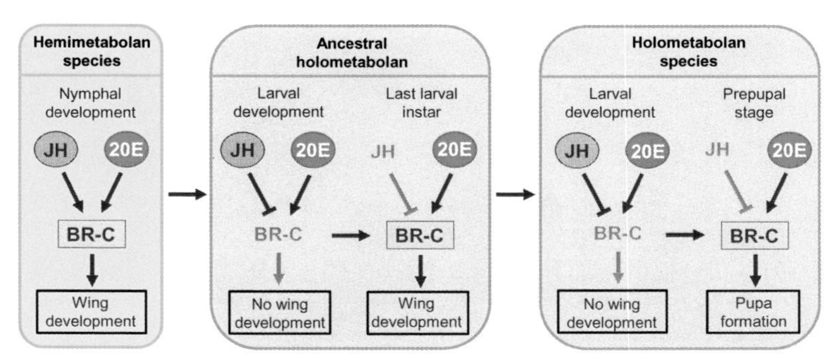

FIGURE 12.5 The Broad-complex hypothesis. In the transition from hemimetaboly to holometaboly, the regulation and functions of Broad-complex (BR-C) would have experienced two main changes. First, a shift of JH action on *BR-C* expression, from stimulatory in hemimetabolan species to inhibitory in holometabolans. Second, an expansion of functions, from controlling wing development along nymphal instars (hemimetabolans) to regulating pupal formation at the end of the larval life (holometabolans). *From Huang et al. (2013), slightly modified, with permission.*

clarify their functions in relation to the pupal-like stages of these insects. Given its small size, the RNAi approach by injecting a dsRNA in vivo can be technically problematic. An imaginative possibility is to use symbiotic or commensal bacteria to synthesize dsRNAs inside the insect host (Whitten, 2019). Symbiont-mediated Tubulin (Tub) knockdown has already been achieved in the thrips *F. occidentalis* (Whitten et al., 2016).

Research on the *B. mori* cell line BM-N has revealed a simple molecular mechanism by which JH, via Kr-h1, represses *BR-C* expression (Kayukawa et al., 2016). The *BR-C* gene of *B. mori* has two transcriptional start sites: the distal promoter and the proximal promoter. In the BM-N cells, the distal promoter is activated by 20E, and JH represses this activation. In contrast, the proximal promoter is constitutively activated irrespective of 20E and JH. Kayukawa et al. (2016) identified a Kr-h1 response element (KBS) located between two EcR response elements (EcREs) in the distal promoter of *BR-C*. This location, and further analysis of the role of 20E and JH in the regulation of the *BR-C* distal promoter, led the authors to propose that following induction of *Kr-h1* expression by JH through the Met-Taiman complex, two molecules of Kr-h1 bind to the KBS upstream of *BR-C*, thereby preventing its 20E-dependent activation (Fig. 12.6) (Kayukawa et al., 2016) (see Chapter 7: Molecular mechanisms regulating hormone production and action). Although the situation in vivo can be more complicated (see Kayukawa et al., 2014), for example, with the contribution of the isoforms expressed through the proximal promoter, the study of Kayukawa et al. (2016) reveals a part of the picture explaining the inhibitory action of JH on *BR-C* expression in the larval period.

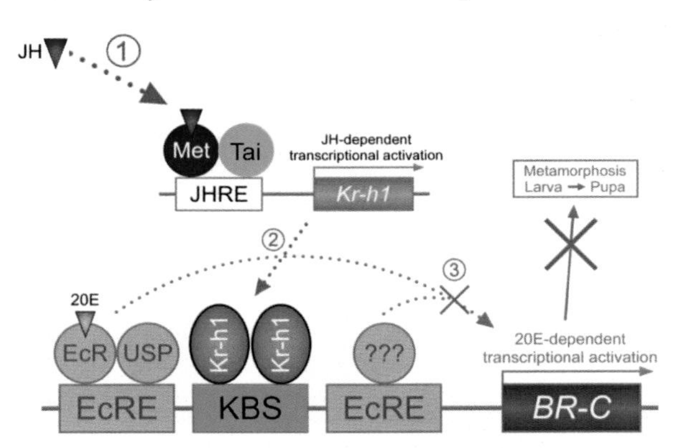

FIGURE 12.6 Mechanism of inhibition of Broad-complex (BR-C) by juvenile hormone (JH) as shown with experiments in *Bombyx mori* cells. The expression of *BR-C* is induced by 20-hydroxyecdysone (20E) through the ecdysone receptor (EcR)/Ultraspiracle (USP); then the 20E-receptor complex binds to the EcR response element in the promoter region of *BR-C*, thus triggering the gene expression. If JH is present, however, it induces the expression of *Krüppel homolog 1* (*Kr-h1*) via the JH receptor complex Methoprene-tolerant (Met)/Taiman (Tai); then two molecules of Kr-h1 bind to the Kr-h1 response element (KBS) in the *BR-C* promoter, which prevents the 20E-induction of *BR-C* expression. *From Kayukawa et al. (2016), with permission.*

Nothing is known about the molecular mechanisms underlying the enhancing effect of JH on *BR-C* expression observed in hemimetabolan nymphs. The mechanism described in *B. mori* suggests, by opposition, that binding of Kr-h1 in the corresponding KBS of the *BR-C* promoter in hemimetabolan species would not prevent the binding of EcR-20E to the EcREs. Kr-h1 binding would rather have a coactivating effect on *BR-C* expression. Nevertheless, the mechanisms involved in the transcriptional regulation of *BR-C* in hemimetabolans and holometabolans are surely more complicated than a crosstalk of only two players, Kr-h1 and EcR. Arguably, the mechanism might involve various EcREs and KBSs, and more or less large transcription initiation complexes, which could be different in hemimetabolan and holometabolan species.

The Broad-complex hypothesis only conjectures an initial general mechanism that might explain both the formation of the larva and the pupa, but it does not enter, it cannot enter yet, into details. From a first inhibitory effect of JH on *BR-C* activity, which would have prevented the development of the wings in juveniles, and led to their formation in the pupa, further effects and functions of JH and BR-C should have emerged successively, as the larva became more divergent with respect to the adult. In the juvenile period, the action of JH upon *BR-C* should coordinate with the regulation of the competence to metamorphose. In the prepupal stage, BR-C should have acquired new functions additional to the formation of the wings, which are required to build the adult body shape. The versatility given by the different BR-C isoforms and alternative gene promoters (Kayukawa et al., 2016) would have helped to develop new tissue-specific functions. These new functions must have emerged progressively, from the less modified species, now represented by the neuropterids, to the more derived ones, such as lepidopterans and dipterans. Certainly, the most appropriate way to find the differential regulatory mechanisms that emerged in the last common ancestor of holometabolans is to compare late-branching hemimetabolans (like hemipterans with juvenile quiescent stages) with early-branching holometabolans (like megalopterans, with larval and pupal stages similar to the adult). That approach, in which almost everything is to be done, could shed crucial new light to the mechanisms that determined the emergence of holometaboly, one of the most successful innovations in insect evolution.

REFERENCES

Aboulafia-Baginsky, N., Pener, M.P., Staal, G.B., 1984. Chemical allatectomy of late *Locusta* embryos by a synthetic precocene and its effect on hopper morphogenesis. J. Insect Physiol. 30, 839–852. Available from: https://doi.org/10.1016/0022-1910(84)90057-X.

Bayer, C., Von Kalm, L., Fristrom, J.W., 1996. Gene regulation in imaginal disc and salivary gland development during *Drosophila* metamorphosis. In: Gilbert, L.I., Tata, J.R., Atkinson, B.G. (Eds.), Metamorphosis. Postembryonic Reprogramming of Gene Expression in Amphibian and Insect Cells. Academic Press, San Diego, CA, pp. 321–361. Available from: https://doi.org/10.1016/B978-012283245-1/50011-7.

Belles, X., Santos, C.G., 2014. The MEKRE93 (Methoprene tolerant-Krüppel homolog 1-E93) pathway in the regulation of insect metamorphosis, and the homology of the pupal stage. Insect Biochem. Mol. Biol. 52, 60–68. Available from: https://doi.org/10.1016/j.ibmb.2014.06.009.

Bergot, B.J., Baker, F.C., Cerf, D.C., Jamieson, G., Schooley, D.A., 1981. Qualitative and quantitative aspects of JH titers in developing embryos of several insect species: Discovery of a new JH-like substance extracted from eggs of *Manduca sexta*. In: Pratt, G.E., Brooks, G.T. (Eds.), Juvenile Hormone Biochemistry. Elsevier, Amsterdam, pp. 33–45.

Berlese, A., 1913. Intorno alle metamorfosi degli insetti. Redia 9, 121–136.

Brüning, E., Lanzrein, B., 1987. Function of juvenile hormone III in embryonic development of the cockroach, *Nauphoeta cinerea*. Int. J. Invertebr. Reprod. Dev. 12, 29–44.

Carpenter, F.M., 1953. The geological history and evolution of insects. Am. Sci. 41, 256–270.

Cohen, S.M., 1993. Imaginal disc development. In: Bate, M., Martinez Arias, A. (Eds.), The Development of *Drosophila melanogaster*. Cold Spring Harbor Laboratory Press, pp. 747–841.

Cole, B.J., 1980. Growth ratios in holometabolous and hemimetabolous insects. Ann. Entomol. Soc. Am. 73, 489–491. Available from: https://doi.org/10.1093/aesa/73.4.489.

Costa, C., Ide, S., Simonka, C.E., 2006. Insectos inmaduros. Metamorfosis e identificación. Monografías del tercer milenio, vol. 5. SEA, Zaragoza.

Daimon, T., Uchibori, M., Nakao, H., Sezutsu, H., Shinoda, T., 2015. Knockout silkworms reveal a dispensable role for juvenile hormones in holometabolous life cycle. Proc. Natl. Acad. Sci. U.S.A. 112, E4226–E4235. Available from: https://doi.org/10.1073/pnas.1506645112.

Downes, W.L., 1987. The impact of vertebrate predators on early arthropod evolution. Proc. Entomol. Soc. Washingt. 89, 389–406.

Emery, I.F., Bedian, V., Guild, G.M., 1994. Differential expression of Broad-Complex transcription factors may forecast tissue-specific developmental fates during *Drosophila* metamorphosis. Development 120, 3275–3287.

Erezyilmaz, D., Riddiford, L., Truman, J., 2004. Juvenile hormone acts at embryonic molts and induces the nymphal cuticle in the direct-developing cricket. Dev. Genes Evol. 214, 313–323. Available from: https://doi.org/10.1007/s00427-004-0408-2.

Erezyilmaz, D.F., Riddiford, L.M., Truman, J.W., 2006. The pupal specifier broad directs progressive morphogenesis in a direct-developing insect. Proc. Natl. Acad. Sci. U.S.A. 103, 6925–6930.

Erezyilmaz, D.F., Rynerson, M.R., Truman, J.W., Riddiford, L.M., 2009. The role of the pupal determinant broad during embryonic development of a direct-developing insect. Dev. Genes Evol. 219, 535–544. Available from: https://doi.org/10.1007/s00427-009-0315-7.

Fernandez-Nicolas, A., Belles, X., 2017. Juvenile hormone signaling in short germ-band hemimetabolan embryos. Development 144, 4637–4644. Available from: https://doi.org/10.1242/dev.152827.

Ferral, N., Gomez, N., Holloway, K., Neeter, H., Fairfield, M., Pollman, K., et al., 2020. The extremely low energy cost of biosynthesis in holometabolous insect larvae. J. Insect Physiol. 120, 103988. Available from: https://doi.org/10.1016/j.jinsphys.2019.103988.

Finn, R.N., Chauvigné, F., Stavang, J.A., Belles, X., Cerdà, J., 2015. Insect glycerol transporters evolved by functional co-option and gene replacement. Nat. Commun. 6, 7814. Available from: https://doi.org/10.1038/ncomms8814.

Friedrich, M., 2003. Evolution of insect eye development: first insights from fruit fly, grasshopper and flour beetle. Integr. Comp. Biol. 43, 508–521. Available from: https://doi.org/10.1093/icb/43.4.508.

Friedrich, M., 2006. Continuity versus split and reconstitution: exploring the molecular developmental corollaries of insect eye primordium evolution. Dev. Biol. 299, 310–329. Available from: https://doi.org/10.1016/j.ydbio.2006.08.027.

Heming, B.S., 2003. Insect Development and Evolution. Comstock Pub. Associates.

Hinton, H.E., 1948. On the origin and function of the pupal stage. Trans. R. Entomol. Soc. Lond. 99, 395–409.

Hinton, H.E., 1963. The origin and function of the pupal stage. Proc. R. Entomol. Soc. Lond. 38, 77–85.

Huang, J.-H., Lozano, J., Belles, X., 2013. Broad-complex functions in postembryonic development of the cockroach *Blattella germanica* shed new light on the evolution of insect metamorphosis. Biochim. Biophys. Acta Gen. Subj. 1830, 2178–2187. Available from: https://doi.org/10.1016/j.bbagen.2012.09.025.

Inoue, Y., Mito, T., Miyawaki, K., Matsushima, K., Shinmyo, Y., Heanue, T.A., et al., 2002. Correlation of expression patterns of homothorax, dachshund, and Distal-less with the proximodistal segmentation of the cricket leg bud. Mech. Dev. 113, 141–148. Available from: https://doi.org/10.1016/S0925-4773(02)00017-5.

Ishimaru, Y., Tomonari, S., Watanabe, T., Noji, S., Mito, T., 2019. Regulatory mechanisms underlying the specification of the pupal-homologous stage in a hemimetabolous insect. Philos. Trans. R. Soc. Lond. Biol. Sci. 374, 20190225. Available from: https://doi.org/10.1098/rstb.2019.0225.

Jindra, M., 2019. Where did the pupa come from? The timing of juvenile hormone signalling supports homology between stages of hemimetabolous and holometabolous insects. Philos. Trans. R. Soc. Biol. Sci. 374, 20190064. Available from: https://doi.org/10.1098/rstb.2019.0064.

Jindra, M., Belles, X., Shinoda, T., 2015. Molecular basis of juvenile hormone signaling. Curr. Opin. Insect Sci. 11, 39–46. Available from: https://doi.org/10.1016/j.cois.2015.08.004.

Kayukawa, T., Murata, M., Kobayashi, I., Muramatsu, D., Okada, C., Uchino, K., et al., 2014. Hormonal regulation and developmental role of Krüppel homolog 1, a repressor of metamorphosis, in the silkworm *Bombyx mori*. Dev. Biol. 388, 48–56. Available from: https://doi.org/10.1016/j.ydbio.2014.01.022.

Kayukawa, T., Nagamine, K., Ito, Y., Nishita, Y., Ishikawa, Y., Shinoda, T., 2016. Krüppel homolog 1 inhibits insect metamorphosis via direct transcriptional repression of Broad-Complex, a pupal specifier gene. J. Biol. Chem. 291, 1751–1762. Available from: https://doi.org/10.1074/jbc.M115.686121.

Kiss, I., Beaton, A.H., Tardiff, J., Fristrom, D., Fristrom, J.W., 1988. Interactions and developmental effects of mutations in the Broad-Complex of *Drosophila melanogaster*. Genetics 118, 247–259.

Konopová, B., Jindra, M., 2008. Broad-Complex acts downstream of Met in juvenile hormone signaling to coordinate primitive holometabolan metamorphosis. Development 135, 559–568. Available from: https://doi.org/10.1242/dev.016097.

Konopová, B., Zrzavý, J., 2005. Ultrastructure, development, and homology of insect embryonic cuticles. J. Morphol. 264, 339–362. Available from: https://doi.org/10.1002/jmor.10338.

Konopová, B., Smykal, V., Jindra, M., 2011. Common and distinct roles of juvenile hormone signaling genes in metamorphosis of holometabolous and hemimetabolous insects. PLoS One 6, e28728. Available from: https://doi.org/10.1371/journal.pone.0028728.

Kukalová-Peck, J., 1991. Fossil history and the evolution of hexapod structures. The Insects of Australia. Melbourne University Press, Carlton, pp. 141–179.

Lameere, A., 1899. [La raison d'être des métamorphoses chez les insectes]. Bull. Ann. Société Entomol. Belgique 43, 619–636.

Lindstedt, C., Murphy, L., Mappes, J., 2019. Antipredator strategies of pupae: how to avoid predation in an immobile life stage? Philos. Trans. R. Soc. B: Biol. Sci. 374, 20190069. Available from: https://doi.org/10.1098/rstb.2019.0069.

Lubbock, J., 1873. On the Origin and Metamorphoses of Insects. Macmillan, London.

Maestro, J.L., Pascual, N., Treiblmayr, K., Lozano, J., Belles, X., 2010. Juvenile hormone and allatostatins in the German cockroach embryo. Insect Biochem. Mol. Biol. 40, 660–665. Available from: https://doi.org/10.1016/j.ibmb.2010.06.006.

Mayhew, P.J., 2007. Why are there so many insect species? Perspectives from fossils and phylogenies. Biol. Rev. 82, 425–454. Available from: https://doi.org/10.1111/j.1469-185X.2007.00018.x.

Minakuchi, C., Tanaka, M., Miura, K., Tanaka, T., 2011. Developmental profile and hormonal regulation of the transcription factors broad and Krüppel homolog 1 in hemimetabolous thrips. Insect Biochem. Mol. Biol. 41, 125–134. Available from: https://doi.org/10.1016/j.ibmb.2010.11.004.

Misof, B., Liu, S., Meusemann, K., Peters, R.S., Donath, A., Mayer, C., et al., 2014. Phylogenomics resolves the timing and pattern of insect evolution. Science 346, 763–767.

Morgan, T.H., Bridges, C., Sturtevant, A.H., 1925. The genetics of *Drosophila*. Bibliogr. Genet. 2, 145.

Nel, A., Roques, P., Nel, P., Prokin, A.A., Bourgoin, T., Prokop, J., et al., 2013. The earliest known holometabolous insects. Nature 503, 257–261. Available from: https://doi.org/10.1038/nature12629.

Nel, P., Bertrand, S., Nel, A., 2018. Diversification of insects since the Devonian: a new approach based on morphological disparity of mouthparts. Sci. Rep. 8, 3516. Available from: https://doi.org/10.1038/s41598-018-21938-1.

Novák, V.J., 1969. Morphogenetic analysis of the effects of juvenile hormone analogues and other morphogenetically active substances on embryos of *Schistocerca gregaria* (Forskål). J. Embryol. Exp. Morphol. 21, 1–21.

Parthasarathy, R., Tan, A., Bai, H., Palli, S.R., 2008. Transcription factor broad suppresses precocious development of adult structures during larval-pupal metamorphosis in the red flour beetle, *Tribolium castaneum*. Mech. Dev. 125, 299–313.

Paulus, H.F., 1989. Das Homologisieren in der Feinstrukturforschung: Das Bolwig-Organ der hoeheren Dipteren und seine Homologisierung mit Stemmata und Ommatidien eines urspruenglichen Facettenauges der Mandibulata. Zool. Beitr. N.F. 32, 437–478.

Piulachs, M.-D., Pagone, V., Belles, X., 2010. Key roles of the Broad-Complex gene in insect embryogenesis. Insect Biochem. Mol. Biol. 40, 468–475. Available from: https://doi.org/10.1016/j.ibmb.2010.04.006.

Polivanova, E.N., 1979. Embryonization of ontogenesis, origin of embryonic moults and types of development in insects. Zool. Zhurnal 43, 1269–1280 (in Russian, English summary).

Poyarkoff, E., 1914. Essai d'une théorie de la nymphe des insectes holométaboles. Arch. Zool. Exp. Gén. 54, 221–265.

Rédei, D., Štys, P., 2016. Larva, nymph and naiad – for accuracy's sake. Syst. Entomol. 41, 505–510. Available from: https://doi.org/10.1111/syen.12177.

Reza, A.M.S., Kanamori, Y., Shinoda, T., Shimura, S., Mita, K., Nakahara, Y., et al., 2004. Hormonal control of a metamorphosis-specific transcriptional factor Broad-Complex in silkworm. Comp. Biochem. Physiol. B: Biochem. Mol. Biol. 139, 753–761. Available from: https://doi.org/10.1016/j.cbpc.2004.09.009.

Rolff, J., Johnston, P.R., Reynolds, S., 2019. Complete metamorphosis of insects. Philos. Trans. R. Soc. B: Biol. Sci. 374, 20190063. Available from: https://doi.org/10.1098/rstb.2019.0063.

Ross, A.J., Jarzembowski, E.A., Brooks, S.J., 2000. The Cretaceous and Cenozoic record of insects (Hexapoda) with regard to global change. In: Culver, S.J., Rawson, P.F. (Eds.), Biotic Response to Global Change, the Last 145 Million Years. Cambridge University Press, Cambridge, UK, pp. 288–302.

Sehnal, F., 1985. Growth and life cycles. In: Kerkut, G.A., Gilbert, L. (Eds.), Comprehensive Insect Physiology, Biochemistry and Pharmacology. Pergamon Press, Oxford, pp. 1–86.

Sehnal, F., Švácha, P., Zrzavý, J., 1996. Evolution of insect metamorphosis. In: Gilbert, L.I., Tata, J.R., Atkinson, B.G. (Eds.), Metamorphosis. Postembryonic Reprogramming of Gene Expression in Amphibian and Insect Cells. Academic Press, San Diego, CA, pp. 3–58.

Shepherd, D., Bate, C.M., 1990. Spatial and temporal patterns of neurogenesis in the embryo of the locust (*Schistocerca gregaria*). Development 108, 83–96.

Smykal, V., Daimon, T., Kayukawa, T., Takaki, K., Shinoda, T., Jindra, M., 2014. Importance of juvenile hormone signaling arises with competence of insect larvae to metamorphose. Dev. Biol. 390, 221–230. Available from: https://doi.org/10.1016/j.ydbio.2014.03.006.

Suzuki, Y., Truman, J.W., Riddiford, L.M., 2008. The role of Broad in the development of *Tribolium castaneum*: implications for the evolution of the holometabolous insect pupa. Development 135, 569–577. Available from: https://doi.org/10.1242/dev.015263.

Švácha, P., 1992. What are and what are not imaginal discs: reevaluation of some basic concepts (Insecta, Holometabola). Dev. Biol. 154, 101–117.

Tanaka, K., Truman, J.W., 2007. Molecular patterning mechanism underlying metamorphosis of the thoracic leg in *Manduca sexta*. Dev. Biol. 305, 539–550. Available from: https://doi.org/10.1016/j.ydbio.2007.02.042.

Thomas, J.B., Bastiani, M.J., Bate, M., Goodman, C.S., 1984. From grasshopper to *Drosophila*: a common plan for neuronal development. Nature 310, 203–207. Available from: https://doi.org/10.1038/310203a0.

Truman, J.W., 1996. Metamorphosis of the insect nervous system. In: Gilbert, L.I., Tata, J.R., Atkinson, B.G. (Eds.), Metamorphosis: Postembryonic Reprogramming of Gene Expression in Amphibian and Insect Cells. Academic Press, San Diego, CA, pp. 283–320.

Truman, J.W., 2019. The evolution of insect metamorphosis. Curr. Biol. 29, R1252–R1268. Available from: https://doi: 10.1016/j.cub.2019.10.009.

Truman, J.W., Riddiford, L.M., 1999. The origins of insect metamorphosis. Nature 401, 447–452. Available from: https://doi.org/10.1038/46737.

Truman, J.W., Riddiford, L.M., 2002. Endocrine insights into the evolution of metamorphosis in insects. Annu. Rev. Entomol. 47, 467–500. Available from: https://doi.org/10.1146/annurev.ento.47.091201.145230.

Truman, J.W., Riddiford, L.M., 2019. The evolution of insect metamorphosis: a developmental and endocrine view. Philos. Trans. R. Soc. B: Biol. Sci. 374, 20190070. Available from: https://doi.org/10.1098/rstb.2019.0070.

Uhlirova, M., Foy, B.D., Beaty, B.J., Olson, K.E., Riddiford, L.M., Jindra, M., 2003. Use of Sindbis virus-mediated RNA interference to demonstrate a conserved role of Broad-Complex in insect metamorphosis. Proc. Natl. Acad. Sci. U.S.A. 100, 15607–15612.

Vea, I.M., Tanaka, S., Shiotsuki, T., Jouraku, A., Tanaka, T., Minakuchi, C., 2016. Differential juvenile hormone variations in scale insect extreme sexual dimorphism. PLoS One 11, e0149459. Available from: https://doi.org/10.1371/journal.pone.0149459.

Wang, Y.-H., Engel, M.S., Rafael, J.A., Wu, H.-Y., Rédei, D., Xie, Q., et al., 2016. Fossil record of stem groups employed in evaluating the chronogram of insects (Arthropoda: Hexapoda). Sci. Rep. 6, 38939. Available from: https://doi.org/10.1038/srep38939.

Whitten, M.M., 2019. Novel RNAi delivery systems in the control of medical and veterinary pests. Curr. Opin. Insect Sci. 34, 1–6. Available from: https://doi.org/10.1016/j.cois.2019.02.001.

Whitten, M.M.A., Facey, P.D., Del Sol, R., Fernández-Martínez, L.T., Evans, M.C., Mitchell, J. J., et al., 2016. Symbiont-mediated RNA interference in insects. Proc. Biol. Sci. 283, 20160042. Available from: https://doi.org/10.1098/rspb.2016.0042.

Zhimulev, I.F., Belyaeva, E.S., Mazina, O.M., Balasov, M.L., 1995. Structure and expression of the BR-C locus in *Drosophila melanogaster*, Diptera: Drosophilidae. Eur. J. Entomol. 92, 263–270.

Zhou, B., Riddiford, L.M., 2001. Hormonal regulation and patterning of the Broad-Complex in the epidermis and wing discs of the tobacco hornworm, *Manduca sexta*. Dev. Biol. 231, 125–137. Available from: https://doi.org/10.1006/dbio.2000.0143.

Zhou, X., Riddiford, L.M., 2002. Broad specifies pupal development and mediates the 'status quo' action of juvenile hormone on the pupal-adult transformation in *Drosophila* and *Manduca*. Development 129, 2259–2269.

Zhou, B., Hiruma, K., Shinoda, T., Riddiford, L.M., 1998. Juvenile hormone prevents ecdysteroid-induced expression of broad complex RNAs in the epidermis of the tobacco hornworm, *Manduca sexta*. Dev. Biol. 203, 233–244.

Zhou, B., Williams, D.W., Altman, J., Riddiford, L.M., Truman, J.W., 2009. Temporal patterns of broad isoform expression during the development of neuronal lineages in *Drosophila*. Neural Dev. 4, 39. Available from: https://doi.org/10.1186/1749-8104-4-39.

Zhu, G.-H., Jiao, Y., Chereddy, S.C.R.R., Noh, M.Y., Palli, S.R., 2019. Knockout of juvenile hormone receptor, Methoprene-tolerant, induces black larval phenotype in the yellow fever mosquito, *Aedes aegypti*. Proc. Natl. Acad. Sci. U.S.A. 116, 21501–21507. Available from: https://doi.org/10.1073/pnas.1905729116.

Epilogue

After reading a scientific monograph, the current reader may be tempted to think that all questions have been answered and that the subject is practically closed. This is obviously false, particularly in science territory, in which the answer to a question opens many other questions, often more exciting than the answered one. This is also the case with the subject of this book. The temptation is not to think that everything is resolved, but to think about which are the most interesting questions still unanswered. And predict on which issues the next and most exciting progress will be achieved. The following list is an attempt in these concerns.

We still don't know what are the enzymes that catalyze the specific epoxidations of *Drosophila melanogaster* and hemipteran juvenile hormones. We also know little about the enzymes that regulate the steps of the ecdysone biosynthetic pathway within the so-called black box. Moreover, different lines of evidence indicate that, in addition to the juvenile hormone nuclear receptor, there is a membrane receptor for this hormone. The identification of these fundamental players is an important pending issue.

The clarification of the role of the juvenile hormone in embryogenesis, especially in hemimetabolan species, is also a pending issue. The information available is incomplete and sometimes appears contradictory. To get clearer results, the more practical approach should be based on gene knockout through genome editing.

What is downstream E93 during adult morphogenesis in hemimetabolan and holometabolan species? What is the role of chromatin modification in the action of E93? ChIP-seq experiments and more specific studies on the action of E93 on chromatin, in order to see which genomic regions are activated or inactivated by E93, appear to be the most immediate approaches to solve the above questions.

High expression of E93 (and hence metamorphosis) cannot be induced before a certain period of juvenile growth has been reached, which has led to the concept of competence to metamorphose. The identification of the factors that confer it and the elucidation of the mechanisms that regulate them are also challenging issues that require further research. Experiments manipulating the size and the hormones involved and perhaps classical approaches of extract purification, biological tests, and chemical elucidation can help to find the answer.

The regulation of postembryonic hemimetabolan development seems to be essentially simple, summarized in the MEKRE93 pathway. However, it is not clear how the formation of the pupal-like stages of thrips, scale insects, and whiteflies, within the paraneopterans, is regulated. Does Broad-complex play a specifying role of those stages as in the holometabolan pupa? Or is its role limited mainly to modulating the development of wing pads, as in other hemimetabolan species? As the tiny size of these insects limits the use of conventional RNAi, the use of symbiotic or commensal bacteria to synthesize dsRNAs inside the insect could be a practical alternative.

Much progress has been made in the knowledge of the mechanisms that control the postembryonic development and metamorphosis in holometabolans, which also follow the MEKRE93 pathway. However, it remains to be resolved how the complex holometabolan life cycles known as hypermetamorphosis are regulated. The obvious first challenge that needs to be overcome is to keep those species in the laboratory throughout the life cycle. Then a practical way to start would be to have transcriptomes of the different stages for analysis.

At the other extreme, the regulation of the oversimplified life cycles of cave beetles, which in their most dramatically modified version are composed of a single quiescent larval instar and a pupal stage, remains to be unveiled. How do the concepts of MEKRE93 pathway, critical size, and competence to metamorphose apply to these species? For them, the initial challenges and research approaches would be the same as for hypermetamorphic species.

Certain observations suggest that juvenile hormone could have morphogenetic roles in ametabolan insects. It would be interesting to study the functions of this hormone in ametabolan postembryonic development and its transduction mechanisms, including the study of Krüppel homolog 1 and E93 roles and possible interactions. The information obtained would allow a better interpretation of the emergence of hemimetaboly. The recently developed gene knockout through genome editing in firebrats paves the way for this functional research.

In hemimetabolan nymphs, *Broad-complex (BR-C)* expression is stimulated by juvenile hormone (JH) while promoting the development of the wings. In contrast, JH represses the expression of *BR-C* in the holometabolan larva. The shift of JH action on *BR-C* might be based on changes in *BR-C* cis-elements, the composition of the transcription initiation complex that activates *BR-C* and/or the use of alternative gene promoters. The elucidation of these changes will bring new light to the study of the evolution of holometaboly.

The role of epigenetics in metamorphosis is in its infancy. It is not just about investigating more epigenetic mechanisms that contribute to regulate metamorphosis. It will also be interesting to investigate the heritability of epigenetic marks. Can they influence the differential success of the juvenile stages with respect to the adult? Epigenetic mechanisms can provide answers to the questions of whether natural selection can operate at different stages of the life cycle, which would be a driver for holometabolan evolution.

While I read the list of pending issues again, I realize that more than being predictions, they are rather questions that I would like to see answered soon, and that the most exciting findings will be those that are unpredictable. In any case, what is easily predictable is that the discoveries in the field of insect metamorphosis in the coming years will be very exciting. With a bit of luck, we can expect that this book will become outdated pretty soon.

X.B.

Index

Note: Page numbers followed by "*f*" refer to figures.

C

D

E

F

I

J

K

L

Q

R

S

Printed in the United States
By Bookmasters